AF334456

Principles and Practice

Regine Eibl • Dieter Eibl • Ralf Pörtner
Gerardo Catapano • Peter Czermak

Cell and Tissue Reaction Engineering

With a Contribution by Martin Fussenegger
and Wilfried Weber

 Springer

Prof. Dr.-Ing. Regine Eibl
Prof. Dr.-Ing. Dieter Eibl
Institute for Biotechnology
Zurich University of Applied Sciences
Department for Life Sciences and Facility
Management
P.O. Box
8820 Wädenswil, Switzerland
e-mail: regine.eibl@zhaw.ch
e-mail: dieter.eibl@zhaw.ch

Prof. Dr. Eng. Gerardo Catapano
Department of Chemical Engineering
and Materials
University of Calabria
87030 Rende (CS), Italy
e-mail: catapano@unical.it

PD Dr.-Ing. Ralf Pörtner
Hamburg University of Technology
(TUHH)
Institute for Bioprocess and Biosystems
Engineering
Denickestr. 15
21073 Hamburg, Germany
e-mail: poertner@tuhh.de

Prof. Dr.-Ing. Peter Czermak
Institute of Biopharmaceutical Technology
University of Applied Sciences
Giessen-Friedberg
Wiesenstrasse 14
35390 Giessen, Germany
and
Department of Chemical Engineering
Kansas State University, Manhattan, USA
e-mail: peter.czermak@tg.fh-giessen.de

ISBN: 978-3-540-68175-5   e-ISBN: 978-3-540-68182-3
DOI: 10.1007/978-3-540-68182-3

Library of Congress Control Number: 2008926595

Cover design: WMX Design GmbH, Heidelberg, Germany

Cover illustration: Alginate-encapsulated human mesenchymal stem cell line (hMSC-TERT) © Institute of Biopharmaceutical Technology, University of Applied Sciences Giessen-Friedberg, Giessen, Germany

Printed on acid-free paper

9 8 7 6 5 4 3 2 1

springer.com

Preface

The completion of the Human Genome Project and the rapid progress in cell biology and biochemical engineering, are major forces driving the steady increase of approved biotech products, especially biopharmaceuticals, in the market. Today mammalian cell products (**"products from cells"**), primarily monoclonals, cytokines, recombinant glycoproteins, and, increasingly, vaccines, dominate the biopharmaceutical industry. Moreover, a small number of products consisting of in vitro cultivated cells (**"cells as product"**) for regenerative medicine have also been introduced in the market.

Their efficient production requires comprehensive knowledge of biological as well as biochemical mammalian cell culture fundamentals (e.g., cell characteristics and metabolism, cell line establishment, culture medium optimization) and related engineering principles (e.g., bioreactor design, process scale-up and optimization). In addition, new developments focusing on cell line development, animal-free culture media, disposables and the implications of changing processes (multi-purpose-facilities) have to be taken into account. While a number of excellent books treating the basic methods and applications of mammalian cell culture technology have been published, only little attention has been afforded to their engineering aspects.

The aim of this book is to make a contribution to closing this gap; it particularly focuses on the interactions between biological and biochemical and engineering principles in processes derived from cell cultures. It is not intended to give a comprehensive overview of the literature. This has been done extensively elsewhere. Rather, it is intended to explain the basic characteristics of mammalian cells related to cultivation systems and consequences on design and operation of bioreactor systems, so that the importance of cell culture technology, as well as differences compared to microbial fermentation technology, is understood.

The book is divided into a part related to mammalian cells in general (part I) and a second part discussing special applications (part II). Part I starts with an overview of mammalian cell culture technology in **Chapter 1**, where we present definitions, the historical perspective and application fields including products; we summarize the most commonly used mammalian cell types as well as their characteristics and discuss resulting process requirements in **Chapter 2**. This chapter also outlines the current understanding of the cell metabolism including mammalian cell culture kinetics and modelling. **Chapters** 3 to 5 cover engineering aspects of mammalian cell culture technology: **Chapter 3** includes a categorization of suitable mammalian cell culture bioreactor types with their descriptions, and highlights the trends for R&D and biomanufacturing, as well as special features of bioreactors for cell therapy and tissue engineering. **Chapters 4 and 5** discuss key process limitation issues and show how they should be handled with respect to optimized mammalian cell bioreactor and process design. In this context we distinguish between suspension cells (non-anchorage dependent growing cells) and adherent cells (anchorage dependent growing cells) and focus on shear stress by aeration as well as mixing, oxygen transport, nutrient transfer, process strategy and monitoring, scale-up factors and strategies.

In part II, special cell culture applications which are predicted to have a high potential for further development are treated (**see Chapter 6 to Chapter 8**). These chapters treat insect cell-based recombinant protein production, bioartificial organs and plant cell-based bioprocessing.

The questions and problems included should enable the reader, especially students, to develop a clear understanding of the fundamentals and interactions presented in each chapter of the book. We hope that this book will be of interest to students of biotechnology who specialize in cell cultivation techniques and biochemical engineering as well as to more experienced scientists and industrial users who work with cell cultures as production organisms.

Finally, we acknowledge our contributing authors, Dr. Wilfried Weber and Prof. Dr. Martin Fussenegger and thank them for their dedication and diligence. We would also like to thank our families for their support during the writing and editing of this book. In addition, we are very grateful to Springer for their keen interest in bringing out this book of quality work.

May 2008

G. Catapano
P. Czermak
R. Eibl
D. Eibl
R. Pörtner

Contents

3   Bioreactors for Mammalian Cells: General Overview 55

D. Eibl and R. Eibl

4   Special Engineering Aspects 83

P. Czermak, R. Pörtner, and A. Brix

Part I
Mammalian Cells

Chapter 1
Mammalian Cell Culture Technology: An Emerging Field

D. Eibl, R. Eibl, and R. Pörtner

Abstract Mammalian cell culture technology has become a major field in modern biotechnology, especially in the area of human health and fascinating developments achieved in the past decades are impressive examples of an interdisciplinary interplay between medicine, biology and engineering. Among the classical products from cells we find viral vaccines, monoclonal antibodies, and interferons, as well as recombinant therapeutic proteins. Tissue engineering or gene therapy opens up challenging new areas. Bioreactors from small- (ml range up to 10L) to large-scale (up to $20\,m^3$) have been developed over the past 50 years for mammalian cell culture-based applications. In this chapter we give a definition of mammalian cells and a brief outline of the historical development of mammalian cell culture technology. Fields of application and products from mammalian cells, as well as future prospects, are discussed.

1.1 Definition and History

Mammalian cell culture technology has become a major field in modern biotechnology, especially in the area of human health, and fascinating developments achieved in the past few decades are impressive examples of an interdisciplinary interplay between medicine, biology and engineering (Kelly et al. 1993; Howaldt et al. 2005). Among the classical products from cells we find viral vaccines, monoclonal antibodies and interferons, as well as recombinant therapeutic proteins

D. Eibl, R. Eibl
Institute of Biotechnology, Zurich University of Applied Sciences, Department of Life Sciences and Facility Management, Wädenswil, Switzerland

R. Pörtner
Hamburg University of Technology (TUHH), Institute of Bioprocess and Biosystems Engineering, Denickestr. 15, D-21073 Hamburg, Germany
poertner@tuhh.de

R. Eibl et al., *Cell and Tissue Reaction Engineering: Principles and Practice,*
© Springer-Verlag Berlin Heidelberg 2009

(Fig. 1.1). Tissue engineering or gene therapy has opened up challenging new areas (Otzturk 2006). "Mammalian cell culture" refers to the cells of a mammalian, isolated from specific tissues (i.e. skin, liver, glands, etc.) and further cultivated and reproduced in an artificial medium (Fig. 1.2) (Butler 2004; Shuler and Kargi 2002). During the cultivation of mammalian cells *in vitro*, outside a living organism, some specific difficulties arise in the extraction of the cells from a "safe" tissue. Slow growth rates with doubling times between 12 and 28 h, low productivity, sensitivity against shear stress due to the lack of a cell wall (Cherry 1993), and complex requirement of growth medium are the challenges in developing techniques for mammalian cell culture. Moreover, many cell lines grow adherent, and a suitable surface for attachment has to be provided for these cells to proliferate. As mammalian cells have originated from multi-cellular organisms, they still hold the genetic program

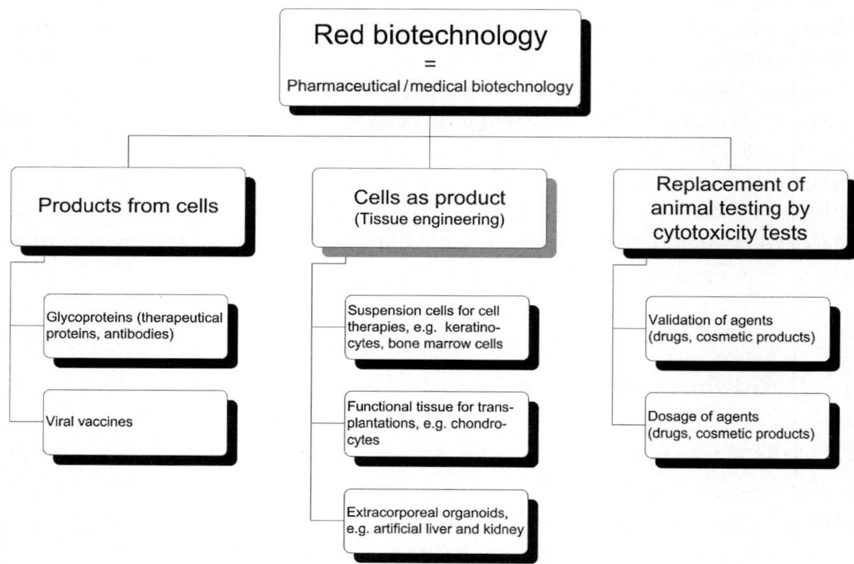

Fig. 1.1 Application fields of mammalian cells

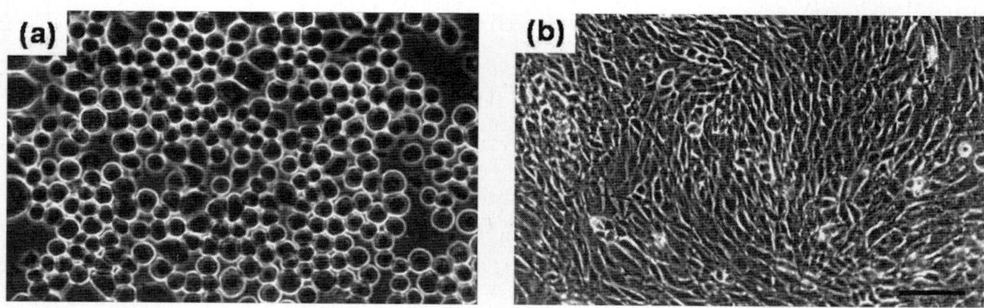

Fig. 1.2 Morphology of (**a**) suspendable and (**b**) adherent mammalian cells (bar approx. 30 μm)

of inducing their own cell death; a process called "apoptosis" or "programmed cell death" (Cotter and Al-Rubeai 1995; Singh et al. 1994; Al-Rubeai and Singh 1998). This can limit culture productivity in biotechnological processes. Another major problem is the finite life-span of primary cells, which die after several doublings *in vitro*. This problem was solved by transforming primary cells into immortal, "established" or "continuous" cell lines.

Considerable effort has been directed towards the development of mammalian cell culture technology, and appreciable progress has been achieved in the past few decades (Glaser 2001; Kretzmer 2002). Media that used to contain up to 10% serum have been continuously improved, and the cultivation in defined serum-free and even chemically defined, protein-free media is now common for most relevant industrial cell lines. Bioreactors have been developed, which provide the required low shear-stress environment by introducing gentle agitation and aeration with slow moving stirrers in tanks, or designing special aerators for air-lift reactors, or separating the cells from stressful conditions like hollow-fibre, fluidized-bed, and fixed-bed reactors. By using specific fermentation techniques for cell culture improvement (Fig. 1.3) mammalian cells can now be cultured in very large volumes (up to 20,000 L) to provide the needed quantity and quality of the desired product. On the industrial scale the adherent cell lines can be cultivated on micro-carriers (e.g. for vaccine production) or adapted to grow in suspension, e.g. cell lines derived from baby hamster kidney cells (BHK) or chinese hamster ovary cells (CHO). International developments have focused on application of disposables (single-use

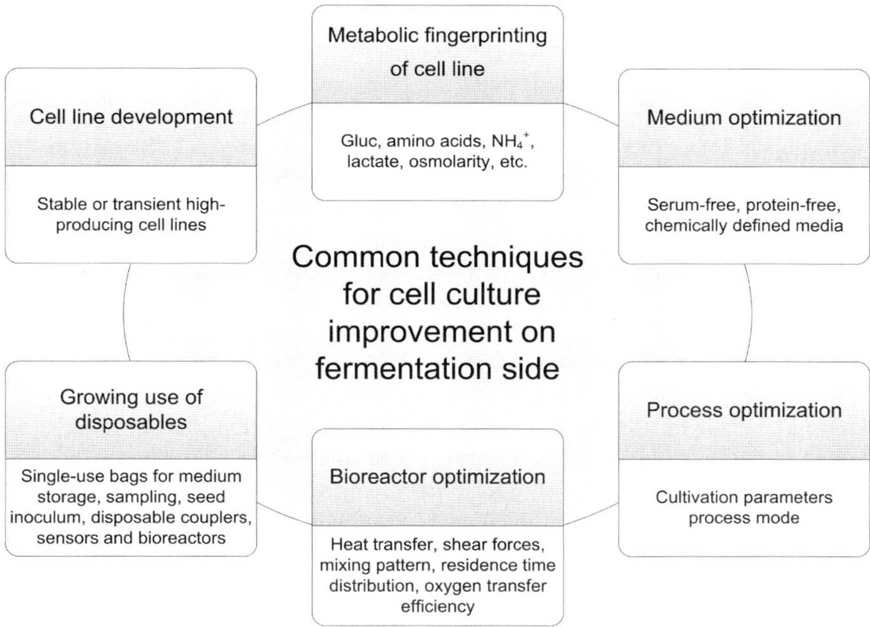

Fig. 1.3 Techniques for cell culture improvement

equipments) in mammalian cell-based processes to reduce the development and process costs, shorten time to market and enable quick process modifications. (Besides disposable bags for storage and sampling, liquid-handling container systems, disposable sensors, filters and couplings, disposable bioreactors have also become increasingly accepted.

Genetic engineering has contributed significantly to the recent progress in this area (Korke et al. 2002; Wurm 2004; Butler 2005). With this technology, functional proteins can be produced by introducing recombinant DNA into cell lines, e.g. chimeric (humanized) antibodies are produced for *in vivo* applications in transfectoma or recombinant CHO cells. New promoters have been developed to enhance productivity, and product titres up to five gram per litre have been reported for some industrial cell lines. By genetically induced proliferation, novel cell lines can now be constructed from primary cells, without losing functionality (Kim et al. 1998; Bebbington et al. 1992).

The following sections provide a basic understanding of the specific requirements of mammalian cells, describe state of the art process technology for cultivation of these cells and give a future perspective. A comprehensive overview on mammalian cell culture technology is given by Ozturk and Hu (2006).

1.2   Fields of Application and Products from Mammalian Cells

A detailed overview of the products from mammalian cells is given in (Bebbington et al. 1992; Griffiths 2000). Among "Products from cells" *viral vaccines* (Cryz et al. 2005) against polio, hepatitis B, measles, and mumps for human use and rubella, rabies, and foot-and-mouth-disease (FMD) for veterinary use are important products. Viral vaccines are produced efficiently by cell-based vaccine technology. For this purpose, primary cells, diploid cells or permanent cell lines (e.g. VERO) and recently, even recombinant cell lines are used. A breakthrough for the large-scale production of viral vaccines with anchorage-dependent cells was the development of microcarriers in the late 1960s which permitted cultivation in stirred tanks on thousand-litre scale. New targets for cell-based vaccines are the human immuno-deficiency virus (HIV), herpes simplex virus, and influenza. Recent developments include genetically engineered or DNA-vaccines.

Monoclonal antibodies (Galfre et al. 2005) have become a valuable tool for diagnostic purposes, as well as therapy. Antibodies synthesized by B-lymphocytes play an important role in the immune system of mammalians. Traditionally polyclonal antibodies were isolated from blood samples. In the 1970s Milstein and Kohler developed a technique to generate hybridoma cells producing monoclonal antibodies (Köhler and Milstein 1975). Due to specific binding, monoclonal antibodies are widely used for diagnosis, as tens of thousands of different monoclonal antibodies are available. The importance of monoclonal antibodies as therapeutic agents has evolved only recently, as immunogenic mouse antibodies were replaced by chimeric, humanized or human antibodies. Areas of application are organ transplantation (OKT3), cancer diagnosis and treatment, rheumatoid arthritis, leukaemia, asthma, and multiple sclerosis. Presently several antibodies are being produced in kilogram quantities (Ozturk 2006). Modern recombinant techniques focus on new antibody formats such as fragmented antibodies (FAb's) or bivalent antibodies, with a broad range of applications.

Glycoproteins are another important group of products produced from mammalian cells. Starting with the production of α-Interferon as an anti-infectious drug by (non-recombinant) Namalwa cells in the late 1970s, a growing number of glycoproteins for treatment of a wide variety of diseases are produced by means of mostly recombinant mammalian cells. Prominent examples are cytokines (e.g. Interferons and Interleukins), hematopoetic growth factors (e.g. Erythropoeitin for treatment of anemia), growth hormones, thrombolytic agents (e.g. tissue plasminogen activator [tPA]), coagulation factors (factor VII, factor VIII, factor IX, etc.), and recombinant enzymes (DNAse) (Ozturk 2006).

Recombinant proteins may be produced by bacterial, yeast, or mammalian cells. From a technological point of view, hosts such as bacteria or yeast have an advantage as regards the growth rate, final cell density and product concentration. Nevertheless, mammalian cells are preferred for those proteins requiring a specific, human-like glycosylation pattern (Andersen and Goochee 1994; Harcum 2006), which is difficult to obtain in other host systems. Another problem for microorganisms is the maximum size of the protein produced which must be below a molecular weight of approximately 30,000 Da. Further, in contrast to extra cellular release of most proteins produced in mammalian cells, products from microorganisms are often accumulated intra-cellularly in 'inclusion bodies'. This requires a more complex down-streaming. Besides this, for mammalian cells, important parameters such as product yield, medium requirements and growth characteristics (suspendable, shear resistant) have been significantly improved.

Proteins produced in the milk of transgenic animals have begun to compete with "classical" mammalian cell culture (Werner 1998; Young et al. 1997; Garner 1998; Wilmut et al. 1997). The proteins are usually expressed in large titres, over one order of magnitude higher than those obtained from cell cultures. The price for the production in transgenic animals will probably continue to decrease significantly due to the development of more efficient reproduction technologies. These new approaches will significantly decrease the time for the development of a product. The disadvantages of transgenic animals are (i) more sophisticated down-streaming (high protein loads), (ii) long development time, (iii) inability to produce proteins that might impair the health of the animal (e.g. insulin), (iv) higher risk of viral contamination and (v) the possibility of prion contamination (scrapies, BSE).

The field of "Cells as products" includes (i) the development of artificial organs (tissue engineering of liver, kidney) and tissues (skin, cartilage, bone), (ii) the expansion of hematopoietic cells for bone marrow transplantation and (iii) gene therapy. The loss and damage of tissues cause serious health problems (Langer 2000; Griffith and Naughton 2002; Petersen et al. 2003). For example, in the United States, annually almost one-half of the costs of medical treatments are spent on implant devices. Worldwide, 350 billion USD are spent on substitution of organs. The substitution of tissues (such as bone or cartilage) or joints with allograft materials includes the risk of infections by viruses (such as HIV, hepatitis C) or a graft rejection. Artificial implants such as those used in knee or hip replacement, have limitations due to their limited lifespan, insufficient bonding to the bone, and allergic reactions caused by material abrasion. These call for new concepts in practical medical applications. Tissue-engineered substitutes generated *in vitro* could provide new strategies for the restoration of damaged tissues. The aim of tissue engineering can be defined as the development of cell-based substitutes to restore, maintain or improve tissue functions. These substitutes should have organ-specific properties with respect to biochemical activity, microstructure, mechanical integrity and bio-stability. Cell-based concepts include the following:

- Direct transplantation of isolated cells
- Implantation of a bioactive scaffold for the stimulation of cell growth within the original tissue
- Implantation of a three dimensional (3D) bio-hybrid structure of scaffold and cultured cells or tissue. Non- implantable tissue structures can be applied as external support devices (e.g. an extra-corporal liver support when a compatible donor organ is not readily available) or engineered tissues can be used as *in vitro* physiological models for studying disease pathogenesis and developing new molecular therapeutics, e.g. drug screening (Pörtner et al. 2005).

Somatic gene therapy implies transfection of a specific gene to cells isolated previously from a patient suffering from a genetic disease (Mulligan 1993; Le Doux and Yarmush 1996; Schlitz et al. 2001). The transfected cells are then re-introduced into the patient. Gene therapy involves transfer of genetic material and encoding therapeutic genes and the sequence necessary for expression to target cells to alter their genetic code for a desired therapeutic effect. Many diseases originate from gene defects. Expression of the transferred genes can result in the synthesis of therapeutic proteins or correction of a gene defect. Gene transfer could also lead to desired apoptosis or inhibition of cell proliferation. Theoretically gene therapy can be applied for repairing single entailed gene defects, acquired gene defects such as chronic infectious diseases, multifactor genetic diseases such as cardiovascular disease and finally cancer. Since the first application of gene therapy in 1990, the number of clinical trials has increased significantly. Gene therapy is now widely used in ongoing clinical trials for the treatment of cancer, infectious diseases such as human immunodeficiency virus (HIV) infection and tissue engineering.

1.3 Future Prospects

Mammalian cell culture products are currently used mainly as medicines or in diagnostics. As medicines some products have a demand of 500 kg/year and generate $1–2 billion in revenue. Among these are tissue plasminogen activator (tPA), erythropeithin (EPO)-products and Remicade® (Centocor) or MAbThera® (Otzturk 2006). The pipeline for new biopharmaceutical therapeutics targets a large number of diseases (e.g. 425 for USA in 2002) (PhRMA 2002), most of them for cancer therapy, infectious diseases, autoimmune diseases, or AIDS/HIV. It can be expected, that at least some of these will find their way to the market. As for some established compounds, the patents have already expired or will expire in the near future. Consequently, there will be a market for biosimilar or generic products. On the other hand, the costs for target identification, clinical trials and process development will increase (approx. \$0.5–1 billion at the current market rates) and only mass produced items will make it finally to the market. New products or medical protocols can be expected for tissue engineering and cell therapy.

1.4 Exercises

- Name industrially relevant products from mammalian cells.
- Investigate the techniques used for the production of viral vaccines.
- Discuss the pros and cons of glycoprotein expression in hosts of bacterial, fungi or mammalian origin.
- Name industrially relevant drugs (glycoproteins) produced by recombinant mammalian cells.
- Give examples of "cells as products".

References

Al-Rubeai M, Singh RP (1998) Apoptosis in cell culture. Curr Opin Biotechnol 9: 152–156

Andersen DC, Goochee CF (1994) The effect of cell-culture conditions on the oligosaccharide structures of secreted glycoproteins. Curr Opin Biotechnol 5: 546–549

Bebbington CR, Renner G, Thomson S, King D, Abrams G, Yarrington GT (1992) High-level expression of a recombinant antibody from myeloma cells using a glutamine synthetase gene as an amplifiable selectable marker. Biotechnol 10: 169–175

Butler M (2004) Animal Cell Culture and Technology – The Basics. Oxford University Press, New York

Butler M (2005) Animal cell cultures: Recent achievements and perspectives in the production of biopharmaceuticals. Appl Microbiol Biotechnol 68: 283–291

Cherry RS (1993) Animal cells in turbulent fluids: Details of the physical stimulus and the biological response. Biotechnol Adv 11: 279–299

Cotter TG, Al-Rubeai M (1995) Cell death (apoptosis) in cell culture systems. Trends Biotechnol 13: 150–154

Cryz STJ, Granstro M, Gottstein B, Perrin L, Cross A, Larrick J (2005) Immunotherapy and Vaccines. In: Ullmann's Encyclopedia of Industrial Chemistry, Wiley-VCH Verlag GmbH & Co. KGaA, Weinheim

Doux J, Yarmush M (1996) Engineering gene transfer technologies: Retroviral-mediated gene transfer. Bioeng Sci 20: 3–8

Galfre GL, Secher DS, Crawley P (2005) Monoclonal antibodies. In: Ullmann's Encyclopedia of Industrial Chemistry, Wiley-VCH Verlag GmbH & Co. KGaA, Weinheim

Garner I (1998) The production of proteins in the milk of transgenic livestock: A comparison of microinjection and nuclear transfer. In: Merten OW et al. (eds) New Developments and New Applications in Animal Cell Technology. Kluwer, Dordecht, pp 745–750

Glaser V (2001) Current trends and innovations in cell culture. Gen Eng N 21: 11

Griffith LG, Naughton G (2002) Tissue engineering-current challenges and expanding opportunities. Science 295: 1009–1014

Griffiths B (2000) Animal cell products, overview. In: Spier RE (ed) Encyclopedia of Cell Technology, Vol. 1. Wiley, New York, pp 70–76

Harcum SW (2006) Protein glycosylation. In: Ozturk SS, Hu WS (eds) Cell Culture Technology for Pharmaceutical and Cell-Based Therapies. Taylor & Francis, New York, pp 113–154

Howaldt M, Walz F, Kempken R (2005) Kultur von Tierzellen. In: Chmiel H (ed), Bioprozesstechnik, Spektrum Akademischer Verlag

Kelly BD, Chiou TW, Rosenberg M, Wang DIC (1993) Industrial animal cell culture. In: Stephanopolous G (ed) Biotechnology, Vol. 3 (Bioprocessing), VCH Verlagsgesellschaft, Weinheim

Kim YH, Iida T, Fujita T, Terada S, Kitayama A, Ueda H, Prochownik EV, Suzuki E (1998) Establishment of an apoptosis-resistant and growth-controllable cell line by transfecting with inducible antisense c-jun gene. Biotechnol Bioeng 58: 65–72

Köhler G, Milstein C (1975) Continuous cultures of fused cells secreting monoclonal antibodies of predifined specifity. Nature 256: 495

Korke R, Rink A, Seow TK, Chung MC, Beattie CW, Hu WS (2002) Genomic and proteomic perspectives in cell culture engineering. J Biotechnol 94: 73–92

Kretzmer G (2002) Industrial processes with animal cells. Appl Microbiol Biotechnol 59: 135–142

Langer R (2000) Tissue engineering. Mol Ther 1: 12–15

Mulligan RC (1993) The basic science of gene therapy. Science 60: 926–932

Otzturk SS (2006) Cell culture technology – an overview. In: Ozturk SS, Hu WS (eds) Cell Culture Technology For Pharmaceutical and Cell-Based Therapies. Taylor & Francis, New York, pp 1–14

Ozturk SS, Hu WS (eds) (2006) Cell Culture Technology For Pharmaceutical and Cell-Based Therapies. Taylor & Francis, New York

Petersen JP, Rücker A, von Stechow D, Adamietz P, Pörtner R, Rueger JM, Meenen NM (2003) Present and future therapies of articular cartilage defects. Eur J Trauma 1: 1–10

PhRMA. Biotechnol New Medicines in Development (2002). Pharmaceutical Research and Manufacturers of America

Pörtner R, Nagel-Heyer ST, Goepfert CH, Adamietz P, Meenen NM (2005) Bioreactor design for tissue engineering. J Biosci Bioeng 100: 235–245

Schlitz AJ, Kühlcke K, Fauser AA, Eckert HG (2001) Optimization of retroviral vector generation for clinical application. J Gene Med 3: 427–436

Shuler ML, Kargi F (2002) Bioprocess Engineering – Basic Concepts. Prentice Hall PTR, Englewood Cliffs, NJ

Singh RP, Al-Rubeai M, Gregory CD, Emery AN (1994) Cell death in bioreactors: a role for apoptosis. Biotechnol Bioeng 44: 720–726

Werner RG (1998) Transgenic technology – a challenge for mammalian cell culture production systems. In: Merten OW et al. (eds) New Developments and New Applications in Animal Cell Technology. Kluwer, Dordecht, pp 757–763

Wilmut I, Schnieke A, McWhir J, Kind A, Campbell K (1997) Viable offspring derived from fetal and adult mammalian cells. Nature 385: 810–813

Wurm FM (2004) Production of recombinant protein therapeutics in cultivated mammalian cells. Nat Biotechnol 11: 1393–1398

Young MW, Okita WB, Brown EM, Curling JA (1997) Production of biopharmaceutical proteins in the milk of transgenic dairy animals. BioPharm 6: 34–38

Complementary Reading

Butler M (1996) Animal Cell Culture and Technology – The Basics. Oxford University Press, New York

Butler M (2005) Animal cell cultures: recent achievements and perspectives in the production of biopharmaceuticals. Appl Microbiol Biotechnol 68: 283–291

Cryz STJ, Granstro M, Gottstein B, Perrin L, Cross A, Larrick J (2005) Immunotherapy and Vaccines. In: Ullmann's Encyclopedia of Industrial Chemistry, Wiley-VCH Verlag GmbH & Co. KGaA, Weinheim

Doux J, Yarmush M (1996) Engineering gene transfer technologies: Retroviral-mediated gene transfer. Bioeng Sci 20: 3–8

Galfre GL, Secher DS, Crawley P (2005) Monoclonal antibodies. In: Ullmann's Encyclopedia of Industrial Chemistry, Wiley-VCH Verlag GmbH & Co. KGaA, Weinheim

Glaser V (2001) Current trends and innovations in cell culture. Gen Eng N 21: 11

Griffith LG, Naughton G (2002) Tissue engineering-current challenges and expanding opportunities. Science 295: 1009–1014

Griffiths B (2000) Animal cell products, overview. In: Spier RE (ed), Encyclopedia of Cell Technology, Vol. 1. Wiley, New York, pp 70–76

Howaldt M, Walz F, Kempken R (2005) Kultur von Tierzellen. In: Chmiel H (ed), Bioprozesstechnik, Elsevier – Spektrum Akademischer Verlag

Langer R (2000) Tissue engineering. Mol Ther 1: 12–15

Ozturk SS, Hu WS (eds) (2007) Cell Culture Technology For Pharmaceutical and Cell-Based Therapies. Taylor & Francis, New York, pp 113–154

Werner RG (1998) Transgenic technology – a challenge for mammalian cell culture production systems. In: Merten OW et al. (eds) New Developments and New Applications in Animal Cell Technology. Kluwer, Dordecht, pp 757–763

Wurm FM (2004) Production of recombinant protein therapeutics in cultivated mammalian cells. Nat Biotechnol 11: 1393–1398

Chapter 2
Characteristics of Mammalian Cells and Requirements for Cultivation

R. Pörtner

Abstract This chapter aims at presenting a general overview on the specific characteristics of mammalian cells to those not familiar with cell biology. At first, differences between mammalian cells, plant cells and microbes are discussed. Then consequences for design of bioreactors and processes are also discussed. Different types of mammalian cells (primary cells, permanent (established) cell lines, hybridom cells) are introduced. Techniques required, to get from primary cells to permanent (established) cell lines and the hybridom technique for production of monoclonal antibodies are also briefly outlined. Culture collections and cell banking are discussed with respect to its purpose and the build-up. Appropriate culture media is an essential requirement for successful mammalian cell culture and is therefore introduced in depth. The chapter also focuses on characteristics of cell growth and metabolism. A short introduction to cell metabolism is followed by a detailed discussion on aspects related to modelling of cell growth and metabolism of mammalian cells.

2.1 Differences Between Mammalian Cells, Plant Cells and Microbes: Consequences of These Differences

Due to their internal complexity living cells are divided into two types – prokaryotic and eukaryotic cells. Prokaryotic cells are simple in structure, with no recognizable organelles. They have an outer cell wall to provide a shape. Just under the rigid cell wall is the more fluid cell membrane. The cytoplasm enclosed within the cell membrane does not exhibit much structure when viewed through electron microscopy. Eukaryotic cells of protozoa, higher plants and mammalians are highly structured. These cells tend to be larger than the cells of bacteria. They have developed specialized packaging and transport mechanisms that may be necessary to support

R. Pörtner
Hamburg University of Technology (TUHH), Institute of Bioprocess and Biosystems Engineering, Denickestr. 15, D-21073 Hamburg, Germany
poertner@tuhh.de

R. Eibl et al., *Cell and Tissue Reaction Engineering: Principles and Practice,*
© Springer-Verlag Berlin Heidelberg 2009

Table 2.1 Culture characteristics (suspension) of microbial, plant cell and mammalian cell culture (t_d: doubling time, vvm: volume of gas per volume of liquid and minute)

Characteristic	Microbial culture	Plant cell culture	Mammalian cell culture
Size	2–10 μm	10–100 μm	10–30 μm
Individual cells	Often	Often aggregates	Sometimes adherent
Inoculation density	Low	High (10%)	High (5–10%)
Growth rates	Rapid (t_d = 1–2 h)	Slow (t_d = 2–7 d)	Slow (t_d = 20–50 h)
Shear sensitivity	Low	Moderate	High
Stability	Stable	Unstable	Unstable
Product accumulation	Intra-/extracellular	Mostly intracellular	Mostly extracellular
Culture medium	Often simple	Often complex	Complex
Temperature	26–36 °C	25–27 °C	29–37 °C
Aeration	Often high (1–2 vvm)	Low (0.1–0.3 vvm)	Low (~ 0.1 vvm)
Foaming	Often high	sometimes foaming	Sometimes foaming
pH-value	3–8	5–6	7.0–7.4
Cell density	(Very) high	Low	Low–middle
Scale-up	Easy	Difficult	Difficult

their larger size. For a deeper understanding of cell physiology special books are available (Butler 2004; Freshney 1994). Here, mainly the consequences of the specific characteristics with respect to cultivation and product formation is discussed.. As can be seen from Table 2.1, the culture characteristics of microbial, plant cell and mammalian cell culture differ significantly.

Some more aspects are relevant for cultivation of mammalian cells. They vary in size (10–30 μm) and shape (spherical, elipsoidal) and have a high tendency to adhere or form aggregates even in suspension cells. This results in homogenisation problems and mass transfer limitations. Moreover, mammalian cells do not have a cell wall, but are surrounded by a thin and fragile plasma membrane that is composed of protein, lipid, and carbohydrate. This structure results in significant shear sensitivity (compare Sect. 3.1). Slow growth rates result in low oxygen uptake rate of the culture. Therefore comparatively low aeration rates are needed.

2.2 Types of Mammalian Cells

In general, mammalian cells relevant for industrial processes can be divided into following groups:

- Primary cells have been isolated from a tissue and then taken into culture (primary culture).
- Permanent or established cell lines originate from a primary culture, but due to some transformation are able (at least theoretically speaking) to divide and proliferate indefinitely. Those cell lines are kept in a cell bank and are very often used as host cells for the expression of recombinant proteins.
- Hybridom cells are the cells, which are obtained through the fusion of lymphocytes and tumour cells and are able to express monoclonal antibodies.

The characteristics of these cell types is discussed briefly in the following pages. For more detail refer to Butler (2004) and Freshney (1994) along with other books.

2.2.1 *From Primary Cells to Permanent (Established) Cell Lines*

In mammalian tissue usually the following cell types can be distinguished (Butler 2004):

- *Epithelial cells*: are those which cover organs (i.e. the skin) or vessels (i.e. veins). Epithelial cells can be taken in culture relatively easily.
- *Fibroblasts*: original from the connective tissue, this consists of fibrous matrix. For example, collagen in cartilage. Fibroblasts in culture exhibit excellent growth characteristics and therefore belong to the most utilised cell groups.
- *Muscle cells*: Muscles are composed by an association of small channels that are built out of myoblasts. The process of building those channels can be repeated by the myoblasts in culture.
- *Nerve cells*: nervous tissue is made of characteristically shaped neurones, which are responsible for the transmission of electric impulses. Neurones are highly differentiated cells and cannot be grown in culture. But some of its characteristics have been observed in cultures of the so-called neuroblasts, i.e. tumour cells originate from nervous tissue.
- *Lymphocytes*: Blood and tissue fluids contain a series of suspended cells, from which some are able to grow in culture. For example, lymphocytes i.e. white blood cells.

The normal procedure for the isolation of a mammalian cell from a tissue involves the following steps: (i) a small piece of tissue is decomposed into single cells or cell clots mechanically, enzymatically [e.g. typsin (Kasche et al. 1980), collagenase] or through a combined procedure. (ii) The cells are then separated from the enzymes by centrifugation and re-suspended in culture medium. (iii) After that, the cells are inoculated into a special glass or plastic vessel with flat bottom. The anchorage dependent cells (compare Fig. 2.1) adhere to the bottom surface and after a lag phase, divide by mitosis. This a cell culture, in which the cells come from a differentiated tissue, is called a primary cell culture. The most critical point in the isolation of primary cells is to avoid external contamination by practising aseptic techniques.

When the primary cells cover the bottom of the culture vessel almost completely, they are enzymatically cleaved (i.e. trypsin) from their support and used to inoculate new cultures. This subculture is the origin of the so called secondary cultures. Some of the cells are stored deeply frozen in liquid nitrogen. They remain as a safety stock, from where,at any needed time, it is possible to get enough "fresh" cells to start a new series of subcultures for mass production. With the primary cells it is possible to repeat subcultures several times. However, the primary cultures have a finite life span and therefore, after a certain number of doublings (from 50 to 100 times), the cells cease to grow and die. A finite growth capacity is a characteristic of all cells derived from normal mammalian tissue. A cell can be considered as normal, when it shows a certain set of characteristics (Hayflick and Moorhead 1961):

- A diploid number of chromosomes (i.e. 46 chromosomes for human cells) with which it is shown that no gross chromosome damage has occurred.
- Adherence: the cells require a surface to grow and attach to (anchorage dependent). The growth phase extends until the cells reach a stage of confluence (contact inhibition).
- A finite life span in culture.
- Non-malignant: the cells are not cancerous, i.e. they do not cause tumour in mice.

In the 1960's normal cells were required for the production of human vaccines. This was to ensure the safety of these products (Butler 2004).

Not all cell types produce exclusively primary cell cultures, and die after a limited number of "passages". Some cells acquire an infinite life span and such a population is usually called a "permanent" or "established cell line". These cells have undergone a "transformation", i.e. they have lost their sensitivity to the growth control mechanisms (Fig. 2.2). Transformed cells can also lose the characteristic

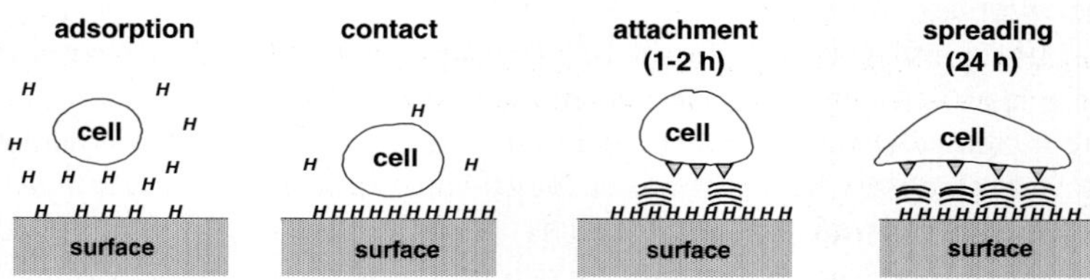

Fig. 2.1 Adhesion of anchorage-dependent cells to a solid substratum (from Butler 2004, modified)

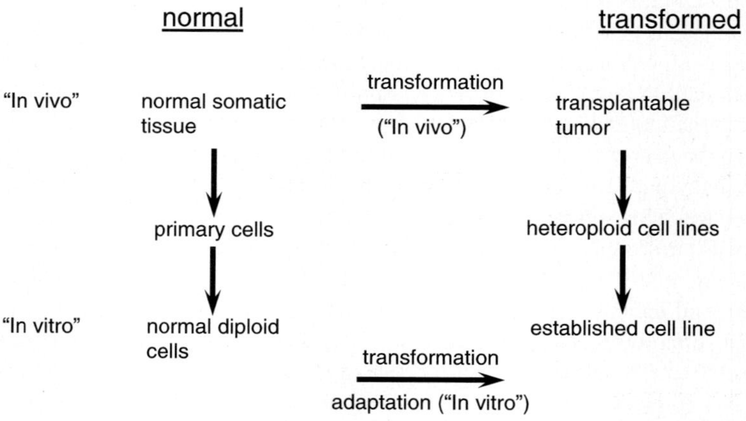

Fig. 2.2 Normal and transformed mammalian cells (adapted from Chmiel 1991)

to grow when adhered to a surface and thus are able to grow in suspension. These transformations are also sometimes reflected in the chromosomes, changing the genotype of the cells. Transformed cells can be easily grown in relatively simple media without addition of expensive growth factors. Initially, transformed cells were identified by chance. But, today there are techniques to cause cell transformation and "immortalisation". For example, the treatment with mutagenic substances, with virus or with oncogenic substances (Butler 2004). Transformation of cells *in vitro* shows some similarities with carcinogenese *in vivo*, but is not identical to it. For instance, not all the transformed cells are malignant. On the other hand, all cells that are isolated from tumours (i.e. HeLa or Namalwa Cells) can be kept in a permanent culture.

The characteristics of normal and transformed cells with respect to cultivation are summarized in Table 2.2. An overview on permanent cell lines important for research and production is given in Table 2.3.

An alternative expression system for recombinant proteins offer insect cells, especially the insect cell-baculovirous expression system (IC-BEVS) (Ikonomou et al. 2003) (compare Part II, Chap. 8). Commonly applied are *Sf*9- or *Sf*21-cells isolated from *Spodoptera frugiperda*. Production of heterologous proteins by the IC-BEVS consists of two stages. Insect cells are first grown to a desired concentration and then infected with a recombinant baculovirus containing the gene coding for the desired protein. In the mean time, several recombinant proteins produced by this technique have already reached the market (e.g. Cervarix, FluBiok, Provenge).

2.2.2 *Hybridom Cells for Production of Monoclonal Antibodies*

Hybridom cells are "artificial" cells produced by the fusion of human or mammalian lymphocytes, which can secret specific antibodies, with a myelom cell. Usually lymphocytes do not survive outside the original organism, i.e. cannot be kept alive in an artificial medium. Köhler and Milstein (1975) were able to solve this problem

Table 2.2 Comparison of "normal" and "transformed" cells

Normal	Transformed
Diploid (46 chromosomes for human cells)	Abnormal number of chromosomes
Non-malignant	Malignant (form tumour in mice)
Finite life-span (50+−10 subcultures max.)	Infinite life-span
Anchorage-dependent (except blood cells)	Non-anchorage-dependent (i.e. suspension culture possible)
Mortal; finite number of divisions	Immortal or continuous cell lines
Contact inhibition; monolayer culture	No contact inhibition; multilayer cultures
Dependent on external growth factor signals for proliferation	May not need an external source of growth factors
Longer retention of differentiated cellular function	Typically loss of differentiated cellular function
Display typical cell surface receptors	Cell surface receptor display may be altered

Table 2.3 Examples of permanent cell lines important for research and production (data from Butler 2004, modified)

Cell Line	Origin	Application
Baby Hamster Kidney (BHK)	Syrian hamster, 1963	Adherent cells, but can be adapted to suspension, foot & mouth disease vaccine, rabies vaccine, recombinant proteins (Factor VIII)
Chinese Hamster Ovary (CHO)	Ovary of Chinese hamster, 1957	Adherent cells, but can be adapted to suspension, recombinant proteins (HBstg, tPA, Factor VIII)
COS	Monkey kidney	Transient protein expression (Edwards and Aruffo 1993)
NAMALWA	Human lymphatic tissue	Alpha-Interferon
HeLa	tumour in a cervical vertebra	Fast growing tumour cell line that was isolated in the beginning of the 50ies.
HEK	Human embryonic kidney, 1977	Transient protein expression
MDCK	rabbit kidney	Adherent cell line with good growth characteristics, animal vaccines
MRC-5	Human embryonic lung cells	"Normal" cells with a finite life span, vaccine production
NS0 and SP2/0	Mouse myelom from B-lymphocytes	Antibody production
3T3	Mouse connective tissue	Suspendable, used in the development of the cell culture technology
WI-38	Human embryonic lung cells	"Normal" cells with a finite life span Vaccine production
Vero	African green monkey kidney, 1962	Established cell line, but with some characteristics of the normal diploid cells.
PERC.6	Human embryonic retina cells, 1998	Immortalized cell line, well characterized, produce high levels of recombinant proteins and viruses (Butler 2005; Maranga and Goochee 2006)

by fusing a mouse myelom cell similar to a tumour cell with a lymphocyte from a mouse spleen, which produced antibodies. The cells generated in this way are hybrid cells (also called "hybridom cells") and have the lymphocyte characteristic to produce antibodies against a certain antigen. They also have the ability from the myelom cells to survive in culture.

Antibodies are glycoproteins that mammalians produce to protect the organism against unknown substances (antigens, e.g. virus, bacteria, foreign tissue) (Milstein 1988). The immune system of the organism reacts by the proliferation of B-Lymphocytes. They belong to the group of white blood corpuscles that produce antibodies against certain groups of the antigen molecules. The B-lymphocytes release the antibodies into the blood and they can then specifically bind to the antigen to induce final removal of the antigen. The antibodies are found in the globuline fraction of blood proteins and are therefore known as immunoglobulins.

The peculiarity of antibodies is binding sites allowing them to bind to specific determined groups on the surface of the antigen (key – keyhole). A very important aspect of the reaction between antibodies and antigens is its specificity. This means the capacity of an antibody to bind specifically to "its" antigen, even under a large number of diverse molecules, i.e. to bind only to the molecule which belongs to the group corresponding for which the antibody was tailored.

There are five major classes of antibodies (IgG, IgA, IgM, IgD and IgE). They are basically differentiated by the structure of the heavy chains. In blood serum mostly IgG antibodies are found. They are composed of two light and two heavy chains, which are bound through disulphide bridges (Fig. 2.3). In addition,we have to distinguish between constant or variable regions. The former shows the same composition and amino acid sequence inside a sub-group and is the characteristic of a certain mammalian, belonging to a major class. A human IgG is only to differentiate from the IgG from a mouse through the constant region. This difference is very important for the application of antibodies (see below). In the variable section, this specifically binds to an antigen. The order of the amino acids changes from antibody to antibody (specific region).

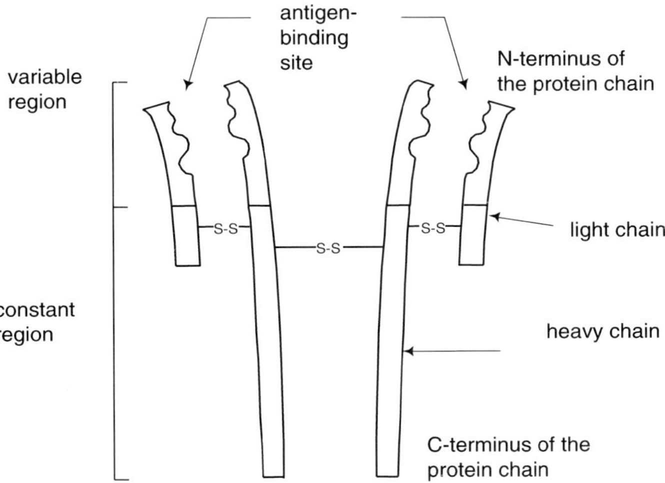

Fig. 2.3 Schematic structure of an antibody (IgG)

The basic technology for production of monoclonal antibodies by hybridoma cells developed by Köhler and Milstein is described briefly in the following (Milstein 1988). At first, the spleen cells from an immunised (i.e. cells that are building antibodies) mouse are fused with the help of polyethylenglycol with myelom cells from a mouse or a rat. The created hybrid cells are separated from the other cells with the help of a selective medium (HAT medium: a mixture of hypoxanthine, aminopterine, thymidine), where only the hybrid cells can survive. Then one should evaluate which kind of antibodies the hybrid cells produce by the isolation of each cell and its cloning. During the first cultivations, the hybrid cells lose part of their chromosomes and finally secret homogenous immunoglobulin. The clones that produce the desired antibody can be frozen for storage. Whenever the antibodies are needed, the hybrid cells are defrozen and proliferated in a suitable medium. The antibodies can be later isolated from the medium. In addition, it is also possible to inject the cells directly in animals, in which large antibodies producing tumours are built and they secrete the antibodies into the blood.

The characteristic of antibodies to selectively bind antigens is widely used by the biological and medical research. For example, the diagnosis and protein purification. Before the introduction of the hybridom technology, the antibodies were isolated from blood serum. As this contains a large variety of different antibodies, these are named "polyclonal" antibodies. The application of polyclonal antibodies is largely affected by the purity of the antibodies and also by the availability and breeding of animals.

A broad therapeutic application of monoclonal mostly murine-derived antibodies was hampered due to a number of side-effects. They are associated with undesirable immune response in humans. This problem could be overcome, to a certain extent, by new techniques to produce humanized or fully human antibodies (Fig. 2.4). Today, monoclonal antibodies or antibody fragments for a number of therapies including transplantation, cancer, infectious disease, cardiovascular disease and inflammation are already approved by the FDA for therapeutic use or they are in the process of getting approval. Examples were given by Dorn-Beinke et al. (2007). To some extent, these antibodies are produced in recombinant host cells (CHO, NS0, etc.) rather than in typical hybridom cells. Larger quantities of therapeutic antibodies are required as for diagnosis or as laboratory use. So, large-scale production capacity will increase during the next few years.

2.2.3   Culture Collections and Cell Banking

A large number of well-characterized cell lines are offered by cell culture collections for many applications. The largest and most well known international mammalian cell culture collections are the American Type Culture Collection (ATCC), the European Collection of Animal Cell Culture (ECACC) as well as the German Resource Centre for Biological Material (DSMZ) (links in references). The application of permanent cell lines for the production of vaccines faced strong opposition,

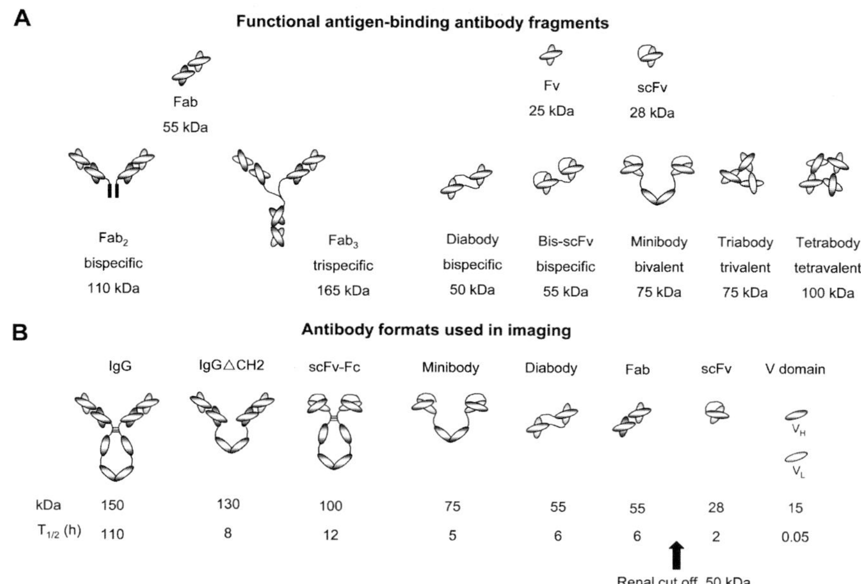

Fig. 2.4 Antibody engineering. Overview of different antibody formats and functional antibody fragments. Modified from Dorn-Beinke et al. (2007), with permission

mainly in the 1960s. This was because of the suspicion that these cell lines, which behave similarly to tumour cells, could infect patients with unknown cancerous substances. Fortunately, no such case has been found till today. In the meantime, more and more permanent cell lines have been applied for the production of therapeutic and diagnosis substances. However, the application of permanent cell lines and, recombinant cell lines, is strictly controlled by supervisory boards (as the FDA in the US, and the EMEA within the EU, links in references). During the entire production cycle of a pharmaceutical, not only the product has to be identical from charge to charge, but also the phenotypical characteristics of the cell line have to be maintained. To assure that, before the beginning of the production, a "master cell bank" has to be created, from which a "master working cell bank" is derived, providing the inoculation material for the production cultures.

2.3 Culture Media

Basically, a growth medium for mammalian cells has to supply all the necessary nutrients required for growth and product formation (Eagle 1959; Butler and Jenkins 1989). Moreover, it should have a certain buffer capacity to stabilize the pH (pH optimum 7.0–7.3) and should provide an appropriate osmolality (approx.

350 mOsm L^{-1}) to avoid a damage of the sensitive cell membrane (Zhanghi et al. 1999; Oh et al. 1995; Ryu and Lee 1997; Zhu et al. 2005; Zengotita et al. 2002). A fundamental component of all media is a salt solution, which provides the ions necessary for life and keeps the osmotic pressure within the desired range. This, contains one or more buffer systems (Na Phosphate buffer, HEPES and/or bicarbonate) for pH-regulation, and, in some cases, a pH indicator (phenol red). The biocarbonate–carbon dioxide buffer system acts similar to the main buffer system in blood *in vivo*:

$$CO_2 + H_2O \rightleftarrows HCO_3^- + H^+$$

As the pK_a-value of this buffering system is 6.3, a concentration of bicarbonate in the medium (approx. 24 mM) maintains an equilibrium with 5% CO_2 in the gas phase (corresponding to a partial pressure of 40 mmHg) (Butler 2004).

In most formulations, glucose is used as the main carbohydrate to provide the energy source as well as a precursor for biosynthesis. Alternative carbohydrates such as fructose can also be added. Amino acids are included as a precursor for protein synthesis. Whereas for most amino acids the concentration is approx. 0.1–0.2 mmol L^{-1}, glutamine is usually included at higher concentrations (2–4 mmol L^{-1}) in order to act as a precursor for the TCA cycle intermediates (Butler 2004). Moreover, the medium contains vitamins, mineral salts and trace elements.

A number of medium formulations have been developed, e.g. Eagle's minimal essential medium (MEM), Dulbecco's enriched (modified) Eagle's medium (DMEM), Ham's F12, and RPMI 1640 among others. Examples of medium compositions can be found in Table 2.4 and in the reference literature (Butler 2004; Freshney 1994; Fletscher 2005). Traditionally, these basic media were supplied with approx. 5–10% serum (e.g. foetal calf serum [FCS] or horse serum [HS]) in order to supply specific growth factors and to protect the cells against shear stress. The disadvantages of media,containing serum are (i) the indefinite composition of such media, (ii) the high serum costs, (iii) the difficulty of product purification, (iv) variations between the charges and (v) the risk of virus contamination (Griffiths 1987).

Serum can be partially substituted by the addition of transferin, insulin, ethanol amine, albumin or eventually fibronectin as adherence factor in a serum-free but still protein containing medium (Barnes and Sato 1980; Gambhir et al. 1999; Kallel et al. 2002). The next step to chemically defined, protein-free media was made possible by replacing these animal derived proteins by iron salts or iron complexes, IGF-1, chemically defined lipid concentrates, precursors or other simulating agents such as fatty acids, biotin, cholin, glycerin, ethanolamin, thiole, hormones and vitamins (Franék and Dolníková 1991a,b; Fassnacht et al. 1997). For chemically defined media, chemical structure of all compounds is known (Table 2.5). Next, addition of peptone or yeast extract can be helpful (Pham et al. 2003; Kim et al. 2004). However, the growth rate and productivity in a serum free medium can decrease (Fassnacht et al. 1998). In addition, the sensitivity to shear stress increases and additives with protecting characteristics, such as Pluronic F68 have to be added (Zhang et al. 1995; Al-Rubeai et al. 1992) (compare Sect. 4.1.3). Immobilization can improve the culture

Table 2.4 Basic formulation of a typical cell culture medium, 1:1 mixture of IMDM* and Ham's F12 (*Iscove's Modification of Dulbecco's Medium)

Salts (mg L^{-1})	Vitamins (mg L^{-1})	Amino acids (mg L^{-1})	Other compounds (mg L^{-1})
CaCl$_2$ (116.60)	biotine (0.0035)	L-alanine (4.45)	D-glucose (3151.00)
CuSO$_4$·5H$_2$O (0.0013)	D-Ca pantothenate (2.24)	L-asparagine·H$_2$O (7.50)	Hypoxantine (sodium salt) (2.39)
Fe(NO$_3$)·9H$_2$O (0.05)	Choline chloride (8.98)	L-arginine·HCl (147.50)	Linolic acid (0.042)
FeSO$_4$·7H$_2$O (0.417)	Folic acid (2.65)	L-aspartic acid (6.65)	Lipolic acid (0.105)
KCl (311.80)	i-inositol (12.60)	L-cysteine HCl·H$_2$O (17.56)	Phenol red (8.10)
MgSO$_4$ 7H$_2$O (100.00)	nicotineamide (2.02)	L-cysteine 2HCl (31.29)	Putrescine 2HCl (0.081)
MgCl$_2$·6H$_2$O (61.00)	Pyridoxal HCl (2.031)	L-glutamic acid (7.35)	Thymidine (0.365)
NaCl (6996.00)	riboflavin (0.219)	L-glutamine (365.00)	Na-pyruvate (55.00)
NaHCO$_3$ (1200.00)	Thiamine HCl (2.17)	Glycine (18.75)	
NaH$_2$PO4·H$_2$O (62.50)	Vitamine B12 (0.68)	L-histidine HCl·H$_2$O (31.48)	
NaH$_2$PO$_4$·7H$_2$O (134.00)		L-isoleucine (54.47)	
ZnSO$_4$·7H$_2$O (0.432)		L-lysine HCl (91.25)	
		L-methionine (17.24)	
		L-phenylalanine (35.48)	
		L-proline (15.25)	
		L-serine (26.25)	
		L-threonine (53.45)	
		L-tryptophane (9.02)	
		L-tyrosine (di-sodium salt) (55.79)	
		L-valine (25.85)	

stability of non-anchorage-dependent mammalian cells grown in serum-free media (Lee and Palsson 1990; Lüdemann et al. 1996). Antibiotics (penicillin, streptomycin) are often added on laboratory scale to prevent contamination. On production scale, antibiotics are avoided and are used only as selection marker. Furthermore, on production scale phenolred is avoided to simplify downstream processing.

Different cell lines require different compositions. Adaptation of a cell line to grow without serum is quite time consuming. Not all cell lines are adapted to serum-free or protein-free media (Sinacore et al. 2000; Link et al. 2004). For cultivation of primary cells, for basic cell culture research or for vaccine production, which are complex processes serum-containing media are common. In industrial production with established optimized cell lines, serum-free, bovine-free and chemically defined media are state of the art technology. Well adapted medium

Table 2.5 Comparison of medium with and without serum

Medium with serum	CD Medium (without serum)
Not chemically defined	Chemically defined
Varying composition of every medium batch	Identic composition of every batch
Potential source of contaminations (viruses, mycoplasms, prions)	Lower contamination risks
More complicated downstreaming	Less complicated downstream
	Often higher biological activity of the product
Lower biological activity of the desired product	Easier validation and product registration
More complicated validation and product registration	
Albumins protect cells against shear stress	Viscosity of the medium is increased by use of Pluronic F68 (cell protection agent)
	Pluronic can hinder gas input

formulations allow cell densities of approx. 5×10^6 to $> 10^7$ cells mL^{-1} and product titer of 3–5 g L^{-1} (Morrow 2007; Voedisch et al. 2005; Wurm 2004).

Sterilisation of medium used during cell culture assures you that media and reagents are not a source of contaminants. Heat sterilization (autoclaving at 121 °C) is not the option for many media, as autoclaving may destroy certain components and biomolecules needed for cell growth. Filtration, therefore, is the preferred method for sterilizing cell culture media and additives. This is done usually by using 0.2 μm filters. Serum can be sterilised by heat inactivation, but is usually supplied sterilized. HTST (high temperature short time) processing is also suitable (Grob et al. 1998).

2.4 Characteristics of Cell Growth and Metabolism

2.4.1 Short Introduction to Cell Metabolism

For optimisation of cell culture processes a basic understanding of the metabolic pathways in the cells is required. This information can be subsequently used to improve the process itself and/or the medium and to improve the cells in appropriate ways. The metabolic pathway of mammalian cells is very complex and flexible. Moreover, the metabolic pattern of mammalian cells alters substantially and becomes highly deregulated *in vitro* (Gódia and Cairó 2006). This is characterized by a high and inefficient consumption of glucose and glutamine as the main sources for carbon, nitrogen, and energy, leading to generation of metabolic waste products such as lactate, ammonia and some amino acids (alanine, glutamic acid, etc.). The main metabolic pathways comprise of:

- Glycolysis
- Pentose phosphate pathway
- Tricarboxylic (TCA) cyle
- Oxidative phosphorylation
- Glutaminolysis
- Metabolism of other amino acids (Gódia and Cairó 2006)

Here the discussion should give an idea of the complexity of cell metabolism only as far as it is necessary to understand the cell culture processes and to formulate kinetic models. For a more detailed description refer to the reference literature (Otzturk 2006; Wagner 1997; Häggström 2000; Rensing and Cornelius 1988). A simplified overview of the cell metabolism is shown in Fig. 2.5.

The complexity of mammalian cell metabolism, and especially the highly deregulated pattern leads to cell specific substrate uptake rates much higher than strictly necessary. It limits the performance of mammalian cells in culture. Different strategies are suggested in references (reviewed by Gódia and Cairó 2006) to reduce the metabolic deregulation of the cells in culture and to obtain more physiologically and metabolically balanced patterns: metabolic engineering, medium redefinition and balance, and optimized bioprocess engineering design based on metabolic requirements. The final aim is to generate more efficient cell culture processes.

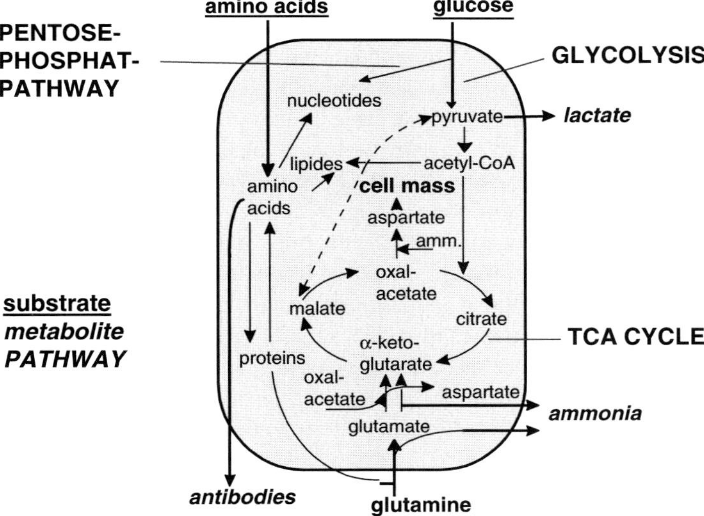

Fig. 2.5 Metabolic pathways in hybridom cells, from Miller et al. (1988), modified

2.4.2 Glucose, Glutamine and Amino Acids as Carbon and Energy Source

As mentioned before, the main substrates for mammalian cells are glucose and glutamine (Duval et al. 1991). Ammonia and lactate, the two main metabolites, can act as an inhibitor for high density cultures, and have to be minimized in the medium with appropriate methods. The metabolic conversion of these compounds follows a combination of different mechanisms and is summarized by Gódia and Cairó (2006): "transport into the cell, gycolysis and glutaminolysis pathways, the TCA cycle, phosphorylative oxidation, generation of cellular energy, regeneration of reducing power, among others".

Glucose is contained in concentrations of 10–25 mmol L^{-1} in the culture medium. As the cytoplasmic membrane is impermeable to polar molecules like glucose, the uptake of glucose is achieved by means of transport molecules located in the plasmatic membrane and is mainly driven by the concentration gradient across the cell membrane (Gódia and Cairó 2006). Glucose mainly provides the energy metabolism for the cells and can be metabolised via glycolysis to pyruvate. This results in a number of intermediates for biosynthesis, 2 mol ATP (adenosine triphosphate) per mol of glucose, and two NADH molecules. Lactate is produced as an inhibitory metabolite in a ratio of about 1.1–1.7 mol lactate/mol glucose. In an oxygen limiting environment (anaerobic metabolism) 2 moles of lactate are produced per mol of glucose. The pyruvate resulting from the glucolysis is oxidatively decarbonised to acetyl-CoA and transformed through the TCA-cycle and through the respiratory chain into water and CO_2. This pathway results in 36 mol ATP per mol glucose. Another important pathway is represented by the pentose-phosphate pathway for building nucleotides (Batt and Kompala 1989; Butler 1987; Reitzer et al. 1979; Zielke et al. 1978). The percentage of consumed glucose that enters the pentose-phosphate pathway can be quite low, 4–8%, but is within the range of 15–27% (compare Gódia and Cairó 2006).

Glutamine (approx. 1–5 mmol L^{-1} in the culture medium) is an important precursor for synthesis of purines, pyrimidines, amino sugars and asparagines (Butler 2004). It is incorporated into the cell by means of different amino acid transport systems (Gódia and Cairó 2006), and it can be converted to glutamate or aspartate or it is metabolised in many ways in the TCA cycle (Reitzer et al. 1979; Zielke et al. 1978). The complete oxidation of glutamine to CO_2 generates 21 moles ATP per mol glutamine; the incomplete oxidation to aspartate generates 12 mol ATP per mol glutamine; the incomplete oxidation to lactate generates 6 mol ATP per mol glutamine. The main metabolite from the transformation of glutamine is ammonia (about 0.7 mol ammonia per mol glutamine).

Amino acids are present in the medium in concentrations ranging from 0.1 to 0.2 mmol L^{-1} and are mainly used for the protein synthesis. They can be grouped under those, that are consumed (arginine, aspartate, cysteine, histidine, isoleucine, leucine, lysine, methionine, phenyl alanine, proline, serine, threonine, tryptophane, tyrosine, valine) and those that are partially produced (alanine, glycine, glutamic acid).

The actual values for the cell specific substrate uptake rates can vary significantly in the culture medium depending on the type of cell, the state of the culture and the substrate concentration. Similar to most microorganisms, cell specific uptake rates for substrates and production rates for metabolites decrease under substrate limitation. The kinetics of these effects is discussed more in detail in Sect. 2.5.2. In addition, a number of phenomena specific for mammalian cells have been observed and are,summarized in the following paragraphs.

For hybridom cells, several authors were able to show a decrease of cell specific substrate uptake and metabolite production rates during batch-culture, even at constant growth rate (Seamans and Hu 1990; Shirai et al. 1992; Pörtner et al. 1994). An example is given in Fig. 2.6, where in repeated-batch experiments the cell specific glucose uptake rate decreased with each cycle, but the growth rate remained constant (Pörtner et al. 1994). Autocrine compounds were assumed to be responsible for this effect (Shirai et al. 1992; Lee et al. 1992). Another reason might be that in cell culture technology uptake and production rates are usually related to the number of cells. This implies a constant cell volume. But, as can be seen from Fig. 2.7,the cell volume decreases during batch culture,almost similar to the cell specific glucose uptake rate (Pörtner 1998; Frame and Hu 1990). The ratio between cell specific uptake rate and cell volume remained constant at $2.7{\times}10^{-13}$ mmol μm^{-3} h^{-1}. The dry weight of a hybridom cell is approx. $1.95{\times}10^{-10}$ mg μm^{-3} (Frame and Hu 1990). Therefore,the mass specific glucose uptake rate during exponential growth phase can be estimated to be approx. 1.4 mmol g^{-1} cells^{-1} h^{-1} during exponential growth (constant growth rate), at least, for the example shown in Figs. 2.6 and 2.7 Similar to glucose, glutamine is also consumed by most mammalian cells at high rates, especially at high glutamine concentrations in the medium (Pörtner et al. 1994; Miller et al. 1988; Glacken et al. 1986; Miller et al. 1989b; Butler and Spier 1984; Jeong and Wang 1995; Linz et al. 1997; Vriezen et al. 1997). Moreover, the cells can react quite sensitively to sudden changes in the culture environment,

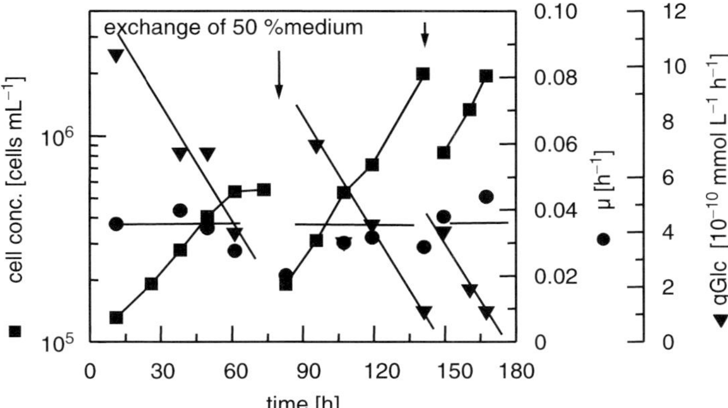

Fig. 2.6 "Repeated-Batch"-culture of hybridom cell line IV F 19.23 Kultur in a 1 L suspension reactor (working volume 600 mL, blade impeller). Concentration of viable cells, cell specific glucose uptake rate q$_{Glc}$ and cell specific growth rate μ vs. culture time (data from Pörtner et al. 1994)

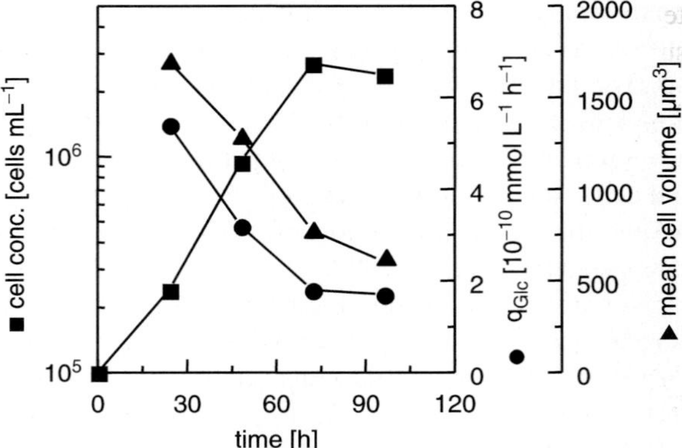

Fig. 2.7 Batch-culture (T-flask) of hybridom cell line IV F 19.23. Concentration of viable cells, cell specific glucose uptake rate q_{Glc} and mean cell volume vs. culture time (data from Pörtner 1998)

e.g. to step changes of substrate concentration provided in continuous chemostat culture (Miller et al. 1989a,b). Data for hybridom cells is given in Sect. 2.5.2.

Similar effects have been observed for Chinese hamster ovary (CHO) and baby hamster kidney (BHK) cells. For CHO cells specific uptake rates up to 3×10^{-10} mmol cells^{-1} h^{-1} were observed in batch experiments and at high glucose concentrations, whereas under glucose limiting conditions in chemostat culture specific uptake rates were in the range of $1.18–1.23 \times 10^{-10}$ mmol cells^{-1} h^{-1} (Hayter et al. 1991). BHK cells showed high specific uptake rates up to 1.08×10^{-10} mmol cells^{-1} h^{-1} in case of high glucose, whereas low values of 0.13×10^{-10} mmol cells^{-1} h^{-1} were found under glucose limitation (Cruz et al. 1999). The cell specific glutamine uptake rate decreased in continuous chemostat culture from 0.33 to 0.16×10^{-10} mmol cells^{-1} h^{-1} when the feed concentration was reduced from 0.52 to 0.14 mmol L^{-1}.

The amount of energy available to cells, produced from glucose or glutamine, varies significantly between cell types and between different growth conditions. The contribution of glutamine is in the range of 40–70% (Reitzer et al. 1979; Zielke et al. 1978). The uptake rate of glucose and glutamine as well as the yield coefficient of lactate and ammonium is also determined by concentration of both compounds in the medium. An increase in the glucose concentration leads to an increased lactate production, a decreased conversion via pentose-phosphate pathway and to lower glutamine and oxygen uptake rates. High glutamine concentrations induce a high glutamine uptake rate, which in turn causes a higher glutamine oxidation rate and a higher production rate of the intermediate metabolites from the TCA-cycle. At the same time, it seems that a higher glutamine concentration facilitates the metabolism of glucose through the pentose-phosphate pathway but makes more difficult the oxidation of glucose in the TCA-cycle (Siano and Muthrasan 1991). When glucose levels are limiting and glutamine is present at non-limiting concentrations, the cell metabolism adapts itself by increasing the glutamine consumption rate, leading to an increased ammonia production and oxygen consumption (Gódia and Cairó 2006).

When glutamine is the limiting substrate, glutamine and glucose consumption seems to follow a parallel profile. Both increase simultaneously at higher glutamine concentrations and decrease at lower glutamine concentrations (Miller et al. 1989b; Vriezen et al. 1997; Linz et al. 1997; Pörtner 1998).

Gódia and Cairó (2006) concluded from the various studies conducted on the complementarily consumption of glucose and glutamine and its dynamic nature, that cell metabolism will adapt to changing conditions,. This will occur also as a function of existing culture conditions at the time when a given perturbation or change takes place. In this context, the observation of multiplicity of steady states made by several authors in continuous culture is very interesting (Altamirano et al. 2001; Europa et al. 2000; Follstad et al. 1999).

2.4.3 *The Effects of Lactate and Ammonia*

The influence of lactate concentration on mammalian cell growth has hardly been examined, as inhibition occurs at concentrations which are not usually reached during growth. In systems with pH control, lactate concentration has an inhibitory effect at concentrations of $40\,\mathrm{mmol\ L^{-1}}$ or higher. For hybridom cells Glacken et al. (1989) found for $70\,\mathrm{mmol\ L^{-1}}$ slight inhibition of cell growth, for $40\,\mathrm{mmol\ L^{-1}}$ none at all. Omasa et al. (1992) observed a decrease in specific growth rate, or increase in specific death rate at lactate concentrations of $56\,\mathrm{mmol\ L^{-1}}$. At $78\,\mathrm{mmol\ L^{-1}}$ the specific growth rate still reached 50% of the original value. The observed effects are primarily due to the increased medium osmolality, as the direct metabolic effects are observed later (Omasa et al. 1992; Ozturk et al. 1992). In conclusion, lactate might inhibit cell growth in systems like fed-batch, where due to extension of the culture period, the cumulative amount of consumed glucose and subsequently the amount of produced lactate increases. It is interesting to note, that some cells (e.g. CHO-cells) are capable of consuming lactate due to metabolic shift in the later stage of a culture.

The accumulation of ammonia in the course of the cultivation leads to growth inhibition even when a process is kept under otherwise well controlled physiological conditions (Ryll et al. 1994). At high ammonia concentrations, the quality of product can significantly decrease (Ryll et al. 1994; Thorens and Vassalli 1986). Ammonia stems not only from the glutamine metabolism (Miller et al. 1988; Glacken et al. 1986), but also from the chemical decomposition of glutamine to pyrrolidon-carboxylic acid and ammonia (Glacken et al. 1986; Ozturk and Palsson 1990; Tritsch and Moore 1962). A summary of possible mechanism for the inhibitory effect of ammonia can be found by Gódia and Cairó (2006).

The effect of ammonia on the growth of hybridom cells has been examined in both, batch and continuous cultures. Static batch cultures were carried out with varying initial ammonia-concentrations in the medium. When results of different authors are compared (Glacken et al. 1986; Lüdemann et al. 1994; Hassel et al. 1991; Reuveny et al. 1986; Doyle and Butler 1990), we can find a corresponding reduction of the specific growth rate to almost 50% of the maximum specific growth rate at overall initial ammonia concentrations of 5–6 mmol L^{-1}. The cultures were usually started at high pH-values (7.4–7.5) to remain in the physiological range (7.0) at the end of the batch.

In continuous chemostat and perfusion cultures performed by Lüdemann et al. (1994) and Matsumura et al. (1990) the total ammonia concentration could be increased up to approx. 18 mmol L^{-1} until it causes a 50%-reduction of growth. In both cultures the pH was controlled at 7.1–7.2. Nayve et al. (1991) found for their hybridom cell-line at a pH of 7.2 a critical ammonia concentration as low as 5 mmol L^{-1}. The authors described this cell line as extremely sensitive to ammonia but it can also be concluded that the decrease in the specific growth rate observed was due to glutamine limitation rather than ammonia inhibition.

McQueen and Bailey (1990) postulated an inhibition mechanism based on the assumption that ammonia generates an acidification of the cytoplasm and thus a decrease in intracellular pH. It seemed that the pH has a significant influence on the effect of ammonia. Therefore Lüdemann et al. (1994) ascribed inhibition by ammonia

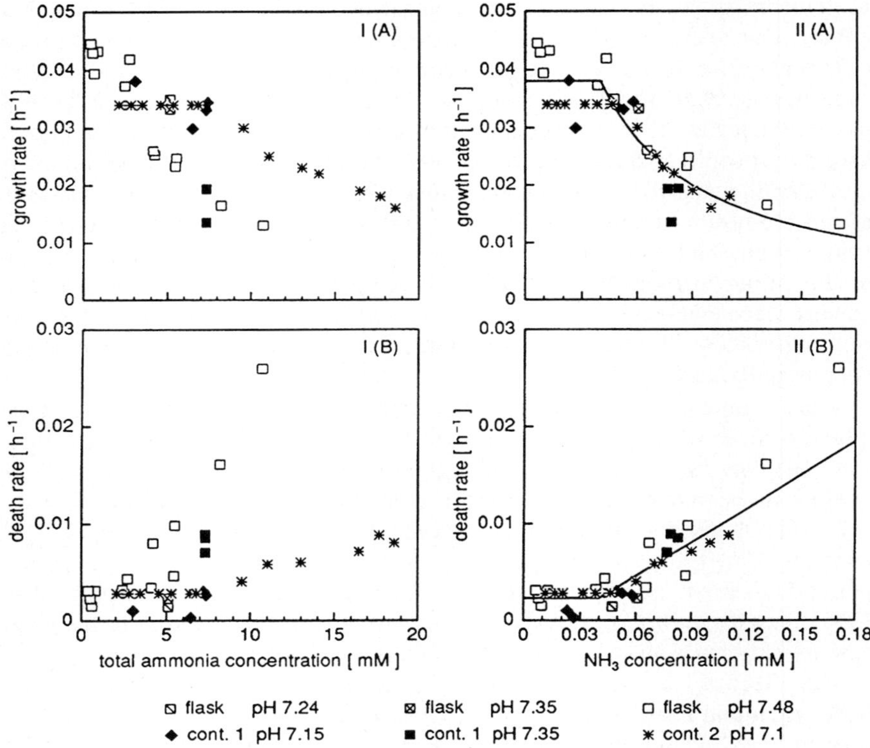

Fig. 2.8 Effect of total ammonia concentration and concentration of NH_3 on cell specific growth rate (A) and cell specific death rate (Lüdemann et al. 1994)

to the undissociated NH_3 where concentration changes considerably with varying pH (Fig. 2.8). According to their equations, the specific growth rate was reduced by 50% at a NH_3-concentration of 0.095 mM L^{-1}, corresponding to a total ammonia concentration of 5.6 mM L^{-1} at pH 7.5. This value coincided very well with those reported by other authors from batch cultures (see above).

Even if usually lactate and ammonia are regarded as the compounds which have an inhibitory effect, there is experimental evidence that other inhibitory substances may also accumulate as a result of cellular metabolic activity (Gódia and Cairó 2006; Gawlitzek et al. 1998).

2.4.4 Oxygen Uptake and Carbon Dioxide Production

The aspects mentioned before were mainly derived under sufficient oxygen supply. As oxygen is an important nutrient, the effect of oxygen level on cell metabolism has been studied quite thoroughly, from very low concentrations to hyperoxic conditions [reviewed by Gódia and Cairó (2006)]. Oxygen is the final electron acceptor in the mitochondrial respiration chain and is directly linked to the generation of energy. With respect to process control, especially the very low solubility of oxygen in the culture medium compared to other nutrients has to be considered. In equilibrium with air the oxygen concentration is just about 0.2 mmol L^{-1} and therefore oxygen supply to the cells is often the limiting factor, especially at high cell densities. The specific effects of oxygen concentration on various cell lines have been discussed extensively by Godia and Cairó (2006). Here some data important for process design are summarized. Table 2.6 gives an overview on cell specific oxygen uptake rates under non-limiting conditions.

According to Henzler and Kauling (1993) optimum conditions for growth, cell concentration, and viability as well as product formation are achieved with 5–80% of air saturation. Below 5% of air saturation significant changes in the metabolic activity of the cells have to be considered (Miller et al. 1987). For hybridom cells a decrease of cell specific glucose and glutamine uptake rates between 10 and 5%

Table 2.6 Range of cell specific oxygen uptake rates q_{O_2} of different cell lines under non-limiting conditions (adapted from (Henzler and Kauling 1993; Zeng and Bi 2006))

Cell line	q_{O_2} [10^{-10} mmol cell^{-1} h^{-1}]
FS-4	0.5
HeLa	5
Skin fibroblasts	0.6
BHK 21	1–2
MRC-5	1.5
Human hybridom	0.2
Mouse hybridom	1–5
Melanoma	0.7–1
CHO	2–8

of air saturation were observed. Below 0.5% the glucose uptake rate increased and
the yield of lactate from glucose approached a value of 2 mol lactate per mol glucose at 0.1% of air saturation.

For bioreactor systems such as stirred tanks or bubble columns operated under
controlled oxygen level (usually between 20 and 50% of air saturation), the impact of
varying oxygen concentration can be obviously ignored (Henzler and Kauling 1993).
In contrast, for bioreactor systems with immobilised cells (fixed bed, fluidized bed,
hollow fibre reactor), where mass transfer limitations are expected, metabolic shifts
due to oxygen limitations have to be taken into account (compare Sect. 4.3).

Carbon dioxide plays an important role in cell culture, as it is one of the end
products of mammalian cell metabolism, and is widely used in pH regulation (compare Sect. 2.3). Attention is focused on the impact of dissolved CO_2 concentration
on cell specific production rate, product quality, growth rate, intracellular pH and
apoptosis (Zanghi et al. 1999; Kimura and Miller 1997). Detrimental effects of elevated CO_2-levels are especially relevant in large scale reactors, operating at high
cell densities (Mostafa and Gu 2003; Matanguihan et al. 2001). Due to this high cell
density, metabolically produced CO_2 will be built up and the CO_2 partial pressure
will increase beyond optimal levels. Whereas pCO_2 levels of approx. 40–50 mmHg
are considered as optimal (DeZengotita et al. 1998), pCO_2-levels up to 170 mmHg
are observed in stirred tanks at high cell densities (Mostafa and Gu 2003). The elevated concentrations are partly due to low mass transfer coefficients in stirred cell
culture reactors (compare Sect. 4.2) (Frahm et al. 2002).

2.5 Kinetic Modelling of Cell Growth and Metabolism

2.5.1 Introduction to Kinetic Modelling for Mammalian Cells

Kinetic models are intended to describe quantitative cell growth and metabolic
activity for better understanding of cell physiology and for optimization and control
of animal cell cultures (Zeng and Bi 2006; Doyle and Griffiths 1998; Adams et al.
2007). Simple models consisting basically of a set of mathematical relationships
among different cellular rates (e.g. death and growth rate, nutrient uptake and
metabolite production rates) and medium compounds, can be used to compute the
time course of culture variables (cell density, concentration of nutrients, metabolites and products, pH, pCO_2, pO_2, etc.) in batch, fed-batch or continuous culture or
to identify kinetic effects like depletion of nutrients (glucose, glutamine and other
amino acids, oxygen) or accumulation of metabolites (e.g. ammonia and lactate) on
cellular growth, death and productivity. More elaborate models describing intracellular metabolic pathways may help to identify limiting metabolic steps during
growth and product synthesis or indicate changes in intracellular metabolic pathways depending on the culture conditions.

As regards the optimization and control of cell culture processes, kinetic models
are required for the design and layout of cell culture reactors, calculation of optimal

process parameters, design of control strategies for optimal and safe operation or observation of the state of the reactor system based on mathematical models (software sensors).

Compared to chemical and mechanical systems, intricate biological systems like mammalian cells have a high complexity. They are capable of compensating for minor changes in environment, for instance, by adaptation of metabolism. This may be an advantage because biological processes do not require a very costly process control, but this is surely a disadvantage, when a growth and production model is to be set up. In general, one can choose between purely descriptive empirical models (unstructured models) which disregard intracellular processes, and the more detailed structured models which are based on a thorough comprehension of cell metabolism (Tsuchiya et al. 1966). The latter is, by its nature, applicable to a large range of different growth conditions, while the former can only be employed for the very process conditions and data range it was derived from (McMeekin et al. 1993). A large number of unstructured and structured kinetic models for the cell growth and cell metabolism have been suggested in the reference literature, especially for hybridom cells and are reviewed extensively (Zeng and Bi 2006; Tziampazis and Sambanis 1994; Pörtner and Schäfer 1996).

In the following paragraphs, aspects relevant for kinetic analysis and setting up an unstructured model based on data provided by Pörtner and coworkers (Pörtner 1998; Pörtner and Schäfer 1996) and suggestions made by Doyle and Griffiths (1998) as well as Adams et al. (2007) for set-up of a kinetic model is discussed. After this, concepts for structured models are introduced. The idea is to give a basic understanding of the different types of modelling rather than to review the available literature.

2.5.2 *Set-Up of an Unstructured Model*

Steps involved in setting up a kinetic model for mammalian cell culture were identified by Doyle and Griffiths (1998):

- "A kinetic analysis of the experimental results with the formation of hypotheses on the nature of the rate-limiting steps;
- The choice of rate expressions describing the influence of these phenomena on the cellular process;
- Evaluation of parameter values;
- And validation of the model with different experimental results."

Following the simplified flow scheme of growth and metabolism of mammalian cell cultures shown in Fig. 2.9, the intended model for a mammalian cell line producing a recombinant protein or a monoclonal antibody should provide equations for:

- Growth rate μ and death rate k_d (substrate limitation, metabolite inhibition)
- Substrate uptake rates q_s (glucose, glutamine, amino acids, oxygen)

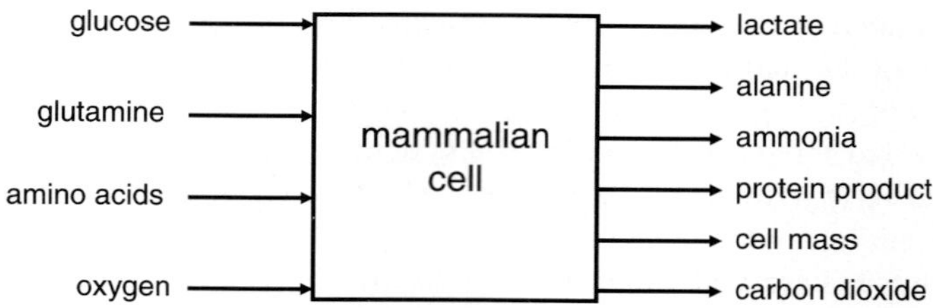

Fig. 2.9 Simplified diagram of material flows of growth and metabolism of mammalian cell cultures

- Metabolite production rate q_p (ammonia, lactate among others)
- Production rate for recombinant protein or monoclonal antibodies q_{MAb}

Even if cell metabolism and the medium composition are very complex, the number of variables considered for modelling is often reduced due to the following variables: cell density (viable, dead), glucose and glutamine as nutrients, ammonia and lactate as limiting metabolites, as well as the desired product (e.g. monoclonal antibodies). At first, these variables can be determined quantitatively using well established analytical methods. Then the main effects on cell growth and cell death can be attributed in most cases to these variables. Nevertheless, this simplified approach might lead to misinterpretation of physiological effects and is discussed at later stage.

Kinetic data can be obtained from experiments performed in different cultivation modes and bioreactor systems. Batch experiments are quite common, as they are easy to perform. Cultures in T- or shake-flask can provide a first intention of the kinetics of a cell line. But often certain culture conditions such as pH or pO_2 are not controlled, these data might lead to wrong conclusions. Batch-cultures performed in bioreactors equipped with appropriate pH- and pO_2-control will result in getting more reliable data. Nevertheless, batch cultures have some limitations with respect to kinetic analysis. As all medium concentrations (nutrients and metabolites) change simultaneously, the precise influence of a single medium component on cell kinetics is difficult to determine. Moreover, the critical phase of the culture, mainly during the decline phase at the end of exponential growth, is usually very short and sampling is required at very short intervals. Chemostat cultures provide more reliable data, even if they require more sophisticated equipment (pumps for feed and harvest, control units) and long-term operation for several weeks or sometimes even for months. But as during "steady state" nutrients and metabolite concentrations can be maintained constant for several days, averaged data allow for a more precise analysis of the influence of medium composition on cellular activity. Examples for different culture modes (batch, fed-batch, chemostat, perfusion, dialysis) as well as the differential equations required to calculate the metabolic parameters are discussed in detail in Sect. 4.4. Finally, when investigating cell kinetics, it is very important not only to control the physicochemical parameters (temperature, pH, pO_2, pCO_2, osmolality among others), but also to define precisely the state of the inoculum, as the "history" of the cells may have a significant impact on the cell metabolism.

Prior to the construction of the kinetic model, one has to perform a detailed analysis of the experimental data in order to identify the main rate-limiting effects. The following discussion is mainly based on data provided and summarized by Pörtner (1998) as well as Pörtner and Schäfer (1996). No attention is be paid to the individual reactor system and culture mode used, as it is the author's opinion that irrespective of the cultivation mode, the growth of a cell line follows the same kinetics (see below), and data gained from batch experiments can therefore be compared with those from continuous culture. The medium composition is considered but the models are not distinguished accordingly, since they ought to hold a certain general applicability. Care was taken to ensure that all data was obtained within the same ranges of pH (7.0–7.2) and at non-limiting dissolved oxygen concentrations, because an unfavourable pH or a very low dissolved oxygen concentration (below 0.5% of air saturation) is known to lead to changes in cell metabolism (see above).

2.5.2.1 Cell Specific Growth and Death Rate

Formal equations for cell specific growth rate, defined as the number of new cells produced per unit of living cells present in the culture medium per unit time, can be derived from a Monod-type equation

$$\mu = \frac{\mu_{max} c_s}{k_s + c_s},\tag{2.1}$$

where μ_{max} is the maximum specific growth rate, c_S is the concentration of the controlling substrate such as glucose, and k_S is the concentration of the controlling substrate where the specific rate is half of the maximum rate (Adams et al. 2007) . If $c_S \gg k_S$ then $\mu \to \mu_{max}$. In this case, the compound S is regarded as the controlling substrate concentration. Thus, the other substrates are abundantly supplied and their concentrations do not affect the cell growth rate. The specific growth rate can also be described with multiple substrate limitation and multiple inhibitors, also as controlling factors:

$$\mu = \mu_{max} \Pi \left(\frac{c_{s,i}}{k_{s,i} + c_{s,i}} \right) \Pi \left(\frac{k_{p,j}}{k_{p,j} + c_{p,j}} \right).\tag{2.2}$$

With the maximal growth rate μ_{max}, usually the growth rate during exponential growth, substrate concentrations c_{Si} of certain nutrients (e.g. glucose or glutamine) and metabolite concentrations c_{Pj} (e.g. lactate or ammonia) and the corresponding kinetic constants $k_{s,i}$ for substrate limitation and $k_{p,j}$ for metabolite inhibition.

The cell specific death rate, k_d, is defined as the number of dying cells per unit of living cells present in the culture medium per unit time (Doyle and Griffiths 1998). Modelling of the cell death rate is usually a problem. Mechanisms leading to cell death are quite complex and two possible ways are apoptosis or necrosis

(Cotter 1994; Franék 1995; Al-Rubeai and Singh 1998). The latter is characterized by sudden swelling of the cells and subsequent disintegration caused mainly by external influences such as shear stress, disruption by bubbles or osmotic pressure. Apoptosis or the so-called "programmed cell death" is caused actively by the cell itself which means that the cell death is under genetic control. Mercille and Massie (1994) found that substrate limitation leads mainly to apoptosis whilst metabolite inhibition mainly causes necrosis. Therefore, a number of different equations were suggested in the literature (Pörtner and Schäfer 1996). While setting up a kinetic model, the following equations will provide a useful starting point for kinetic analysis of cell death:

$$k_d = k_{d,\mathrm{min}} + k_{d,\mathrm{max}} \Pi\left(\frac{k_{d,s,i}}{k_{d,s,i} + c_{s,i}} \right) \Pi\left(\frac{c_{P,j}}{k_{d,p,j} + c_{P,j}} \right), \tag{2.3}$$

$$k_d = d_0 e^{d_1/\mu}, \tag{2.4}$$

with a minimal death rate $k_{d,\mathrm{min}}$, a maximal death rate $k_{d,\mathrm{max}}$ and the kinetic constants $k_{d,s,i}$ for substrate limitation and $k_{d,p,j}$ for metabolite inhibition. The parameters d_0 and d_1 are fit-parameters. If these equations do not turn out to be appropriate for a specific cell line then refer to the reference literature (Zeng and Bi 2006; Doyle and Griffiths 1998; Pörtner and Schäfer 1996).

To give a better understanding of cell growth kinetic, data for a hybridom cell line on cell specific growth rate and cell specific death rate are provided in Fig. 2.10. In this case, glutamine was regarded as the limiting substrate, as under all culture conditions, glucose and all amino acids were still available in non-limiting concentrations. Moreover, ammonia and lactate did not reach inhibiting concentrations, as discussed above. Even if data from batch or fed-batch cultures scatter considerably compared to those from continuous chemostat cultures, the data obtained for different culture modes can be described by a single kinetic expression. In this case, for the cell specific growth rate a typical monod-equation can be applied,. For the cell specific death rate the equation similar to (2.3) can be applied.

Even if most kinetic growth models are based on glucose or glutamine as the main carbon source, they need not be growth limiting. Pörtner and Schäfer (1996) analysed data and kinetic models from various authors with respect to maximum growth rates μ_{max}, K_s-values found for glucose and/or glutamine, and the medium used (Table 2.7). While comparing the data one has to take into account that some authors (Glacken et al. 1989; Pörtner et al. 1996; Dalili et al. 1990; Frame and Hu 1991b) formulated the specific growth rate as a function of only one substrate. But others set up a complex model comprising not only of substrate limitation by glucose and/or glutamine but also inhibition by the metabolites lactate and ammonia. The respective constants were then gained by parameter identification in the overall model (Miller et al. 1988; Bree et al. 1988; de Tremblay et al. 1992; de Tremblay et al. 1993).

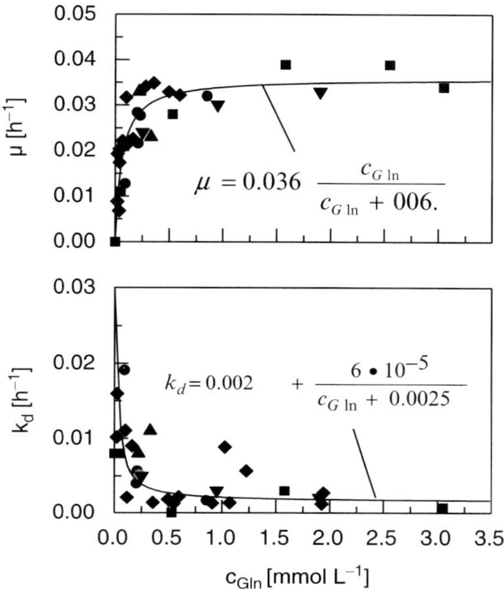

$$\mu = 0.036 \; \frac{c_{G\,ln}}{c_{G\,ln} + 006.}$$

$$k_d = 0.002 \; + \; \frac{6 \cdot 10^{-5}}{c_{G\,ln} + 0.0025}$$

Fig. 2.10 Cell specific growth μ and death rate k_d vs. glutamine concentration c_{Gln} for the hybridom cell line IV F 19.23 (Prof. Kasche, TUHH) (■ flask batch, ● chemostat, ▲ dialysis cont., ▼ dialysis batch, ◆ fed-batch)

As can be seen from Table 2.7, the k_s-values for glutamine or glucose are low when the feed concentration of glucose, glutamine or concentration of both was comparatively low. For models which show a pure glutamine dependence the conclusion drawn was that the respective k_s-values are within the range of 0.06–0.15 mmol L⁻¹. Differences seen for the maximum specific growth rates can be mainly related to the different medium compositions and fractions of serum. The maximum specific growth rates of all authors listed in Table 2.7 are also within the same range for media with serum concentrations between 10 and 20%, while disregarding the value determined by Bree et al. (1988). Pörtner et al. (1996) gained kinetic parameters from a different medium formulation (IMDM/Ham's F12) with a comparably small fraction of serum (3%), and Kurokawa et al. (1993) used serum-free medium resulting in relatively low maximum specific growth rates. Bree et al. (1988) fitted their parameters on the basis of only a single batch run. The values obtained must be viewed sceptically as with means of parameter identification they found for the maximum specific growth rate a value of 0.125 h⁻¹. However, according to the data published, the specific growth rate reached only 0.03 h⁻¹ during the exponential growth phase.

Models from different authors describing the cell specific death rate were analysed by Pörtner and Schäfer (1996) by transforming these models into a relationship $k_d = f(\mu)$, as shown in Fig. 2.11. There is obviously the general trend that the specific death rate varies between 0.001 and 0.006 h⁻¹ at high specific growth rates and increases

Table 2.7 Maximum specific growth rate μ_{max}, Monod-constants k_{Glc} and k_{Gln} for glucose and glutamine, respectively, growth medium composition and initial concentrations of glucose (Glc) and glutamine (Gln) for various authors and culture mode; DMEM: -L: low glucose (adapted from Pörtner and Schäfer 1996)

Reference	μ_{max} [h^{-1}]	k_{Glc} [mmol L^{-1}]	k_{Gln} [mmol L^{-1}]	Medium	Glc [mmol L^{-1}]	Gln [mmol L^{-1}]	Mode
De Tremblay et al. (1992)	0.045	1	0.3	DMEM, 10% FCS	25	4	
Frame and Hu (1991a)	0.063	0.034	–	DMEM-L, 10% FCS	5.56	4	Chemostat
Bree et al. (1988)	0.125	–	0.8	DMEM, 5% FCS	25	4	Batch (1)
Kurokawa et al. (1993)	0.033	0.28	–	RDF, serum-free	Var.	Var.	
Miller et al. (1988)	0.063	0.15	0.15	DMEM, 10% FCS	22	4.8	Chemostat
Dalili et al. (1990)	0.056	–	0.06	RPMI, 20% FCS	13.8	1.5	
Linardos et al. (1991)	~0.04	–	–	DMEM, 1.5% FBS	25	4.8	
Gaertner and Dhurjati (1993)	0.043	–	–	DMEM, 10% FCS	Var.	Var.	Batch
Glacken et al. (1989)	0.055	–	0.15	DMEM, 0.75–1% FCS	25	4	Batch, fed-batch
Pörtner et al. (1996)	0.036	–	0.06	IMDM/Ham's F12, 3%HS/FCS	17.5	4	Batch, fed-batch, chemostat, dialysis

drastically with decreasing specific growth rate. It should be noted that cell-lines which exhibit rather similar growth behaviour (data not shown) seem to differ as for their death kinetics. Here, factors so far not taken into account during cultivation or pre-cultivation might have a considerable influence on the cell metabolism.

The inhibitory effect of ammonia and lactate is already discussed in Sect. 2.4.3. Especially for ammonia the kinetics of cell growth inhibition is studied in depth. As discussed earlier the inhibitory effect of ammonia can be linked to the concentration of undissociated NH_3, rather than linking it to the concentration of total ammonia. An example was shown in Fig. 2.8. In this case the following kinetic expressions describe the relationship between the cell specific growth rate μ as well as the death rate μ_d and the concentration of NH_3 above a threshold of $c_{cr,NH3} = 0.04\,\mathrm{mmol}\,L^{-1}$ (Lüdemann et al. 1994):

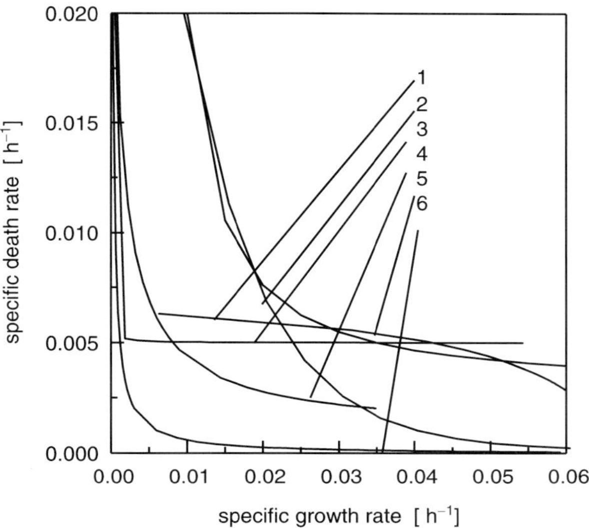

Fig. 2.11 Cell specific death rate vs. cell specific growth rate at substrate limitation – literature survey: 1 Frame and Hu (1991), 2 Glacken et al. (1989), 3 Dalili et al. (1990), 4 Pörtner et al. (1996), 5 Linardos et al. (1991), 6 Bree et al. (1988)

$$\mu = \mu_{max} \frac{k_{p,NH_3}}{c_{NH3} - c_{cr,NH_3} + k_{p.NH_3}} \qquad (2.5)$$

$$k_d = k_{d,min} \frac{c_{NH_3} - c_{cr,NH_3} + k_{d,p.NH_3}}{K_{d,p,NH_3}} \qquad (2.6)$$

with $\mu_{max} = 0.036\,h^{-1}$; $k_{p,NH3} = 0.055\,mmol\,L^{-1}$, $k_{d,min} = 0.002\,h^{-1}$; $k_{d,p,NH3} = 0.02\,mmol\,L^{-1}$. Below the threshold value, cell growth does not seem to be affected by ammonia.

For lactate inhibition critical values are discussed in Sect. 2.4.3. Only little information is available in the literature about inhibitory constants ($k_{p,i}$, $k_{d,p,j}$). Most of this data was obtained from multi-substrate and – metabolite kinetics rather than from studies on lactate inhibition. The reason may be, that under most culture conditions ammonia accumulation is more critical than lactate accumulation.

2.5.2.2   Cell Specific Substrate Uptake Rates (Glucose and Glutamine)

In the literature, specific uptake rates for glucose and glutamine are described as a function of either the specific growth rate (Miller et al. 1988; Dalili et al. 1990; Frame and Hu 1991b; Linardos et al. 1991; Bree et al. 1988; Harigae et al. 1994; Hiller et al. 1991), the substrate concentration (Glacken et al. 1989; Kurokawa et al. 1993; Pörtner et al. 1996) or as combination of both (de Tremblay et al. 1992). The main inhibiting metabolites lactate and ammonia are generally ignored throughout these correlations and seem to be negligible due to our own observations (data not published). A survey of correlations for both the specific glucose and glutamine uptake rate is given in the literature (Zeng and Bi 2006; Pörtner and Schäfer 1996).

The relationship between the cell specific substrate uptake rate and the specific growth rate can often be found following the maintenance-energy model (Pirt 1985).

$$q_s = \frac{1}{Y_{x,s}} \mu + m \tag{2.7}$$

where $Y_{x,s}$ denotes the yield coefficient of cells on substrate and m stands for the maintenance term. Miller et al. (1988) modelled by means of (2.7) the specific glucose and glutamine uptake rate with the maintenance term equal to zero for the latter. This linear correlation fitted well in the range of higher specific growth rates. However, for lower specific growth rates the uptake rate for both glucose and glutamine was lower than expected. Linardos et al. (1991) observed the same pattern and thus extended (2.7) by a term including the specific death rate. The necessity of the additional term might be questioned on the ground that the data cover a rather small range of specific growth rates and may scatter within the analytical error. Harigae et al. (1994) also modelled their results obtained from perfusion culture strictly due to (2.7). In contrast to Miller et al. (1988), the uptake rates for glucose and glutamine could be fitted well within the range of low specific growth rates, whereas at high specific growth rates, i.e. in the vicinity of the maximum specific growth rate, the values increased considerably and ceased being linear. Data obtained by other authors (Ray et al. 1989; Robinson and Memmert 1991; Hiller et al. 1991) follow this same trend depending on the respective maximum specific growth rates (Pörtner and Schäfer 1996). Frame and Hu (1991b) explained this deviation from linearity by introducing a critical specific growth rate beyond which the yield of cells on substrate is increased. Moreover, they added glucose-consumption term for the synthesis of antibodies.

Apart from the maintenance-energy-model, the saturation-type model offers alternatively the modelling of the specific substrate uptake rate as a function of the limiting substrate.

$$q_s = q_{s,max} \frac{c_s}{c_s + k_s}. \tag{2.8}$$

where q_s is the cell specific substrate uptake rate, c_s the substrate concentration, k_s the limitation-constant, and $q_{s,max}$ maximal cell specific substrate uptake rate. An example is given in Fig. 2.12. This model approach seems to be more accurate about the error of the independent variables.

Pörtner and Schäfer (1996) calculated for kinetic expressions reported for some hybridom cell lines from the literature the specific glucose and glutamine uptake

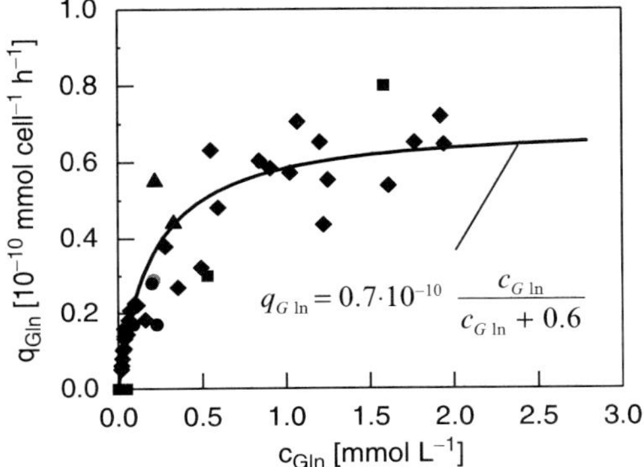

$$q_{G\,ln} = 0.7 \cdot 10^{-10} \; \frac{c_{G\,ln}}{c_{G\,ln} + 0.6}$$

Fig. 2.12 Cell specific glutamine uptake rate q_{Gln} vs. glutamine concentration c_{Gln} for the hybridom cell line IV F 19.23 (■ flask batch, ● chemostat, ▲ dialysis cont., ◆ fed-batch)

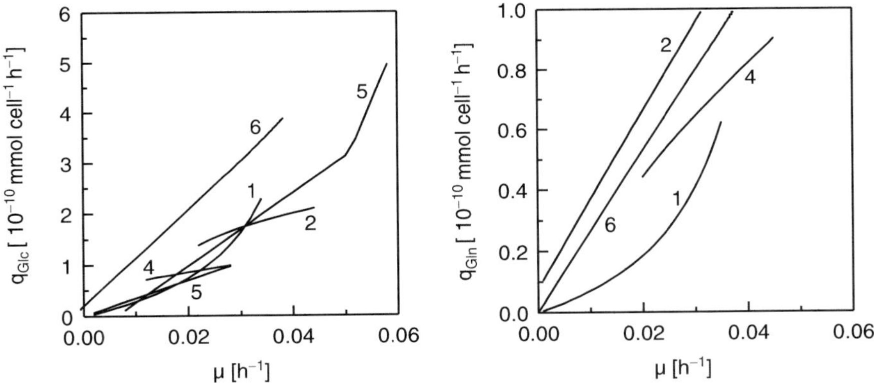

Fig. 2.13 Cell specific glucose q_{Glc} and glutamine uptake rate q_{Gln} vs. cell specific growth rate μ – literature survey: 1 Pörtner et al. (1996), 2 Linardos et al. (1991), 3 Bree et al. (1988), 4 Harigae et al. (1994), 5 Frame and Hu (1991), 6 de Tremblay et al. (1992)

rates as a function of the specific growth rate (Fig. 2.13). Those correlations not giving $q = f(\mu)$ were transformed accordingly. The uptake rates for glucose of the hybridom cell lines investigated, were found to be within a relatively narrow range. For glutamine, they differed considerably between the cell lines. The maintenance term for glucose and glutamine (uptake rate at zero growth rates) turned out to be very low, if not negligible. It should be pointed out that Fig. 2.13 clearly demonstrates how a small range of data allows a variety of different kinetic interpretations, all of which might be valid within this range. However, relating this kinetics to other data found beyond this range and the analytical margin of error may prove apparent phenomena to be mere scatter.

The models for substrate uptake rates discussed so far are valid only for growth rates below the maximal growth rate. Several authors found a decline of cell specific substrate uptake and metabolite production rates during exponential growth, e.g. maximal growth rate (compare Sect. 2.4.2). Equations describing this effect can be found in the reference literature (Pörtner et al. 1994; Zeng and Bi 2006; Pörtner and Schäfer 1996).

2.5.2.3 Yield Coefficients for Lactate and Ammonia

Lactate stems mainly from the glucose-metabolism and to a small extent from utilisation of glutamine. Typical values for hybridom cells are in the range of 1.1–1.7 mol lactate/mol glucose. The yield can be found constant (Frame and Hu 1991a; de Tremblay et al. 1992), increasing (Pörtner 1998; Miller et al. 1988), and declining (Harigae et al. 1994), with decreasing specific growth rate. Several publications describe a dependency of the yield coefficient for lactate on the glucose and or glutamine concentration. Pörtner et al. (1995) observed an increasing yield of lactate from glucose with increasing glutamine and glucose concentration. During fed-batch culture Kurokawa et al. (1994) found an increasing yield of lactate with increasing glucose concentration independent of the glutamine concentration. Moreover, increasing yields of lactate from glucose were observed concurring with decreasing yields of ammonia from glutamine. Reitzer et al. (1979) observed in studies with HeLa cells a higher lactate production at higher glucose concentrations. Zielke et al. (1978) found in studies with human fibroblasts that the lactate production increased at high glutamine concentrations. Until now it is not clear whether it is glucose or the glutamine which affects the lactate yield from glucose.

Ammonia stems mainly from the glutamine metabolism. The apparent yield of ammonia from glutamine can be found varying between 0.46 and 0.85 (Pörtner 1998; Harigae et al. 1994; Hiller et al. 1993; Miller et al. 1988; Bree et al. 1988; Pörtner et al. 1996; de Tremblay et al. 1992; Linardos et al. 1991). Its dependence on the specific growth rate seems to be of less significance compared to the yield of lactate from glucose. Only Hiller et al. (1993) observed a slight increase of the ammonia yield with increasing growth rate.

The yield coefficients for lactate and ammonia have not been studied in depth. Still many questions regarding the mechanisms influencing these coefficients have to be investigated.

2.5.2.4 Cell Specific Production Rate of Monoclonal Antibodies

The production of a majority of recombinant protein products or monoclonal antibodies by mammalian cells falls into the following categories (Adams et al. 2007):

- "Product formation results in cell destruction

- Product formation is initiated by induction or infection, and, after a relatively short production stage, cell destruction occurs
- Product formation is initiated by infection, but it does not result in immediate cell destruction
- Product formation is primarily during the growth stage
- Production formation is "constitutive" and an extended production period can be achieved.

Instead of using more detailed models discussed earlier in this section" the product formation equation can be set up differently depending on the categories described above correlating to growth.

Here production of monoclonal antibodies by means of hybridom cells is discussed in more detail, extensive literature data is available in this case. The cell specific antibody productivity differs significantly between different cell lines with respect to the level of productivity as well as the influence of culture conditions. In most cases the cell specific antibody production rate was related to the specific growth rate and it was found to increase or decrease with increasing growth rate or was independent from it [reviewed by Zeng and Bi (2006), Pörtner and Schäfer (1996)]. Increased antibody production during the death phase of batch cultures was reported. Increased release of intracellular antibodies (Musielski et al. 1994) and a higher antibody production rate due to stress conditions (Linardos et al. 1991), respectively, was concluded. Serum seems to have a regulatory as well as a stimulating effect on the antibody productivity. Dalili et al. (1990) and Gaertner and Dhurjati (1993) observed an increase in antibody productivity with increasing serum concentration. On the other hand, it was shown that in optimized serum- or protein-free media the antibody productivity can reach the same value as in serum containing medium (Lüdemann et al. 1996). The influence of substrate concentrations (glucose, glutamine) has been found to vary between cell lines. In some cases, a constant (Pörtner et al. 1996; Hiller et al. 1991) or decreasing antibody productivity was observed at decreasing substrate concentrations (Linardos et al. 1991; Dalili et al. 1990; Frame and Hu 1991b). Kurokawa et al. (1993) reported about a significant increase of the antibody productivity with decreasing glucose- and glutamine concentrations. Toxic metabolites such as ammonia and lactate seem to have a minor influence on the antibody productivity (Omase et al. 1992; Lüdemann et al. 1994). Whether the concentration of oxygen has any impact is not yet clear (compare Tziampazis and Sambanis 1994).

A number of unstructured models for the cell specific antibody production rate were summarized by Pörtner and Schäfer (1996). In Fig. 2.14, the relationship between the cell specific antibody production rate and the specific growth rate is shown for data from various authors and models that could be transformed correspondingly. The broad variety of different approaches reflects the different behaviour of the cell lines investigated. Alternatively, cell cycle models for antibody productivity were developed (reviewed by Zeng and Bi 2006). Hybridom cells produce monoclonal antibodies mainly in the late G1-phase of the cell cycle. It is assumed that with decreasing growth rate the cells remain in G1-phase for a longer time and therefore an increase of the antibody productivity with increasing death

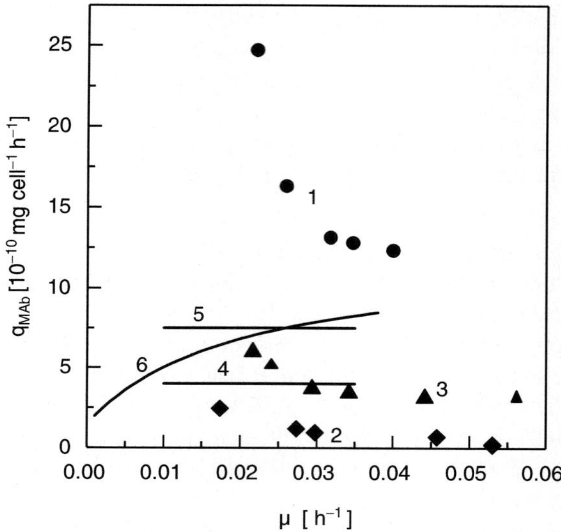

Fig. 2.14 Cell specific antibody production rate q_{MAb} vs. growth rate μ – literature survey: 1 Leno et al. (1992), 2 Ray et al. (1989), 3 Miller et al. (1988), 4 Pörtner et al. (1996), 5 Hiller et al. (1991), 6 de Tremblay et al. (1988)

rate was postulated. Moreover, a number of structured models are proposed. These are discussed in the following.

2.5.3 Structured Models

Besides unstructured models a number of structured models are developed based on knowledge of intracellular structure, bioreactions and their regulation mechanisms. The large number of structured models have been reviewed by Zeng and Bi (2006). Among these are (i) "chemically or metabolically" structured models, (ii) cell cycle models especially for antibody production as well as (iii) a cybernetic approach. In the following paragraphs an example is introduced briefly to give a basic idea of the methodology.

A typical chemically or metabolically structured model was suggested by Batt and Kompala (1989) for hybridom cells. Basically the biomass is divided into several chemical components (pools) such as cell membrane (lipids), protein, and nuclei acids. The stoichiometric relationships (balance equations) and rate equations for the compounds form the basis for modelling the entire cell (Zeng and Bi 2006). The main problem in establishing a model with this approach is the formulation of reliable kinetic rate expressions. Batt and Kompala (1989) defined "metabolic pools" for amino acids, proteins, nuceotides and lipids. As metabolites ammonia, lactate and monoclonal antibodies were considered. The antibodies were regarded as part of the protein pool. The interactions between different pools as well as the extracellular environment were described by saturation type kinetic expressions. The model consists of 18 dynamic equations with 49 coefficients describing the earlier mentioned metabolic pools, extracellular products (ammonia, lactate, MAb), substrate concentrations (glucose, glutamine, amino acids) as well as cell growth (growth and death rate). The model parameters were adjusted by using the data of Miller et al. (1988).

Normally, structured models are considered superior to unstructured models, especially for extrapolation to conditions outside the experimental range. Care has to be taken, as the high complexity of mammalian cell metabolism can hardly be described mathematically. The underlying mechanisms used for model formation are mostly a working hypothesis and do not explain the metabolic reactions of the cells (Zeng and Bi 2006). Usually structured models are quite complex incorporating a huge number of variables and kinetic parameters, that are difficult to determine and are accessible only by parameter adaptation.

2.5.4 Conclusions for Set-Up of a Kinetic Model

The variety of models set up in recent years for growth and metabolism of mammalian cells, especially hybridom cell has become prominent. A careful comparison of the available data and models showed that cell culture has considerable error, either analytical or caused by metabolic changes, but also has certain unpredictability due to lack of understanding of the real metabolism. For simplicity and easy handling it is common to choose one or two growth limiting substrates which are easy to measure, if possible, on-line. The majority of the models cited in literature deal with the cell specific growth rate as a function of the glucose and/or glutamine concentration, since these are the main substrates. Nevertheless, they need not be growth limiting as expressed by models incorporating a "threshold term" (zero growth at residual glucose and/or glutamine concentrations). Thus, a careful medium composition must be chosen to obtain distinct data to get accurate kinetic correlations.

A number of models were formulated as multi-component-kinetics for which limitation and inhibition constants of glucose, glutamine, lactate or ammonia, respectively, whereas in some cases it was determined by a small data. Especially inhibition constants for lactate and ammonia determined in this way differ significantly from those reported in publications that investigated inhibition kinetics separately. Until now research is not conducted to investigate superimposition of substrate limitation and metabolite inhibition.

The comparison of the model equations shows very clearly that for model set up a data range too small can lead to wrong conclusions and model equations should not be therefore used outside the data range. For proper model design care has to be taken to cover the entire range of process conditions. Static batch cultures can be used to determine the maximal specific growth rate but they give only little information on the relationship between growth rate and death rate and the substrate or metabolite concentration. Continuous cultures (chemostat) yield reliable data due to steady state conditions. However, at low dilution rates, which are necessary to obtain very low substrate concentrations, they are not stable. Data at very low substrate concentrations can be drawn from fed-batch cultures.

The diversity of the unstructured models cited above mostly reveal their descriptive character compared to the structured models. Due to this limitation, empirical models can still serve as a valuable tool for process design. For understanding the cell metabolism they might have been over-rated in the past. For most parameters a considerable range of scatter was observed besides the analytical error. This reflects deviations of the cell metabolism between different cultures or changes after long culture periods. Since high performance processes (high antibody concentration and yield, efficient medium use) are usually run at low substrate and high metabolite concentrations, these deviations in cell metabolism might lead to a wrong prediction of process parameters. Experiences with applications of "a priori" predictions of process parameters e. g. for fed-batch cultures show the importance of this problem (Pörtner et al. 1996; de Tremblay et al. 1992). For further improvements it is essential to adapt model parameters during the process by parameter identification or to apply knowledge based control strategies (fuzzy control, neural networks). Examples are discussed in Sect. 4.4.

The models discussed so far were mostly formulated to simulate the kinetic behaviour in "balanced" batch growth or under steady states of continuous cultures. The dynamic behaviour of cells can hardly be described. Here new kinetic models are required, incorporating further information from metabolic flux analysis or genomic and proteomic data (Zeng and Bi 2006).

2.6   Questions and Problems

- Explain the differences between mammalian cells, plant cells and microbes with respect to cultivation and product formation.
- Name characteristic properties of mammalian cells.
- Explain briefly different types of mammalian cells.
- Describe the principal scheme for isolation of primary cells from a tissue.
- Name features of normal cells and three features of transformed cells.
- Explain the principal steps involved in generating a hybridom cell line for production of monoclonal antibodies introduced by Milstein and Köhler.
- Sketch and mark the schematic structure of an IgG-antibody
- Explain the set-up and purpose of cell banking.
- Cell culture medium: What are the main compounds in cell culture medium? What are the main metabolites from mammalian cells?
- Discuss pros and cons of serum- and protein-free medium.
- Explain briefly the fundamentals of cell metabolism (main substrates, pathways and metabolites).
- Explain the mechanism of ammonium toxicity.

- What is the reasonable range for cell specific oxygen uptake rate of mammalian cells?
- Name variables used for modelling the kinetic of mammalian cells.
- Give a general equation describing the relationship between cell specific growth rate, substrate limitation and product inhibition.
- Describe the relationship between cell specific substrate uptake rate and specific growth rate.
- Give a typical example for the yield of lactate from glucose as well as for the yield of ammonia from glutamine.
- Explain "structured" kinetic models.

List of Symbols

c_P	product concentration
c_S	substrate concentration
d_0, d_1	fit parameter
i	index for substrate
j	index for product
k_d	cell specific death rate
$k_{d,max}$	maximal cell specific death rate
$k_{d,min}$	minimal cell specific death rate
$k_{d,P}$	kinetic constant for effect of metabolite on cell death
$k_{d,S}$	kinetic constant for effect of substrate on cell death
k_p	metabolite inhibition constant
k_S	substrate limitation constant
m	maintenance term
NH3	index for undissoziated ammonia
q_P	cell specific production rate
q_S	cell specific substrate uptake rate
$q_{S,max}$	maximal cell specific substrate uptake rate
$Y_{x/s}$	yield coefficient cell number per mol substrate
μ	cell specific growth rate
μ_{max}	maximal cell specific growth rate

References

Adams D, Korke R, Hu WS (2007) Application of stoichiometric and kinetic analyses to characterize cell growth and product formation. In: Pörtner R (ed) Animal Cell Biotechnology – Methods and Protocols. Humana Press, Clifton, UK

Al-Rubeai M, Emery AN, Chalder S (1992) The effect of Pluronic F-68 on hybridoma cells in continuous culture. Appl Microbiol Biotechnol 37: 44–45

Al-Rubeai M, Singh R (1998) Apoptosis in cell culture. Curr Opin Biotechnol 9: 152–156

Altamirano C, Illanes A, Casablancas A, Gámez X, Cairó JJ, Gódia F (2001) Analysis of CHO cells metabolic redistribution in a glutamate-based defined medium in continuous culture. Biotechnol Prog 17: 1032–1041

American Type Culture Collection (ATCC). http://www.atcc.org

Barnes D, Sato G (1980) Serum-free cell culture: an unifying approach. Cell 22: 649–655

Batt CB, Kompala SK (1989) A structured modeling framework for the dynamics of hybridoma growth in continuous suspension cultures. Biotechnol Bioeng 34: 515–531

Bohlen H (1993) Medikamente nach Maß – Monoklonale Antikörper für die Medizin der Zukunft. Bild der Wissenschaften 1: 35–39

Bree MA, Dhurjati P, Geoghegan RF, Robnett B (1988) Kinetic modelling of hybridoma cell growth and immunoglobulin production in a large-scale suspension culture. Biotechnol Bioeng 32: 1067–1072

Butler M (1987) Growth limitations in microcarrier cultures. Adv Biochem Eng 34: 57–84

Butler M (2004) Animal Cell Culture and Technology – The Basics (2nd edition). Oxford University Press, New York

Butler M (2005) Animal cell cultures: Recent achievements and perspectives in the production of biopharmaceuticals. Appl Microbiol Biotechnol 68: 283–291

Butler M, Jenkins H (1989) Nutritional aspects of the growth of animal cells in culture. J Biotechnol 12: 97–110

Butler M, Spier RE (1984) The effects of glutamine utilisation and ammonia production on the growth of BHK cells in microcarrier culture. J Biotechnol 1: 187–196

Chmiel H (1991) Bioprozesstechnik, UTB

Cotter TG (1994) Programmed to die: Cell death via apoptosis. In: Spier RE, Griffiths JB, Berthold W (eds) Animal cell technology: products of today, prospects for tomorrow. Butterworth-Heinemann, pp 175–182

Cruz HJ, Moreira JL, Carrondo MJT (1999) Metabolic shifts by nutrient manipulation in continuous cultures of BHK cells. Biotechnol Bioeng 66: 104–113

Dalili M, Sayles GD, Ollis DF (1990) Glutamine-limited batch hybridoma growth and antibody production: experiment and model. Biotechnol Bioeng 36: 74–82

de Tremblay M, Perrier M, Chavarie C, Archambault J (1992) Optimization of fed-batch culture of hybridoma cells using dynamic programming: single and multi feed case. Bioproc Eng 7: 229–234

de Tremblay M, Perrier M, Chavarie C, Archambault J (1993) Fed-batch culture of hybridoma cells: comparison of optimal control approach and closed loop strategies. Bioproc Eng 9: 13–21

DeZengotita VM, Kimura R, Miller WM (1998) Effects of CO2 and osmolality on hybridoma cells: growth, metabolism and monoclonal antibody production. Cytotechnol 28: 213–227

Dorn-Beinke A, Nittka ST, Neumaier M (2007) Technology and production of murine monoclonal and recombinant antibodies and antibody fragments. In: Pörtner R (ed) Animal Cell Biotechnology – Methods and Protocols. Humana Press, Clifton, UK

Doyle A, Griffiths JB (1998) Cell and Tissue Culture: Laboratory Procedures in Biotechnology. Wiley, New York

Doyle C, Butler M (1990) The effect of pH on the toxicity of ammonia to a murine hybridoma cell line. J Biotechnol 15: 91–100

Duval D, Demangel C, Munier-Jolain K, Miosecc S, Geahel I (1991) Factors controlling cell proliferation and antibody production in mouse hybridoma cells: I. Influence of amino acid supply. Biotechnol Bioeng 38: 561–570

Eagle H (1959) Amino acid metabolism in mammalian cell cultures. Science 130: 432–437

Edwards CP, Aruffo A (1993) Current applications of COS cell based treatment expression system. Curr Opin Biotechnol 4: 558–563

EMEA: European Medicine Agency. http://www.emea.eu.ing

Europa AF, Gambhir A, Fu PC, Hu WS (2000) Multiple steady states with distinct cellular metabolism in continuous culture of mammalian cells. Biotechnol Bioeng 67: 25–34

European Collection of Animal Cell Culture (ECACC). http://www.ecacc.org.uk

Fassnacht D, Rössing S, Franék F, Al-Rubeai M, Pörtner R (1998) Effect of *Bcl-2* expression on hybridoma cell growth in serum-supplemented, protein-free and diluted media. Cytotechnol 26: 119–226

Fassnacht D, Rössing S, Ghaussy N, Pörtner R (1997) Influence of non-essential amino acids on apoptotic and necrotic death of mouse hybridoma cells in batch cultures. Biotechnol Lett 19: 35–38

FDA: U.S. Food and Drug Administration. http://www.fda.gov

Fletscher T (2005) Designing culture media for recombinant protein production. BioProcess 1: 30–36

Follstad BD, Balcarcel RR, Stephanopolous G, Wang DIC (1999) Metabolic flux analysis of hybridoma continuous culture steady state multiplicity. Biotechnol Bioeng 63: 675–683

Frahm B, Blank HC, Cornand P, Oelßner W, Guth U, Lane P, Munack A, Johannsen K, Pörtner R (2002) Determination of dissolved CO_2 concentration and CO_2 production rate of mammalian cell suspension culture based on off-gas measurement. J Biotechnol 99: 133–148

Frame KK, Hu WS (1990) Cell volume measurement as an estimation of mammalian cell biomass. Biotechnol Bioeng 36: 191–197

Frame KK, Hu WS (1991a) Kinetic study of hybridoma cell growth in continuous culture. I. A model for non-producing cells. Biotechnol Bioeng 37: 55–64

Frame KK, Hu WS (1991b) Kinetic study of hybridoma cell growth in continuous culture. II. Behavior of producers and comparison to nonproducers. Biotechnol Bioeng 38: 1020–1028

Franék F (1995) Starvation-induced programmed death of hybridoma cells: prevention by amino acid mixtures. Biotechnol Bioeng 45: 86–90

Franék F, Dolníková J (1991a) Hybridoma growth and monoclonal antibody production in iron-rich protein-free medium: Effect of nutrient concentration. Cytotechnol 7: 33–38

Franék F, Dolníková J (1991b) Nucleosomes occurring in protein-free hybridoma cell cultures. Evidence for programmed cell death. FEBS Lett 248: 285–287

Freshney R (1994) Culture of Animal Cells. Wiley, New York

Gaertner JG, Dhurjati P (1993) Fractional factory study of hybridoma behavior. 2. Kinetics of nutrient uptake and waste production. Biotechnol Prog 9: 309–316

Gambhir A, Zhang C, Europa A, Hu WS (1999) Analysis of the use of fortified medium in continuous culture of mammalian cells. Cytotechnol 31: 243–254

Gawlitzek M, Valley U, Wagner R (1998) Ammonium ion and glucosamine dependent increase of oligosaccharide complexity in recombinant glycoproteins secreted from cultivated BHK-21 cells. Biotechnol Bioeng 57: 518–528

German Resource Centre for Biological Material (DSMZ): http://www.dsmz.de

Glacken MW, Fleischacker RJ, Sinskey AJ (1986) Reduction of waste product excretion via nutrient control: possible strategies for maximizing product and cell yields in cultures of mammalian cells. Biotechnol Bioeng 28: 1376–1389

Glacken MW, Huang C, Sinskey AJ (1989) Mathematical description of hybridoma culture kinetics. III Simulation of fed-batch bioreactors. J Biotechnol 10: 39–66

Gódia F, Cairó JJ (2006) Cell metabolism. In: Ozturk SS and Hu W-S (eds) Cell Culture Technology for Pharmaceutical and Cell-Based Therapies. Taylor & Francis, New York

Griffiths JG (1987) Serum and growth factors in cell culture media: an introduction review. Dev Biol Stand 66: 155–160

Grob D, Lettenbauer C, Eibl R, Meier HP (1998) Vergleich verschiedener Verfahren und Apparate zur Sterilisation von Zellkulturmedien. In: iba (ed) Proceedings zum 9. Heiligenstädter Kolloquium

Häggström L (2000) Animal cell metabolism. In: Spier R (ed) Encyclopedia of Cell Technology. Wiley, New York, pp 392

Harigae M, Matsumura M, Kataoka H (1994) Kinetic study on HBs-MAb production in continuous cultivation. J Biotechnol 34: 227–235

Hassel T, Gleave S, Butler M (1991) Growth inhibition in animal cell culture – the effect of lactate and ammonia. Appl Biochem Biotech 30: 29–41

Hayflick L, Moorhead PS (1961) The serial cultivation of human diploid cell strains. Exp Cell Res 25: 585

Hayter PM, Curling EMA, Baines AJ, Jenkins N, Salmon I, Strange PG, Tong JM, Bull AT (1991) Chinese hamster ovary cell growth and interferon production kinetics in stirred batch culture. Appl Microbiol Biotechnol 34: 559–564

Henzler HJ, Kauling OJ (1993) Oxygenation of cell cultures. Bioproc Eng 9: 61–75

Hiller GW, Aeschlimann AD, Clark DS, Blanch HW (1991) A kinetic analysis of hybridoma growth and metabolism in continuous suspension culture on serum-free medium. Biotechnol Bioeng 38: 733–741

Hiller GW, Clark DS, Blanch HW (1993) Cell retention-chemostat studies of hybridoma cells – analysis of hybrdoma growth and metabolism in continuous suspension culture on serum-free medium. Biotechnol Bioeng 42: 185–195

Ikonomou L, Schneider YJ, Agathos SN (2003) Insect cell culture for industrial production of recombinant proteins. Appl Microbiol Biotechnol 62: 1–20

Jeong YH, Wang SS (1995) Role of glutamine in hybridoma cell culture: effects on cell growth, antibody production, and cell metabolism. Enzyme Microb Technol 77: 45

Kallel H, Jouini A, Majoul S, Rourou S (2002) Evaluation of various serum and animal protein free media for the production of a veterinary rabies vaccine in BHK-21 cells. J Biotechnol 95: 195–204

Kasche V, Probst K, Maass J (1980) The DNA and RNA content of crude and crystalline trypsin used to trypsinize animal cell cultures: kinetics of trypsinization. Eur J Cell Biol 22: 388

Kim JS, Ahn BC, Lim BP, Choi YD, Jo EC (2004) High-level scu-PA production by butyrate-treated serum-free culture of recombinant CHO cell line. Biotechnol Prog 20: 1788–1796.

Kimura R, Miller WM (1997) Glycosylation of CHO-derived recombinant tPA produced under elevated pCO_2. Biotechnol Prog 13: 311–317.

Köhler G, Milstein C (1975) Continuous cultures of fused cells secreting monoclonal antibodies of predifined specifity. Nature 256: 495

Kurokawa H, Ogawa T, Kamihira M, Park YS, Iijima S, Kobayashi T (1993) Kinetic study of hybridoma metabolism and antibody production in continuous culture using serum-free medium. J Ferment Bioeng 76(2): 128–133

Kurokawa H, Park YS, Iijima S, Kobayashi T (1994) Growth characteristics in fed-batch culture of hybridoma cells with control of glucose and glutamine concentration. Biotechnol Bioeng 44: 95–103

Lee GM, Kaminski MS, Palsson BO (1992) Observations consistent with autocrine stimulation of hybridoma cell growth and implications for large-scale antibody production. Biotechnol Lett 14(4): 257–262

Lee GM, Palsson BO (1990) Immobilization can improve the stability of hybridoma antibody productivity in serum-free media. Biotechnol Bioeng 36: 1049–1055

Linardos TI, Kalogerakis N, Behie LA (1991) The effect of specific growth rate and death rate on monoclonal antibody production in hybridoma chemostat cultures. Can J Chem Eng 69: 429–438

Link T, Backstrom M, Graham R, Essers R, Zorner K, Gatgens J, Burchell J, Taylor-Papadimitriou J, Hansson GC, Noll T (2004) Bioprocess development for the production of a recombinant MUC1 fusion protein expressed by CHO-K1 cells in protein-free medium. J Biotechnol 110: 51–62

Linz M, Zeng AP, Wagner R, Deckwer WD (1997) Stoichiometry, kinetics and regulation of glucose and amino acid metabolism of recombinant BHK cell line in batch and continuous cultures. Biotechnol Prog 13: 453–463

Lüdemann I, Pörtner R, Märkl H (1994) Effect of NH3 on the cell growth of a hybridoma cell line. Cytotechnol 14: 11–20

Lüdemann I, Pörtner R, Schaefer C, Schick K, Šrámková K, Reher K, Neumaier M, Franék F, Märkl H (1996) Improvement of the culture stability of non-anchorage-dependent animal cells grown in serum-free media through immobilization. Cytotechnol 19: 111–124

Maranga L, Goochee CF (2006) Metabolism of PER.C6 cells cultivated under fed-batch conditions at low glucose and glutamine levels. Biotechnol Bioeng 94: 139–150

Matanguihan R, Sajan E, Zachariou M, Olson C, Michaels J, Thrift J, Konstantinov K (2001) Solution to the high dissolved CO_2 problem in high-density perfusion culture of mammalian cells. In: Animal Cell Technology: From Target to Market. Kluwer, The Netherlands, pp 399–402

Matsumura M, Nayve R, Shimoda M, Motoki M, Kataoka H (1990) Growth inhibition of hybridoma cells by ammonia and its selective removal. In: Kataoka H, Märkl H (eds), Proceedings of the German–Japanese Workshop on Animal Cell Culture Technology, pp 14–28

McMeekin TA, Olley JN, Ross T, Ratkowsky DA (1993) Predictive Microbiology: Theory and Application. Research Studies Press, Taunton, UK

McQueen A, Bailey JE (1990) Effect of ammonium ion and extracellular pH on hybridoma metabolism and antibody production. Biotechnol Bioeng 35: 1065–1077

Mercille S, Massie B (1994) Induction of apoptosis in nurtrient-deprived cultures of hybridoma and myeloma cells. Biotechnol Bioeng 44: 1140–1154

Miller WM, Wilke CR, Blanch HW (1987) Effects of dissolved oxygen concentration on hybridoma growth and metabolism in continuous culture. J Cell Phys 132: 524–530

Miller WM, Wilke CR, Blanch HW (1988) A kinetic analysis of hybridoma growth and metabolism in batch and continuous suspension culture. Effect of nutrient concentration, dilution rate and pH. Biotechnol Bioeng 32: 947–965

Miller WM, Wilke CR, Blanch HW (1989a) Transient responses of hybridoma cells to nutrient additions in continuous culture: I. Glucose pulse and step changes. Biotechnol Bioeng 33: 477–486

Miller WM, Wilke CR, Blanch HW (1989b) Transient responses of hybridoma cells to nutrient additions in continuous culture: II. Glutamine pulse and step changes. Biotechnol Bioeng 33: 487–499

Milstein C (1988) Monoklonale Antikörper. In: Immunsystem, 2. Auflage, Spektrum der Wissenschaft Verlagsgesellschaft, Heidelberg

Morrow KJ (2007) Improving protein production strategies. GEN 28: 37–39

Mostafa SS, Gu X (2003) Strategies for improved dCO2 removal in large-scale fed-batch cultures. Biotechnol Prog 19: 45–51

Musielski H, Rüger K, Zwanzig M, Lehmann K (1994) Monoclonal antibodies released from viable hybridoma cells at different stages of growth. In: Spier RE, Griffiths JB, Berthold W (eds) Animal Cell Technology: Products of Today, Prospects for Tomorrow. Butterworth-Heinemann, London, pp 485–492

Nayve R, Masamichi M, Matsumura M, Kataoka H (1991) Selective removal of ammonia from animal cell culture broth. Cytotechnol 6: 121–130

Oh SKW, Chua FKF, Choo ABH (1995) Intracellular responses of productive hybridomas subjected to high osmotic pressure. Biotechnol Bioeng 46: 525–535

Omase T, Higashiyama KI, Shioya S, Suga KI (1992) Effects of lactate concentration on hybridoma culture in lactate-controlled fed-batch operation. Biotechnol Bioeng 39: 556–564

Otzturk SS (2006) Cell culture technology – an overview. In: Ozturk SS, Hu WS (eds) Cell Culture Technology for Pharmaceutical and Cell-Based Therapies. Taylor & Francis, New York

Ozturk SS, Palsson BO (1990) Chemical decomposition of glutamine in cell culture media: Effect of media type, pH, and serum concentration. Biotechnol Prog 6: 121–128

Ozturk SS, Riley MR, Palsson BO (1992) Effects of ammonia and lactate on hybridoma growth, metabolism and antibody production. Biotechnol Bioeng 39: 418–431

Pham PL, Perret S, Doan HC, Cass B, St-Laurent G, Kamen A, Durocher Y (2003) Large-scale transient transfection of serum-free suspension-growing HEK293 EBNA1 cells: peptone additives improve cell growth and transfection efficiency. Biotechnol Bioeng 84: 332–342.

Pirt SJ (1985) Principles of Microbe and Cell Cultuvation. Blackwell, Oxford

Pörtner R (1998) Reaktionstechnik der Kultur tierischer Zellen. Shaker, Aachen

Pörtner R, Bohmann A, Lüdemann I, Märkl H (1994) Estimation of specific glucose uptake rates in cultures of hybridoma cells. J Biotechnol 34: 237–246

Pörtner R, Lüdemann I, Bohmann A, Schilling A, Märkl H (1995) Evaluation of process strategies for efficient cultivation of hybridoma cells based on mathematical models. In: Beuvery EC et al. (eds) Animal Cell Technology: Developments Towards the 21st Century, Kluwer, The Netherlands. pp 829–833

Pörtner R, Schäfer TH (1996) Modelling hybridoma cell growth and metabolism – A comparison of selected models and data. J Biotechnol 49: 119–135

Pörtner R, Schilling A, Lüdemann I, Märkl H (1996) High density fed-batch cultures for hybridoma cells performed with the aid of a kinetic model. Bioproc Eng 15: 117–124

Ray NG, Karkare SB, Runstadler PW (1989) Cultivation of hybridoma cells in continuous cultures: Kinetics of growth and product formation. Biotechnol Bioeng 33: 724–730

Reitzer LJ, Wice BM, Kennell D (1979) Evidence that glutamine, not sugar, is the major energy source for cultured HeLa cells. J Biol Chem 254: 2669–2676

Rensing L, Cornelius G (1988) Grundlagen der Zellbiologie, UTB

Reuveny S, Velez D, Mamillan JD, Miller L (1986) Factors affecting cell growth and monoclonal antibody production in stirred reactors. J Immunol Meth 86: 53–59

Robinson DK, Memmert KW (1991) Kinetics of recombinant immunoglobulin production by mammalian cells in continuous culture. Biotechnol Bioeng 38: 972–976

Ryll T, Valley U, Wagner R (1994) Biochemistry of growth inhibition by ammonium ions in mammalian cells. Biotechnol Bioeng 44: 184–193

Ryu JS, Lee GM (1997) Effect of hypoosmotic stress on hybridoma cell growth and antibody production. Biotechnol Bioeng 55: 565–570.

Seamans TC, Hu WS (1990) Kinetics of growth and antibody production by a hybridoma cell line in a perfusion culture. J Ferment Bioeng 70: 241–245

Shirai Y, Hashimoto K, Takamatsu H (1992) Growth kinetics of hybridoma cells in high density culture. J Ferment Bioeng 73: 159–165

Siano SA, Muthrasan R (1991) NADH Fluorescence and oxygen uptake responses of hybridoma cultures to substrate pulse and step changes. Biotechnol Bioeng 37: 141–159

Sinacore MS, Drapeau D, Adamson SR (2000) Adaptation of mammalian cells to growth in serum-free media. Mol Biotechnol 15: 249–257.

Thorens B, Vassalli P (1986) Chloroquine and ammonium chloride prevent terminal glycosylation of immunoglobulins in plasma cells without effecting secretion. Nature 321: 618–620

Tritsch GL, Moore GE (1962) Spontaneous decomposition of glutamine in cell culture media. Exp Cell Res 28: 360–364

Tsuchiya HM, Fredrickson AG, Avis R (1966) Dynamics of microbial cell populations. Adv Chem Eng 6: 125–206

Tziampazis E, Sambanis A (1994) Modelling cell culture processes. Cytotechnol 14: 191–204

Voedisch B, Menzel C, Jordan E, El-Ghezal A, Schirrmann T, Hust M, Jostock T (2005) Expression rekombinanter Proteinpharmazeutika. transkript 7: 47–51

Vriezen N, Romein B, Luyben KCHAM, van Dijken JP (1997) Effects of glutamine supply on growth and metabolism of mammalian cells in chemostat culture. Biotechnol Bioeng 54: 272–286

Wagner R (1997) Metabolic control of animal cell culture processes. In: Hauser H, Wagner R (eds) Mammalian Cell Biotechnology in Protein Production. Walter de Gruyeter, Berlin, pp 193

Wurm FM (2004) Production of recombinant protein therapeutics in cultivated mammalian cells. Nat Biotechnol 22: 1393–1397

Zanghi J, Schmelzer A, Mendoza R, Knop R, Miller W (1999) Bicarbonate concentration and osmolality are key determinants in the inhibition of CHO cell polysialylation under elevated pCO_2 or pH. Biotechnol Bioeng 65: 182–191

Zeng AP, Bi JX (2006) Cell culture kinetics and modeling. In: Ozturk SS, Hu WS (eds) Cell Culture Technology for Pharmaceutical and Cell-Based Therapies. Taylor & Francis, New York

Zengotita de VM, Schmelzer AE, Miller WM (2002) Characterization of hybridoma cell responses to elevated pCO_2 and osmolality: Intracellular pH, cell size, apoptosis, and metabolism. Biotechnol Bioeng 77: 369–380

Zhang Z, Chisti Y, Moo-Young M (1995) Effects of the hydrodynamic environment and shear protectants on survival of erythrocytes in suspension. J Biotechnol 43: 33–40

Zhu MM, Goyal A, Rank DL, Gupta SK, Vanden Boom T, Lee SS (2005) Effects of elevated pCO2 and osmolality on growth of CHO cells and production of antibody-fusion protein B1: a case study. Biotechnol Prog 21: 70–77

Zielke HR, Oznard PT, Tildon JT, Sevdalain DA, Cornblath M (1978) Reciprocal regulation of glucose and glutamine utilization by cultured human diploid fibroblasts. J Cell Phys 95: 41–48

Complementary Reading

Al-Rubeai M, Singh R (1998) Apoptosis in cell culture. Curr Opin Biotechnol 9: 152–156

Butler M (2004) Animal Cell Culture and Technology – The Basics (2nd edition). Oxford University Press, New York

Doyle A, Griffiths JB (1998) Cell and Tissue Culture: Laboratory Procedures in Biotechnology. Wiley, New York

Fletscher T (2005) Designing culture media for recombinant protein production. BioProcess 1: 30–36

Freshney R (1994) Culture of animal cells. Wiley, New York

Häggström L (2000) Animal cell metabolism. In: Spier R (ed) Encyclopedia of Cell Technology. Wiley, New York, pp 392

Henzler HJ, Kauling OJ (1993) Oxygenation of cell cultures. Bioproc Eng 9: 61–75

Ozturk SS, Hu WS (eds) (2006) Cell Culture Technology For Pharmaceutical and Cell-Based Therapies. Taylor & Francis, New York

Pörtner R (1998) Reaktionstechnik der Kultur tierischer Zellen. Shaker, Aachen

Pörtner R (ed) (2007) Animal Cell Biotechnology – Methods and Protocols. Humana Press, Clifton, UK

Tziampazis E, Sambanis A (1994) Modelling cell culture processes. Cytotechnol 14: 191–204

Wurm FM (2004) Production of recombinant protein therapeutics in cultivated mammalian cells. Nat Biotechnol 22: 1393–1397

Chapter 3
Bioreactors for Mammalian Cells: General Overview

D. Eibl and R. Eibl

Abstract For the development and manufacturing of biotechnological medicines, the in vitro cultivation of animal cells has now become an accepted technology. In fact, about 50% of all commercial biotechnological products used for in vivo diagnostic and therapeutic purposes today are made using procedures based on animal cells. In addition to products from cells that are glycoproteins (drug products, e.g., cytokines, growth hormones, hematopoietic growth factors and antibodies, and viral vaccines, see Chaps. 1 and 2), cells as products for regenerative medicine, namely cellular therapies and tissue engineering, have been successfully investigated in clinical trials and introduced on the market.

Bioreactors from small (milliliter range up to 10 L) to large scale (above 500 L) have been developed over the past 50 years for animal-cell-culture-based applications. Suitable cell and tissue culture types displaying similar characteristics and specific differences (Chap. 2) have resulted in a variety of bioreactor types and their modifications, manufactured from plastics, glass, or steel. Although no universal bioreactor suitable for all cell and tissue culture types exists, it is obvious that conventional stirred bioreactors from stainless steel are the gold standard and dominate both in R&D and manufacturing.

This chapter aims to present a general overview of suitable bioreactor types for animal cells. On the basis of differentiation between static and dynamic bioreactors as well as methods of power input and primary pressure, we attempt to categorize the most frequently used cell culture bioreactor types, explain their typical working principles, and deduce possible fields of application. Furthermore, cell culture bioreactor trends for R&D and manufacturing, and special features of bioreactors for 3D tissue formation and stem cell cultivation are summarized.

D. Eibl, R. Eibl
Institute of Biotechnology, Zurich University of Applied Sciences, Department of Life Sciences and Facility Management, Wädenswil, Switzerland
dieter.eibl@zhaw.ch

R. Eibl et al., *Cell and Tissue Reaction Engineering: Principles and Practice,*
© Springer-Verlag Berlin Heidelberg 2009

3.1 Technical Terminology: Bioreactor/Fermentor, Bioreactor Facility

A bioreactor is defined as a closed system (vessel/bag or apparatus) in which a biochemical reaction involving a biocatalyst (in this case animal cells) takes place (Chmiel 2006). In the process, the biocatalyst is converted into the desired product, which is biomass (seed inoculum, cells for transplantation) or an expressed protein in general. It should be pointed out that the term "fermentor" exclusively characterizes a bioreactor operating with fast growing microorganisms (e.g., bacteria and fungi) in American English. Consequently, the term "bioreactor" is used here for cultivation of mammalian cells. However, in Europe it is customary to use the terms "bioreactor" and "fermentor" for cultivation systems involving living cells (Krahe 2000; Stadler 1998).

The primary role of a bioreactor is to provide containment with sustainable conditions for cell growth and/or product formation. Its type and configuration considerably influence cultivation results (which frequently aim for medium to high biomass concentrations/cell densities and/or product titers in grams per liter range) and as a consequence process efficiency (Wurm 2004, 2005). In general, a cell culture bioreactor has to meet the following demands:

- Guaranteed cell-to-cell contact and a surface for cell detachment in case of anchorage-dependent growing cells
- Homogeneous and low-shear mixing and aeration
- Sufficient turbulence for effectual heat transfer
- Adequate dispersion of air and gas
- Avoidance of substrate segregation
- Measurability of process variables and key parameters
- Scale-up capability
- Long-term stability and sterility
- Ease of handling
- Reasonable maintenance.

Traditional small-scale bioreactors (e.g., petri dishes, T-flasks, culture bags, shake flasks, spinner and roller bottles) are characterized by simple design and a low level of instrumentation and control. For this reason, in contrast to more sophisticated bioreactors (self-contained bioreactor facilities), they usually require external equipment such as incubators or shakers to ensure an appropriate physical and chemical environment for the cells. Typically, a bioreactor facility consists of a bioreactor vessel, a measurement and control unit, a drive, a heating and cooling circuit, an air inlet and outlet duct, sampling and harvest systems, sterile couplers, fittings, piping and safety installations (e.g., safety valves), a burst disc, etc. Moreover, systems for cleaning-in-place (CIP) and sterilization-in-place (SIP), vessels for precursors as well as feed and dosage, and systems for cell retention are also available.

3.2 Suitable Bioreactor Types for Mammalian Cell Cultures

3.2.1 Categorization, Functional Principle, Possible Fields of Application

Bioreactor development, which is reflected in a multitude of bioreactor types and modifications, was driven by increased knowledge of cell biology, cultivation techniques, biochemical engineering, and bioreactor design in combination with heightened interest in cell-culture-based products. Devising a categorization of all the bioreactor types that are available on the market today is becoming increasingly difficult. Thus, different approaches for general bioreactor categorization are found in reported studies. In the main, categorization is made according to distribution of the biocatalyst, form and fixation of the biocatalyst, its metabolic and growth type, bioreactor mode, bioreactor configuration, continuous phase, reaction kinetics, power input, and primary pressure (Kim et al. 2002; Kunz et al. 2005; Menkel 1992; Moser 1981; Pàca, 1987; Schügerl 1980; Voss 1991).

We recommend categorizing the most widely used animal-cell-culture bioreactors (including basic small-scale culture systems) in a way that takes mass and energy transfer as well as characterization of power input into account. As depicted in Fig. 3.1, we generally distinguish cell culture bioreactors in dynamic systems, which are characterized by mechanical, hydraulic, or pneumatic power input, from those in static systems. Existing hybrid systems (Auton et al. 2007) that can be operated as combined pneumatically driven airlift bioreactors/stirred bioreactors are not considered here. In dynamic systems, the generated power input of mechanically, hydraulically, or pneumatically driven bioreactors is responsible for heat transfer, mixing, homogenization, dispersion, and suspension, and therefore cell growth and product formation. However, in bioreactors with unenforced power input (static systems), cell growth and product formation are exclusively effected by conduction and reaction processes within the reaction system and by interaction with the environment.

3.2.1.1 Static Bioreactors with Unenforced Power Input

Static systems characterized by unenforced power input are petri dishes (Fig. 3.2a), multiwell plates (plates having a number of wells at the bottom, Fig. 3.2f), T-flasks (plastic screw-capped tissue culture flasks ranging up to 300 cm², Fig. 3.2b), and gas-permeable bags from polypropylene or Teflon (Fig. 3.2d), including their modifications.

They are ideally suited for the in vitro expansion of anchorage-dependent growing and suspension cells. Multitray cell culture systems, also called *multisurface plate units* (multilayer stack with up to 40 trays and a maximum size of 24,000 cm², Fig. 3.2c, e.g., NUNC Cell Factory), and simple membrane flask bioreactors

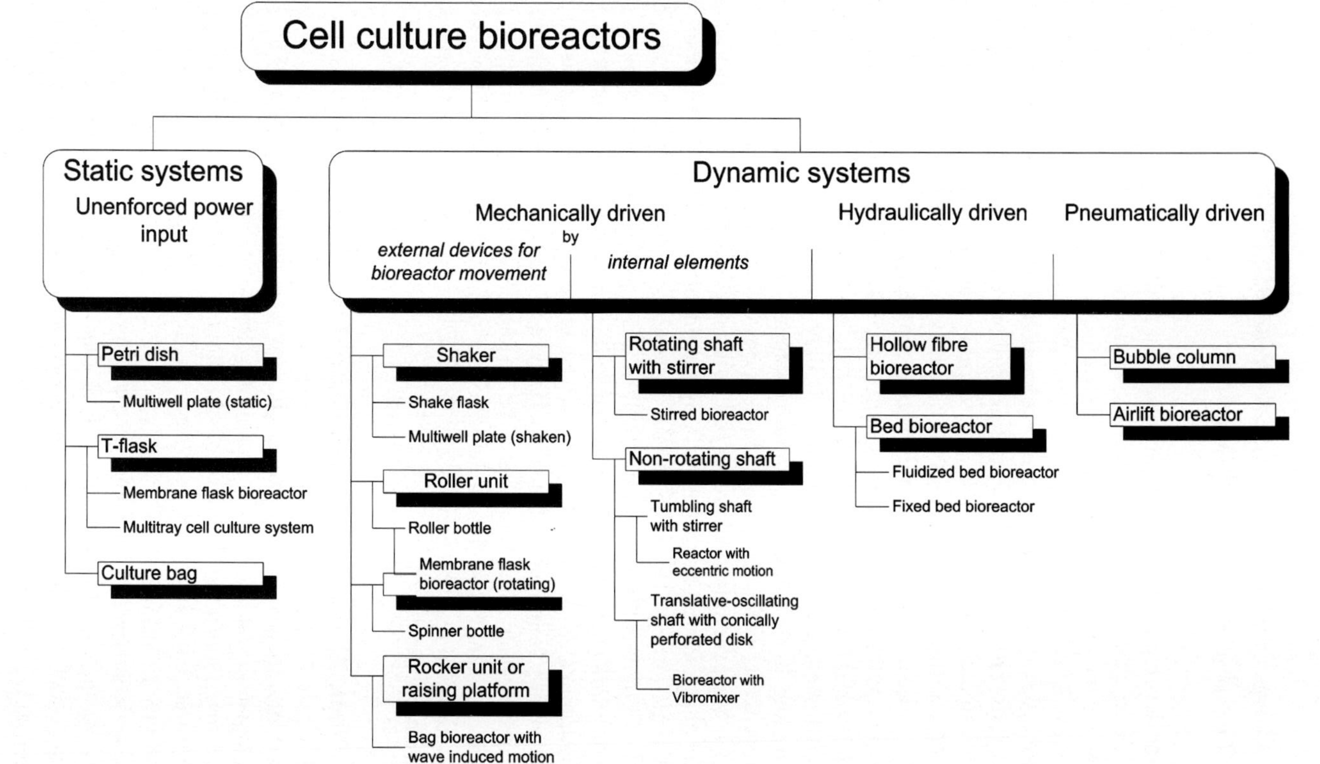

Fig. 3.1 Cell culture bioreactor categorization

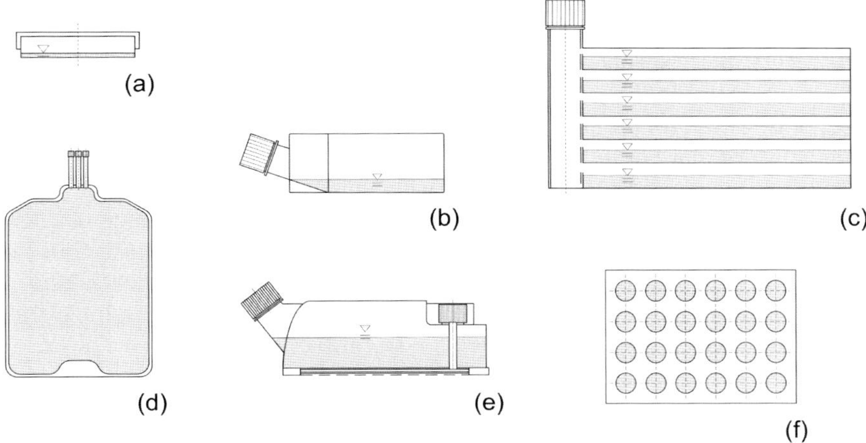

Fig. 3.2 Basic scheme of static cell culture bioreactors: (**a**) Petri dish, (**b**) T-flask, (**c**) Multitray cell culture system, (**d**) Culture bag, (**e**) Static membrane flask bioreactor (CELLine), (**f**) Multiwell plate

(two-compartment systems, Fig. 3.2e, e.g., CELLine) clearly provide larger surface areas (although they originate from the T-flask) and represent typical cell production systems for suspended and anchorage-dependent growth of animal cells. Here, we would like to add that even today animal as well as human vaccine manufacturers use multitray cell culture systems (DePalma 2002; Falkowitz et al. 2006; Schwander and Rasmusen 2005).

In membrane flask bioreactors, cells are kept in a cell compartment, designed as a membrane bag, which has been placed in a large flask containing the culture medium. Cells are fed by nutrients from the outer medium compartment, which pass over the dialysis membrane, while secreted proteins are accumulated in the cell compartment. Efficient gas transfer is ensured by the silicone membrane on the outside of the bioreactor. Despite its small culture volume, this type of bioreactor enables high cell densities above 10^7 cells/mL to be achieved and therefore high levels of protein (Baumann 2004; McArdle 2004). In this regard, membrane flask bioreactors are systems of choice for screening experiments and clinical sample production (Bruce et al. 2002; Eibl et al. 2003; Eibl and Eibl 2006a; Scott et al. 2001; Trebak et al. 1999).

3.2.1.2 Dynamic Bioreactors

Among dynamic cell culture bioreactors, there are bioreactors mechanically driven by external elements, bioreactors mechanically driven by internal elements having rotating or non-rotating shafts with stirrers or disks, hydraulically driven systems, and pneumatically driven systems.

Bioreactors Mechanically Driven by External Devices

With only a few exceptions, mechanically driven bioreactors with an external device for bioreactor movement (shaker platform, roller unit, spinner unit) are used exclusively for cell propagation. The most famous exception is commercial vaccine production in roller bottles (Aunins 2000; Guardia and Hu 1999). Roller bottles (Fig. 3.3b) are cylindrical vessels (2.5 L total volume or 1,750 cm^2 maximum area) that revolve slowly to bathe the cells attached to the inner surface in the medium. While the growing of anchorage-dependent animal cells is preferably performed in roller bottles and their modifications (e.g., MiniPerm, Fig. 3.3c), shake flasks (Fig. 3.3a) and spinner bottles (bottles with a central magnetic stirrer shaft and side arms for the addition and removal of cells, Fig. 3.3d) have been principally employed in the cultivation of suspension cells (Cowger et al. 1999; Falkenberg 1998; Heidemann et al. 1994; Lindl 2002; Müller 2001; Schneider 2004).

If it is necessary to increase oxygen transfer and reduce shear stress for animal suspension cells, bag bioreactors (Fig. 3.3e) with wave induced motion (WIM) are preferred to shake flasks, spinner or roller bottles, and stirred cell culture bioreactors with surface or membrane aeration (Eibl and Eibl 2006a,b; Eibl and Eibl 2007). In rocking or shaking bag bioreactors with WIM (e.g., BioWave, Wave Bioreactor, BIOSTAT CultiBag, AppliFlex, CELL-tainer, Tsunami-Bioreactor, Optima and OrbiCell), mass and energy transfer is manually adjusted via the rocking angle, rocking rate, and filling level (Chap. 5). These reactors have a plastic cultivation chamber, which has been designed as a bag. The gas-permeable and surface-aerated bag is fixed by a clamp arrangement and moved on the rocker unit. Among bag bioreactors with WIM, the BioWave and Wave Reactor have a leading position owing to maximum scale, the availability of scale-up criteria, the hydrodynamic expertise, and the convincing results of oxygen transport efficiency investigations. Moreover, different studies have revealed their potential for seed inoculum production, process development, and small- and middle-volume production of bioactive agents. For example, the application of both reactor systems for monoclonal antibody production, resistin secretion with HEK 293 cells, insect-cell-based protein expression by using the baculovirus system, secreted alkaline phosphatase (SEAP) production with Chinese hamster ovary (CHO) cells, and production of recombinant adeno-associated virus and lentivirus has been described (Eibl and Eibl 2003, 2004, 2005, 2006b; Flanagan 2007; Hami et al. 2003; Negrete and Kotin 2007; Pierce and Shabram 2004; Singh 1999; Tang et al. 2007; Weber et al. 2001). In addition, IDT (Impfstoffwerk Dessau-Tornau) is already using the BioWave for the commercial manufacture of a vaccine for minks (Hundt et al. 2007).

Bioreactors Mechanically Driven by Internal Elements

Independent of scale, production organism type, and product, stirred cell culture bioreactors (Fig. 3.3g), in which power input for mass and heat transfer is controlled mechanically, dominate. They are basically equipped with aeration devices, an

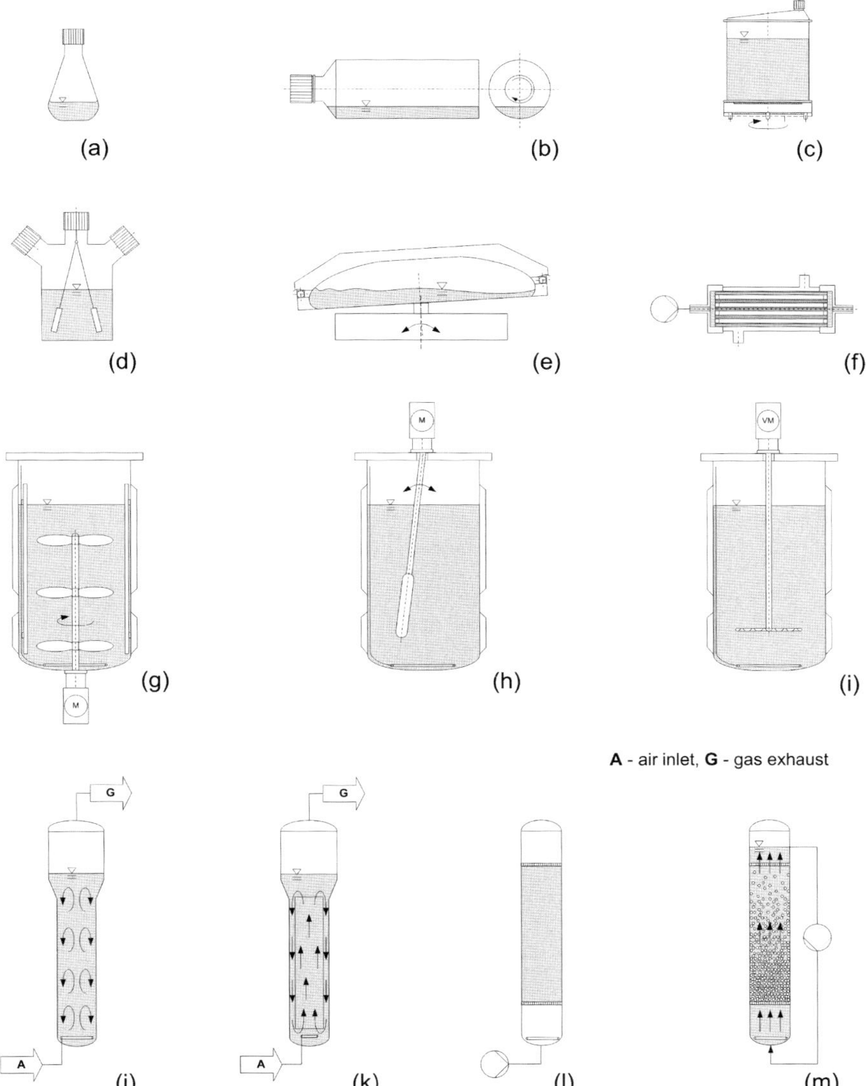

Fig. 3.3 Basic scheme of dynamic cell culture bioreactors: (**a**) Shake flask, (**b**) Roller bottle, (**c**) Rotating membrane flask bioreactor (MiniPerm), (**d**) Spinner flask, (**e**) Rocking bag bioreactor with wave induced motion, (**f**) Hollow fiber bioreactor, (**g**) Stirred bioreactor, (**h**) Bioreactor with eccentric motion stirrer, (**i**) Bioreactor with Vibromixer, (**j**) Bubble column, (**k**) Airlift bioreactor, (**l**) Fixed bed bioreactor, (**m**) Fluidized bed bioreactor

impeller, and baffles (Nienow 2006), and are applied in seed inoculum production, screening experiments, optimization purposes, and manufacturing processes. Prominent commercial product examples are factor VIII, tPA, interferon, herceptin, and avastin (Griffiths 2000; Kretzmer 2002; Ozturk 2006; Walsh 2003, 2005; Wurm 2005). The

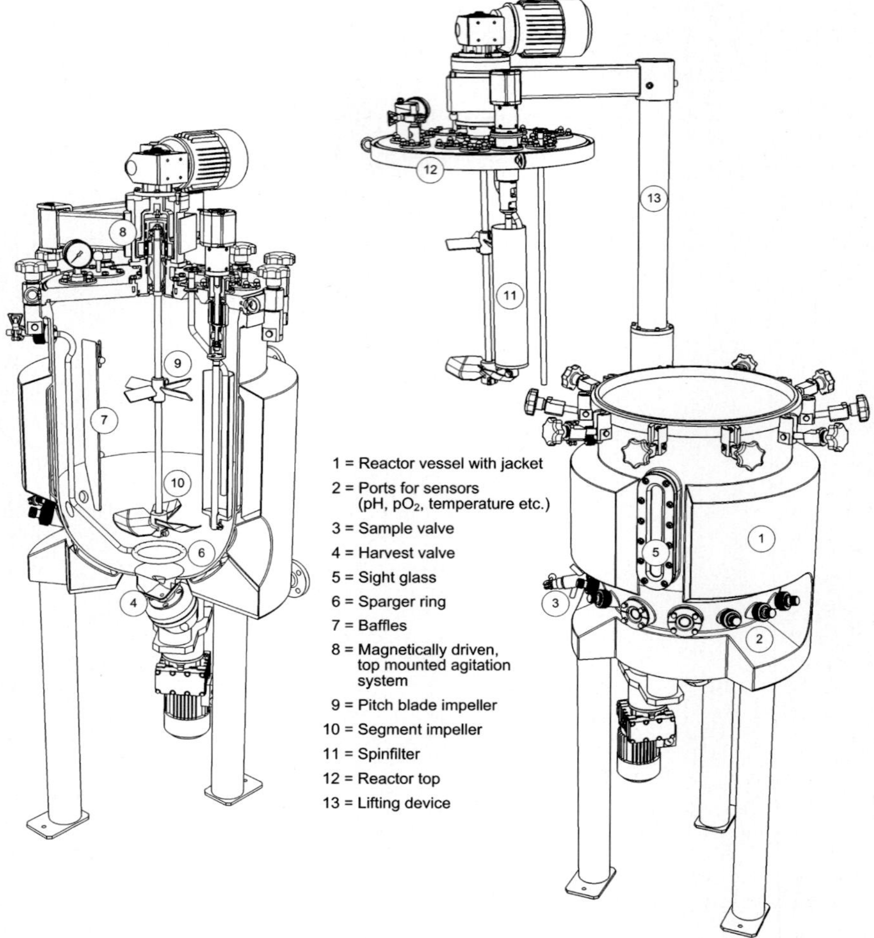

Fig. 3.4 Typical setup of stirred cell culture bioreactor (reproduced with kind permission of ZETA AG)

typical setup of stirred bioreactors for applications with animal cell cultures growing in suspension is schematically drawn in Fig. 3.4.

General design criteria were derived from stirred bioreactors for microbial culture and modified to meet the requirements of the sensitive mammalian cells, as outlined in Table 3.1.

It can be seen that stirred bioreactors for bacteria or yeasts growing in suspension primarily differ from those for plant and mammalian suspension cells iin vessel size, vessel geometry, power input, oxygen transfer coefficient, impeller type, tip speed, and sparger design. For more detailed information and explanations, the reader is referred to Chaps. 4 and 5.

Table 3.1 Comparison of commercial stirred reactors for microbial cells, mammalian cells, and plant suspension cells (Curtis 1999; Eibl and Eibl 2004; Kieran et al. 2000; Nienow 2006; Varley and Birch 1999; Zhong 2001)

Feature	Microorganisms	Mammalian cells	Plant cells
Size (total volume)	> 100 m^3	≤ 25 m^3	≤ 100 m^3
Vessel geometry (overall L:D)	3:1 ratio	2:1 ratio	2:1 ratio
$k_L a$	> 20 h^{-1}	≤ 15 h^{-1}	≤ 15 h^{-1}
P/V	4–10 kW m^{-3}	30–50 W m^{-3}	20 W m^{-3}–2 kW m^{-3}
Impellers	Radial flow impellers (turbine impellers)	Axial flow impellers (marine-type impellers)	Axial flow impellers (marine-type impellers)
	Impellers with distributed power input (e.g., InterMig impeller)	Combinations of axial and radial flow impellers	Combinations of axial and radial flow impellers
		Impellers with distributed power input (e.g., InterMig impeller)	Impellers with distributed power input (e.g., InterMig impeller)
Tip speed	< 20 ms^{-1}	0.3–2 ms^{-1}	≤ 2.5 ms^{-1}
Sparger design	Bubble aeration: ring, pipe, plate, frit	Bubble aeration: ring, pipe, plate, frit	Bubble aeration: ring, pipe, plate, frit
		Bubble-free aeration: tubes of silicone, external aeration (bypass, spinfilter, Vibromixer)	Bubble-free aeration: tubes of silicone, external aeration (bypass, spinfilter, Vibromixer)
Standard instrumentation	Temperature, agitation speed, pressure, pH, dissolved oxygen, dissolved carbon dioxide, liquid level, weight, air flow rate		
Product example	*E. coli*-based Humulin	CHO cell-based Herceptin	*T. chinensis*-based Paclitaxel (Part II, Chap. 3)

Whereas stirred cell culture bioreactors are typically equipped with dynamic seals (e.g., magnetic coupling, single or double mechanical seals), the tumbling or vibrating shafts of bioreactors with eccentric motion (Fig. 3.3h) and bioreactors with Vibromixer (Fig. 3.3i) are sealed with static components, for example, membranes, bellows, or metal compensators. As a consequence, these bioreactor types are considered to be safer (containment) compared to stirred systems equipped with single or double mechanical seals. In addition, lower shear stress and foaming are reported with tumbling motion, together with upward as well as downward movement of the conically perforated disk attached to a translative-oscillating shaft (Blasey et al. 2000; Griffiths 1999; Krahe 2000; Lehmann et al. 1998). In spite of these advantages, bioreactors with Vibromixer and, in particular, bioreactors with eccentric motion have played a subordinate role in mammalian cell cultivation to date.

Hydraulically Driven Systems

Like membrane flask bioreactors, hollow fiber reactors (Fig. 3.3f) support high cell density growth with a tissue-like architecture in animal cell cultivation. However, hollow fiber reactors operating in perfusion mode (Chap. 5) belong to the group of popular hydraulically driven systems, where energy input is produced by special double-phase pumps. In most of their applications, the cells are grown in the extracapillary space outside the thousands of fibers that have been potted into a cylindrical cultivation module. By the fibers' molecular weight cut-off, the passage of macromolecules through the fiber wall is affected as oxygen enriched medium flows continuously through the fibers, the intracapillary space (Goffe et al. 1995). Typically, cell-secreted proteins are retained in the extracapillary space (Labecki et al. 1996; Marx 1998). Oxygen enrichment is accomplished by a separate module from silicone tubes or membranes, that is, the "oxygenator". Hollow fiber bioreactors can be universally used for both cells growing in suspension and adherent animal cells. However, besides probable mass transfer limitations, there is an absence of process monitoring in the immediate cell environment. Furthermore, there is a risk of product contamination by cell fragments and cell lysis products as well as of destruction of sensitive proteins by the high residence time of cells and products in the same bioreactor module. But the main disadvantage of hollow fiber bioreactors is their small culture volume, which ranges only from 2.5 to 1000 mL (Brecht 2004). Taking all these disadvantages into account, it becomes obvious that the most important application of hollow fiber bioreactors is for fast, flexible, small-scale production of antibodies for diagnostic and research purposes (Chu and Robinson 2001; Dowd et al. 1999; Jackson et al. 1999).

Finally, bed bioreactors (Fig. 3.3l–m) are further representatives of hydraulically driven cell culture bioreactors. Bed bioreactors are directly linked to the use of animal cells for cultivations in an immobilized form. Owing to the characteristics of a bed containing cells that have been immobilized on microcarriers (small porous particles, usually spheres from 100 to 300 ∝m in size, see Sect. 4.3.), we distinguish between packed bed or fixed bed bioreactors and fluidized bed bioreactors (Groot 1995; Looby et al. 1990; Nilsson et al. 1986).

As suggested by its name, the fixed bed bioreactor has highly densely packed carrier material, which forms the fixed bed. A typical fixed bed bioreactor (Fig. 3.3l) is composed of a cylindrical bioreactor chamber filled with carriers of porous glass or macroporous materials, a gas exchanger, a medium storage tank and a pump that circulates the culture medium between the bioreactor, and the medium storage tank. Medium channeling in the bed, blockage of pores due to high cell densities, poor gas transfer and detachment, and cell washout (elutriation) limit its applications. The main advantages of this reactor are the low surface shear rate, the absence of particle–particle abrasion, and the increased space–time yield. Hence, fixed bed bioreactors (up to 100 L medium volume) are capable of providing high cell densities in cultivations with animal suspension cells secreting proteins, and with animal anchorage-dependent growing cells used for virus production (Kang et al. 2000; Kompier et al. 1991; Meuwly et al. 2005; Wang et al. 1992; Whiteside and Spier 2004).

When packed bed bioreactors are operated in upflow mode, the bed expands at high liquid flow rates and follows the motion of the microcarriers to which the cells have been attached. This is the working principle of the ordinary fluidized bed bioreactor (Fig. 3.3m). Fluidization of the microcarriers is achieved when the flow of fluid through the bed is high enough to compensate for their weight. With respect to optimum mass and heat transfer, fluidized bed bioreactor operation aims to provide a fluidized bed to ensure movement of all particles (microcarriers and cells) and avoid their sedimentation or flotation. In order to meet these conditions, an optimized incident flow is required along with a narrow range of size, form, and density of the microcarriers. In the cultivation of animal suspension and adherent cells, fluidized bed bioreactors in laboratory and production scale (maximum 100 L total volume) have been proven to be efficient if high cell density, antibody, recombinant protein, or virus titer is the focus (Born et al. 1995; Dean et al. 1987; Lundgren and Blüml 1998; Wang et al. 2002).

Pneumatically Driven Systems

Fluidized bed bioreactors usually differ from pneumatically driven systems owing to the fact that the latter do not specifically require the use of immobilized cells as they were developed for free suspension cells. In bubble columns (Fig. 3.3j) and airlift bioreactors (Fig. 3.3k), mass and heat transfer is mostly achieved by direct sparging of a tall column with air or gas that is injected by static gas distributors (diffuser stones, nozzles, perforated plates, diffuser rings) or dynamic gas distributors (slot nozzles, Venturi tubes, injectors, or ejectors). While the ascending gas bubbles cause random mixing in bubble columns, fluid circulation in airlift bioreactors is obtained by a closed liquid circulation loop, which permits highly efficient mass transfer and improved flow and mixing. In airlift bioreactors, this circulation loop results from the mechanical separation of a channel for gas/liquid upflow (riser) and a channel for downflow (downcomer), and their connection at the top and bottom of the column. Hence, fluid flow in airlift bioreactors is driven by density difference between sparged fluid in the riser and nonsparged fluid in the downcomer. Both airlift bioreactors with an external loop (fluid circulation through the external loop) and with an internal loop (columns with cylindrical draught tube or simple baffles generate an internal loop) are available.

Mainly because of their relatively simple mechanical configuration, bubble columns and airlift bioreactors are characterized by low cost. In comparison to stirred bioreactors, the lower energy requirement, minimized problems of long-term sterility (no moving parts, shaft and mechanical seals), and ease of scale-up can be highlighted as additional benefits of bubble columns and airlift bioreactors. However, if there are strong variations in biomass concentration, viscosity, surface tension, ionic concentration, inadequate mixing, foaming, flotation, and bubble coalescence are potential disadvantages. Moreover, bursting gas bubbles in bubble columns and airlift bioreactors can increase shear or hydrodynamic stress acting on cells (Sect. 4.1). Nevertheless, an out-dated opinion that these systems are low-shear devices owing to the lack of mechanical agitation can still be found. Furthermore, use of suitable sparger types or enriched air or pure oxygen and the availability of protective additives in the culture medium preclude excessive cell damage under conventional cultivation conditions (Christi 2000; Guardia and Hu 1999; Meier et al. 1999). In fact, it seems that the homogeneity of shear forces demonstrated in airlift bioreactors is probably responsible for the success of shear-sensitive cultures in this bioreactor type (Merchuk and Gluz 1999).

Most reports on the use of pneumatically driven cell culture bioreactors are based on airlift bioreactors with a concentric draft tube (Fig. 3.3k) and typical aspect ratios in the range of 6:1 to 14:1 (Christi 2000; Fenge and Lüllau 2006). They are considered to be suitable for animal cells (e.g., CHO, baby hamster kidney (BHK), hybridomas, mouse myeloma cell line NS0) up to a maximum batch size of 2000 L (Kretzmer 2002; Petrossian and Cortessis 1990; Varley and Birch 1999; Wang et al. 2005).

Finally, we would like to mention that all the dynamic cell culture bioreactor systems presented for the cultivation of animal suspension cells can also be used to grow adherent animal cells provided that they are operated with membranes or microcarriers. The maximum reported size of a bioreactor vessel for cultivating animal cells on microcarriers is 6000 L (http://www5.amershambiosciences.com. Cited 26 August 2006). It is used by Baxter Biosciences for influenza production based on Vero cells growing on Cytodex microcarriers in stirred bioreactors. The fundamentals and applications of microcarrier cultivations are discussed comprehensively in Sect. 4.3.

3.2.2 Bioreactor Trends and the Increasing Acceptance of Disposables

In order to describe the existing trends for cell culture bioreactors, which are summarized in Fig. 3.5, the three phases of development for a biotechnological product are considered. It is generally demanded that development time is short, costs are low, and flexibility of the facility as well as process safety are high. Phase 1 includes know-how transfer from basic research and technological development, whereas phase 2 represents the development of good manufacturing practice

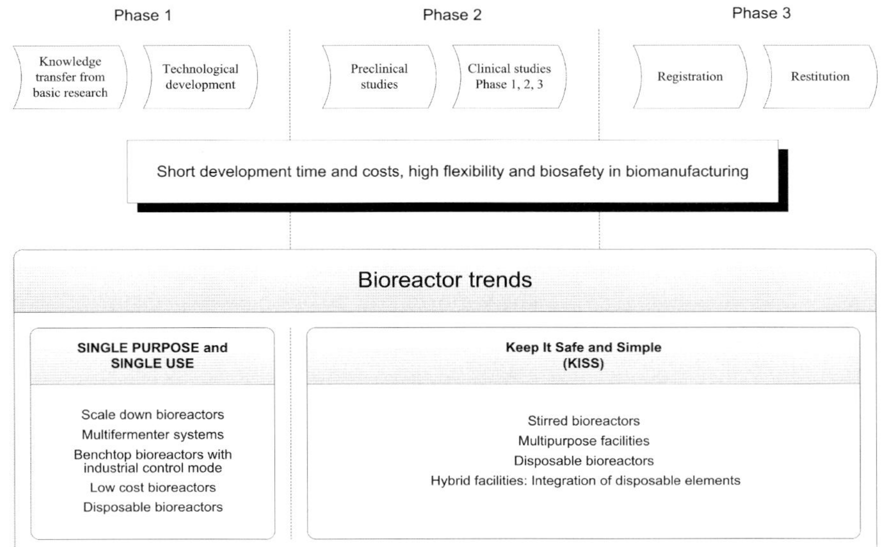

Fig. 3.5 Cell culture bioreactor trends in R&D and manufacturing

(GMP) sample production consisting of preclinical studies and clinical studies. Phase 3 is the actual GMP manufacturing. Because different factors have an impact on bioreactor design and its periphery in every phase of product development, it makes sense to distinguish between bioreactor trends in R&D and bioreactor trends in manufacturing. The dividing line between R&D and manufacturing may be drawn between phases 1 and 2. In phase 1, it is not necessary to work under GMP-compliant conditions; however, phases 2 and 3 require consideration of GMP standards.

As a consequence of cost and time, a change of bioreactor type should be avoided at this point. The change of the bioreactor type is theoretically possible here but implicates high costs. In research there is a tendency to apply single purpose and single use facilities where possible. The focus lies on the following:

- Scale down bioreactors, representing small-scale laboratory bioreactors for cell line screening, medium optimization, and basic studies, which operate with microwell plates, tubes, spinner flasks, or shake flasks and allow online measurement and regulation of relevant process parameters (DeJesus et al. 2004; Kumar et al. 2004; Muller et al. 2006; Puskeiler et al. 2005; Wurm 2004);
- Multifermenter systems equipped with laboratory stirred bioreactors or bubble columns, generating a large amount of process data within a short time and possibly contributing to savings in intermediate steps during process scale-up (Frison and Memmert 2002; Meis et al. 2005; Puskeiler and Weuster-Botz 2004);
- Bench top bioreactors with industrial control mode, characterized by easy handling and an analogous control unit comparable to those of bioreactors for manufacturing (Eibl and Eibl 2004; Glaser 2004);

- Low cost bioreactors in laboratory scale with minimal equipment and therefore low investment costs (Eibl and Eibl 2004; Glaser 2004);
- Disposable bioreactors (membrane bioreactors, bag bioreactors with WIM, bag bioreactor with stirrer or Vibromixer, pneumatically driven bag bioreactors, hybrid bag bioreactors, orbitally shaken bioreactors) with containers or bags manufactured from different types of polymeric material and meeting USP Class VI as well as ISO 10993 specifications (Auton et al. 2007; Aldridge 2004, 2005, 2007; Brecht 2007; Castillo and Vanhamel 2007; DePalma 2004; Eibl and Eibl 2005, 2007; Fries et al. 2005; Glaser 2004; Houtzager et al. 2005; Langhammer et al. 2007; Odum 2005; Ozturk 2007; Kunas 2005; Morrow 2007; Pierce and Shabram 2004; Pörtner 2006; Sinclair and Monge 2004; Thermo Fisher Scientific 2007, Wurm 2007).

Table 3.2 Disposable bioreactor vs. stainless steel bioreactor

Disposable		Stainless steel	
Pros	Short set-up times, presterile → Ready to use	*Cons*	Extensive installation
	No sterilization and cleaning time → Reduction of con-tamination risk and validation efforts		Hygienic design of bioreactor and extensive CIP/SIP cycles to reduce contamination risks → Increased capital investment
	High flexibility and simplicity → Reduction of staff		Inflexibility by defined facility design → Specially trained staff
	Shorter production turnaround times		Higher production turnaround times
Cons	New technology → Intensified logisti-cal effort and per-sonnel training	*Pros*	Proven technology and long-term experi-ence
	Secretion of leacha-bles and extracta-bles → Potential inhibition of cell growth in CD minimal medium		Correct material usage → Harmless opera-tion, no leachables and extractables
	Culture volume limita-tions		Unlimited scalability
	Limited number of disposable sensors for process moni-toring		Proven sensors for bio-reactors from stain-less steel

As soon as GMP conditions are regulated, bioreactors will have to be simple and safe (KISS principle: Keep It Safe and Simple) as in the case of traditional, high-quality finished stirred bioreactors made of stainless steel, which operate primarily in fed-batch mode and are equipped with standard instrumentation. Multipurpose production facilities making high demands on process validation and cleaning are currently popular in this context (Eibl and Eibl 2004; Howaldt et al. 2006; Kranjac 2004; Stadler 1998). On the other hand, the use and acceptance of disposable cultivation technology, namely disposable bioreactors and disposable elements (couplers, filters, storage bags, mixers, sampling systems, sensors, etc.), to set up hybrid manufacturing facilities are increasing (Boehm 2007; DePalma 2006; Glaser 2005, 2006; Haughney and Aranha 2003; Huang 2005; Morrow 2006). Whereas considerable advantages result from the use of disposables and disposable bioreactors in particular (compare Table 3.2), there are some initial limitations concerning the potential secretion of leachables and extractables from plastic layers as well as scalability (Altaras et al. 2007). However, intensive research is being carried out to overcome these limitations and finally to reduce stainless steel in biomanufacturing facilities. Fully controlled stirred bag bioreactors up to 2000 L culture are available, and in the next few years volumes up to 5000 L are expected (DePalma 2005; Langhammer et al. 2007; Ozturk 2007). Along with the increasing protein titers ($> 4\,\mathrm{g}\,\mathrm{L}^{-1}$) that are mainly a result of the improvements in animal cell-line engineering and culture medium optimization over the last 20 years, it is assumed that the way for completely disposable bioproduction trains is paved. Nevertheless, because of the principal advantages of proven performance and virtually unlimited scalability, conventional stirred cell culture bioreactors are not expected to be completely replaced by disposable bioreactors in the future.

3.3 Special Case: Bioreactors for Patient-Specific Therapies Based on Functional Tissue and Stem Cells

Whereas in vitro cultivation of animal cells to produce proteins and viruses is a field in which biotechnologists have long experience, in vitro cultivation of allogenic and autologous cells for patient-specific therapies represent a newly emerging discipline. It is aimed at large-scale cell production for cellular or gene therapies, generation of functional 3D tissue and constructs, and development of organ modules (Gomes and Reis 2004; Liszewsky 2006). Such extracorporeal organoids (also called *bioartificial organs*) functioning as bridge-to-transplant devices or assisting in bridge-to-native organ regeneration will be introduced in Part II/Chap. 2 of this book. They are not considered here. Subsequently, we will focus on bioreactors for growing functional 3D constructs (e.g., skin, bone, blood vessel, cartilage) prior to implantation, and bioreactors for large-scale expansion and differentiation of stem cells (embryonic stem cells, adult stem cells).

3.3.1 Bioreactors for Growing 3D Tissues

The infancy of tissue engineering is evidenced by the small number of registered commercial products (Table 3.3) that allow successful treatment of skin diseases and cartilage and bone injuries (Bock et al. 2003, 2005; CorCell 2003; Senker and Mahdi 2003; Transplant News 9/30/2005).

In growing 3D tissues, the ultimate cultivation goal is high cell and tissue density while guaranteeing tissue-specific differentiation and maintaining the functionality of in vitro-generated tissue. For this reason the consecutive realization of three cultivation steps, namely cell proliferation, cell adhesion on a scaffold, and cell differentiation (tissue formation), either in different bioreactors or within a single bioreactor, is the general rule. Consequently, the specifications of bioreactors for 3D constructs (Fig. 3.6) are different from those of bioreactors for the animal-cell-based production of seed inoculum and therapeutic agents discussed in the previous

Table 3.3 Selected tissue-engineered commercial products

Trade name	Product for	Manufacturing company	Country
Apligraf	Treatment of chronic wounds and skin diseases	Organogensis Inc.	USA
Dermagraft		Advanced Tissue Sciences	USA
CellActiveSkin		IsoTis	NL
BioSeedS, BioSeedM, MelanoSeed		BioTissue Technologies	D
Epicel		Genzyme Biosurgery	USA
Carticel	Knee injuries (ACT)	Genzyme Biosurgery	USA
ACI, MACI, MACI-A		Verigen	D
Chondrotransplant		Co.don	D
BioSeedC		BioTissue Technologies	D
HYALOGRAFT C		Fidia Advanced Biomaterials	Italy
Chondrotransplant disc	Herniated disc (ADCT)	Co.don	D
BioSeed-Oral Bone	Application in jawbone surgery	BioTissue Technologies	D
Osteotransplant	Several applications in bone surgery	Co.don	D
Skeletex	Bone repair	CellFactors	UK
OsteoCel	Bone repair and replacement	Osiris Therapeutics	USA

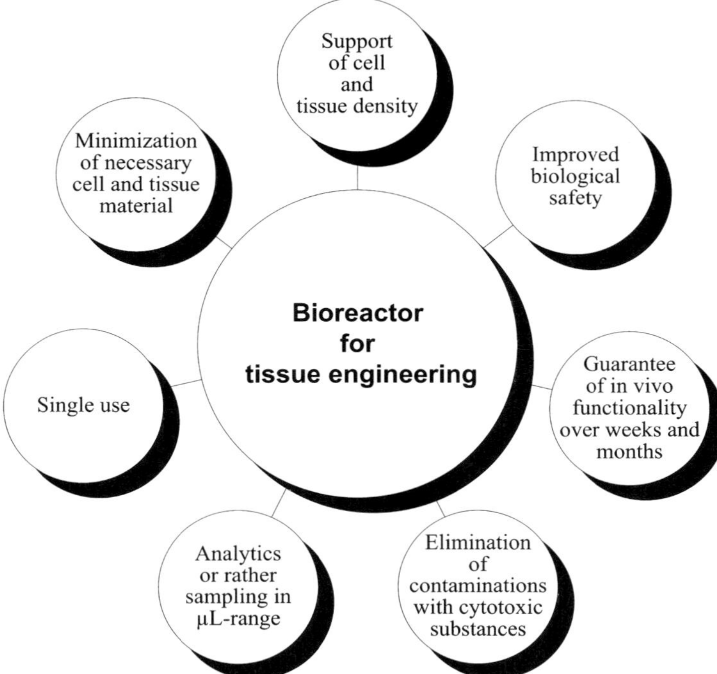

Fig. 3.6 Specifications of bioreactors for tissue-engineered products

sections of this chapter. Primarily, in vivo-like biochemical and biomechanical environment supporting cell and tissue density has to be provided by the bioreactor for proliferation and maturation. In many cases, a culture volume in milliliter range is sufficient for in vitro culture of tissue constructs. Because the functionality of the tissue generated in vitro has to be maintained over weeks and even months, long-term sterility and stability as well as high bioreactor instrumentation become stringent. Above all, disposable design is a necessity for commercial tissue-engineered products.

Cell proliferation, first step in 3D tissue engineering, is usually accomplished in typical static cell expansion devices including petri dishes, multiwell plates, T-flasks, or gas-permeable blood bags (Minuth et al. 2003; Purdue et al. 1997; Safinia et al. 2005). Higher total cell numbers have been achieved in mechanically driven roller bottles and spinner flasks operating with microcarriers (Pörtner and Giesse 2007). The generation of 3D tissue structures is promoted by equipping the roller bottles, spinner flasks, or perfusion chambers with scaffolds, which provide the template for tissue development, and degrade or are resorbed at defined rates (Martin et al. 2004; Ratcliffe and Niklason 2002). More complex bioreactors providing a controlled environment (e.g., temperature, humidity, pH) to direct cellular responses toward specific tissue structures have shown to be more favored to simpler static and dynamic cultivation systems. Therefore, milliliter-scale operating membrane bioreactors as well as fixed bed and fluidized bed bioreactors, where cells are immobilized on carriers or scaffolds, have been investigated for the 3D culture of hepatocytes (Jasmund and Bader 2002; Kullig and Vacanti 2004; Legallais et al. 2000; Gerlach 1996), skin cells (Prenosil and Villeneuve 1998), cardiovascular cells (Barron et al. 2003), and cartilage cells (Darling and Athanasiou 2003). Above all, tissue-specific bioreactors, which simulate in vivo environments better, have been suggested. These systems use mechanical strain (constant, cyclic or dynamic) in vitro, either in the form of shear stress generated by fluid flow, hydrodynamic pressure, or as direct mechanical stress applied to the scaffold (Altman et al. 2002; Bilodeau and Mantovani 2006; Gokorsch et al. 2004; Hoerstrup et al. 2000; Lyons and Pandit 2004). In rotary bioreactors characterized by low shear, high mass transfer rate, and simulated microgravity, a wide range of tissues (e.g., bones and cartilage, cardiac tissue, liver tissue) have been successfully grown (Guo et al. 2006; Marolt et al. 2006; Ohyabu et al. 2006; Okamura et al 2007; Song et al. 2006; Suck et al. 2006). Additional tissue engineering bioreactors have been developed with respect to specific tissue formation over the last years. Detailed reviews are summarized by Chaudhuri and Al-Rubeai (2005).

Further bioreactor developments in growing 3D tissues incorporate fully automated systems in order to reduce laborious manual work, risk of contaminations, and costs. Such autologous clinical tissue engineering systems based on a holistic approach from biopsy to implantation would enable qualified hospitals to perform all procedures for manufacturing the functional tissue within a single closed apparatus (Larcher et al. 2006; Martin et al. 2004). To that end, these new innovative bioreactor systems would eliminate the need for large and expensive GMP tissue manufacturing facilities.

3.3.2 Bioreactors for Large-Scale Expansion and Differentiation of Stem Cells

In case of stem cell cultivation for a cellular or gene therapy, there is normally the requirement for larger amounts of cells than in the manufacturing of 3D tissue. Taking specific end application into account, cultivation of stem cells in clinically relevant numbers results in bioreactors with capacities between 1 and 1000 L (Tzanakakis and Verfaillie 2006). The stem cells have to be expanded while sustaining their pluripotent (embryonic stem cells, multipotent adult progenitor cells) or multipotent (adult stem cells: hematopoietic stem cells, mesenchymal stem cells, gastrointestinal stem cells, epidermal stem cells, hepatitic stem cells) potential. The directed in vitro differentiation results in a mixed population of different cell lineages. Usually, the desired cell type represents only a small proportion of the produced stem cell population. The selection of the appropriate bioreactor is strongly affected by the stem cell type and its characteristics, and the purpose of cultivation (expansion and/or differentiation).

As in cell culture expansion, the bioreactor type for stem cell expansion is largely dependent on whether the cells are growing as single cells in suspension, in monolayers, or in cell aggregates. Large cell densities combined with large cell amounts can be achieved in membrane reactors, stirred reactors, rotary reactors, fixed bed reactors, and fluidized bed reactors partially operating with microcarriers (Cabrita et al. 2003; Chen et al. 2006; Nielsen 1999). However, it was reported that too high a cell density could result in loss of cell properties for some stem cell types (Ulloa-Montoya et al. 2005). In this case, the expansion procedure is aimed at low cell densities and rather frequent passaging.

According to Ulloa-Montoya, Verfaille, and Hu (2005), and Tzanakis and Verfaille (2006), the key in stem cell differentiation is the cell microenvironment mainly affected by the bioreactor, the differentiation medium with growth factors, and the culture mode. The ideal differentiation system should be scalable and fully controlled. It is recommended to differentiate pluripotent stem cells by culturing them as adherent monolayers at high cell densities or by growing them as embryoid bodies in suitable bioreactors (e.g., stirred bioreactors, rotating bioreactors).

3.4 Conclusions

Among the existing static and dynamic bioreactor types that have been described and characterized in this chapter, traditional stirred bioreactors with maximum volumes of $25\,m^3$ for animal cells have become the systems of choice in commercial biomanufacturing processes. They are characterized by their reusability as well as by costly hygienic design, and require specially trained staff for their trouble-free operation, which includes labor-intensive cleaning, sterilization, and validation procedures in GMP-compliant or otherwise regulated processes. This increases process time, process costs, development time, and, therefore, time-to-market for new products. For R&D and manufacturing of niche products in low- and middle-volume scale, disposable bioreactors and certain bag bioreactors, particularly those with culture volumes above 100 L, are alternative solutions. In spite of the growing acceptance of disposable bioreactors, their engineering aspects are insufficiently characterized. Furthermore, to date scale-up criteria have only been published for disposable wave-mixed systems (Chap. 5). On the other hand, knowledge of their advantageous features (easy handling, flexibility, high biosafety, improved process efficiency, and manufacturing costs) has enhanced the development of disposable bioreactors during recent years. Mechanically driven bag bioreactors of 5,000 L capacity are expected in the future. Further developments in biomanufacturing focus on scale down bioreactors, well-instrumented multifermenter systems, benchtop bioreactors with industrial mode, and low cost bioreactors.

On the other hand, there is an urgent need for the development of complex disposable bioreactors for functional tissue-engineered products. To this day, commercially available tissue-engineered products that meet GMP standards are manually manufactured in simply designed cultivation systems using expensive isolator technology. Encouraging results have been shown by prototypes of suitable bioreactors performing the different processing phases to generate 3D tissue via cell disintegration to cell propagation as well as differentiation within a single closed and automated system, and bioreactors applying controlled mechanical forces.

The rapid progress in stem cell research has driven in vitro generation, maintenance, and differentiation investigations of stem cells. Well-characterized disposable bioreactors with capacities between 1 and 1000 L will pave the way for personalized medicine.

3.5   Questions and Problems

1. Explain the terms bioreactor, fermentor, bioreactor facility, conventional bioreactor, and dispoable bioreactor.
2. Discuss the advantages and disadvantages of disposable bioreactors.
3. Explain the working principles of disposable wave-mixed bioreactors, disposable hollow fiber reactors, stirred bioreactors, bubble columns, and airlift bioreactors.
4. Summarize the existing trends for cell culture bioreactors used in R&D and commercial manufacturing.
5. What is implied by "KISS" and how is it realized in biomanufacturing based on mammalian cells?
6. Why do higher demands on bioreactor and manufacturing processes result from the cultivation of CHO cells growing in suspension than from *Escherichia coli, Pichia pastoris, Sf-9* cells, or tobacco suspension cells? Give reasons for your answer.
7. Compare cultivation aims in tissue engineering and mammalian-cell-based protein expression. Deduce general demands and design criteria for a bioreactor which is (a) suitable for mammalian-cell-based protein secretion and (b) suitable for the production of tissue-engineered products.
8. Draw up a scheme for a bioreactor that allows virus production with adherent growing HEK 293 cells.
9. Which bioreactors are suitable to expand stem cells? Give examples and reasons.

List of Abbreviations and Symbols

ACI	autologous chondrocyte implantation
ACT	autologous chondrocyte transplantation
ADCT	autologous disc chondrocyte transplantation
BHK	baby hamster kidney

CD	chemically defined
CHO	chinese hamster ovary
CIP	cleaning-in-place
D	vessel inside diameter
E. coli	*Escherichia coli*
EPO	erythropoietin
GMP	good manufacturing practice
HEK	human embryogenic kidney
k_La	oxygen transfer coefficient
L	vessel inside height
MACI	matrix coupled autologous chondrocyte implantation
NS0	mouse myeloma cell line
P/V	specific power input
R&D	research and development
rpm	revolution per minute
SIP	sterilization-in-place
Sf-9	*Spodoptera frugiperda* (subclone 9)
T. chinensis	*Taxus chinensis*
tPA	tissue plasminogen activator
V	volume of the medium
3D	three-dimensional
°C	degree centigrade

References

Aldridge S (2004) Biomanufacturing faces new set of challenges. GEN 24:1–9

Aldridge S (2005) New biomanufacturing opportunities & challenges. GEN 25:1–16

Aldridge S (2007) Techniques for cell culture improvement. GEN 27:38–39

Altaras GM, Eklund C, Ranucci C, Maheswari G (2007) Quantification of interaction of lipids with polymer surfaces in cell culture. Biotechnol Bioeng 96:999–1007

Altman GH, Lu HH, Horan RL, Calabro T, Ryder D, Kaplan DL (2002) Advanced bioreactor with controlled application of multi-dimensional strain for tissue engineering. J Biomech Eng 124:742–749

Aunins J (2000) Viral vaccine production in cell culture. In: Spier RE (ed) Encyclopedia of cell technology, vol 2. Wiley, New York, pp 1182–1217

Auton KA, Bick JA, Taylor IM (2007) Strategies for the culture of CHO-S cells. GEN 27:42–43

Barron V, Lyons E, Stenson-Cox C, McHugh PE, Pandit A (2003) Bioreactors for cardiovascular cell and tissue growth. Ann Biomed Eng 31:1017–1030

Baumann F (2004) CeLLine Zwei-Kompartment-Bioreaktoren: Einführung in die Technologie und Applikation. Workshop, Disposable Kultivierungstechnologie, Wädenswil, Switzerland, October 5–6

Bilodeau K, Mantovani D (2006) Bioreactors for tissue engineering: Focus on mechanical constraints. A comparative review. Tissue Eng 12:2367–2383

Blasey HD, Brethon B, Hovius R, Vogel H, Tairi AP, Lundström K, Rey L, Bernard AR (2000) Large scale transient 5-HT3 receptor production with the Semikli Forest Expression System. Cytotechnol 32:199–208

Bock AK, Ibarreta D, Rodriguez-Cerezo E (2003) Human Tissue Engineered Products: Today's Markets and Future Prospects (Report EUR 21000 EN). Europäische Kommission, Brüssel Luxembourg

Bock AK, Rodriguez-Cerezo E, Hüsing B, Bührlen B, Nusser M (2005) Human Tissue Engineered Products: Potential Socio-Economic Impacts of a new European regulatory Framework for Authorisation, Supervision and Vigilance (ISI-B-48-05). Europäische Kommission, Brüssel Luxembourg

Boehm J (2007) Linking stainless and single-use systems. GEN 27:42–43

Born C, Biselli M, Wandrey C (1995) Production of monoclonal antibodies in a pilot scale fluidized bed bioreactor. In: Beuvery EC, Griffiths Zeijlemaker WP (eds) Animal Cell Technology: Development Towards the 21st Century. Kluwer, Dordrecht, pp 683–686

Brecht R (2004) Bedeutung von disposable Hohlfasermodulen für die Herstellung von Biopharmazeutika. Workshop, Disposable Kultivierungstechnologie, Wädenswil, Switzerland, October 5–6

Brecht R (2007) Disposable Bioreactor Technologies: Challenges and Trends in cGMP Manufacturing. Presented at the BioProduction 2007, Berlin, Germany, October 29–31

Bruce MP, Boyd V, Duch C, White JR (2002) Dialysis-based bioreactor systems for the production of monoclonal antibodies – alternatives to ascites production in mice. J Immun Meth 264:59–68

Cabrita GJM, Ferreira BS, da Silva CL, Goncaalves, Almeida-Porada G, Cabral JMS (2003) Hematopoietic stem cells: From the bone to the bioreactor. Tibtech 21:233–240

Castillo J, Vanhamel S (2007) Cultivating anchorage-dependent cells. GEN 27:40–41

Chaudhuri JB, Al-Rubeai (2005) Bioreactors for Tissue Engineering. Kluwer, Dordrecht

Chen X, Xu H, Wan C, McCaigue M, Li G (2006) Bioreactor expansion of human adult bone marrow-derived mesenchymal stem cells. Stem Cells (online) doi:10.1634/stemcells.2005–0591

Chmiel H (2006) Bioreaktoren. In: Chmiel H (ed) Bioprozesstechnik. Elsevier, München, pp 195–215

Christi Y (2000) Animal-cell damage in sparged bioreactors. Tibtech 18:420–432

Chu L, Robinson DK (2001) Industrial choices for protein production by large-scale cell culture. Curr Opin Biotechnol 12:180–187

CorCell (2003) Expansion of umbilical cord blood stem cells. http://www.corcell.com. Cited 14 January 2006

Cowger NL, O'Connor KC, Hammond TG, Lacks DJ, Navar GL (1999) Characterization of bimodal cell death of insect cells in a rotating-wall vessel and shaker flask. Biotechnol Bioeng 64:14–26

Darling EM, Athanasiou KA (2003) Articular cartilage bioprocesses and bioreactors. Tissue Eng 9:9–26

Dean RC Jr, Karkare SB, Ray NG, Runstadler PW Jr, Venkatasubramanian K (1987) Large-scale culture of hybridoma and mammalian cells in fluidized bed bioreactors. Ann NY Acad Sci 506:129–146

DeJesus M, Girard P, Bourgeois M, Baumgartner G, Jacko B, Amstutz H, Wurm FM (2004) TubeSpin satellites: A fast track approach for process development with animal cells using shaking technology. Biochem Eng 17:217–223

DePalma A (2002) Options for anchorage-dependent cell culture. GEN 22:58–62

DePalma A (2004) Biomanufacturers face cell culture decisions. GEN 24:52–53

DePalma A (2005) Integration disposables into biomanufacturing. GEN 25:50–56

DePalma A (2006) Bright sky for single-use bioprocess products. GEN 26:50–57

Dowd JE, Weber I, Rodriguez B, Piret JM, Kwok KE (1999) Predictive control of hollow fiber bioreactors for the production of monoclonal antibodies. Biotechnol Bioeng 63:484–492

Eibl D, Eibl R (2005) Einwegkultivierungstechnologie für biotechnische Pharma-Produktionen. BioWorld 3: Supplement Swiss BioteCHnet

Eibl D, Eibl R (2004) Reaktorsysteme für die Herstellung von biopharmazeutischen Proteinwirkstoffen: Quo vadis? In: iba (ed) Proceedings zum 12. Heiligenstädter Kolloquium, Heiligenstadt, Germany, pp 271–279

Eibl R, Rutschmann K, Lisica L, Eibl D (2003) Kosten reduzieren durch Einwegreaktoren? BioWorld 5:22–23

Eibl R, Eibl D (2006a) Disposable bioreactors for pharmaceutical research and manufacturing. Proceedings of the Second International Conference on Bioreactor Technology in Cell, Tissue Culture and Biomedical Applications, Saariselkä, March 27–31, pp 12–21

Eibl R, Eibl D (2006b) Disposable bioreactors for inoculum production and protein expression. In: Pörtner R (ed) Animal Cell Biotechnology. Humana Press, Totowa, pp 321–335

Eibl R, Eibl D (2007) Disposable bioreactors for cell culture based-bioprocessing, Achema Worldwide News:8–10

Falkenberg FW (1998) Production of monoclonal antibodies in the miniPerm bioreactor: Comparison with other hybridoma culture methods. Res Immunol 6:560–570

Falkowitz K, Staggert J, Wedege V (2006) A system approach to improving yields in a disposable bioreactor. BioProcess Int 6:56–59

Fenge C, Lüllau E (2006) Cell culture bioreactors. In: Ozturk SS, Hu WS (eds) Cell Culture Technology for Pharmaceutical and Cell-based Therapies. CRC Press, New York, pp 155–224

Flanagan N (2007) Disposables reach out to new markets. GEN 27:38–39

Fries S, Glazomitsky K, Woods A, Forrest G, Hsu A, Olewinski R, Robinson D, Chartrain M (2005) Evaluation of disposable bioreactors. BioProcess International 10(Supplement):36–44

Frison A, Memmert K (2002) Fed-batch process for mab production. GEN 22:64–66

Gerlach JC (1996) Development of a hybrid liver support system: A review. Int J Artif Organs 19:645–654

Glaser V (2004) Bioreactor and fermentor technology trends. GEN 24:48–50.

Glaser V (2005) Disposable bioreactors become standard fare. GEN 25:80–81

Glaser V (2006) Whither steel & glass in a disposable age. GEN 26:62–65

Goffe RA, Shi JY, Ngyen AKC (1995) Medium additives for high performance cell culture in hollow fiber bioreactors. In: Beuvery EC, Griffiths JB, Zeijlemaker WP (eds) Animal Cell Technology: Development Towards the 21st Century. Kluwer, Dordrecht, pp 187–191

Gokorsch S, Nehring D, Grottke C, Czermak P (2004) Hydrodynamic stimulation and long term cultivation of nucleus pulposus cells: A new bioreactor system to induce extracellular matrix synthesis by nucleus pulposus cells dependent on intermittent hydrostatic pressure. Int J Artif Organs 27:962–970

Gomes MW, Reis RL (2004) Tissue engineering: Key elements and some trends. Macromol Biosci 4:737–742

Griffiths JB (1999) Mammalian cell culture reactors, Scale-up. In: Flickinger MC, Drew SW (eds) Encyclopedia of Bioprocess Technology: Fermentation, Biocatalysis and Bioseparation, vol 3. Wiley VCH, New York, pp 1594–1607

Griffiths JB (2000) Animal cell products, overview. In: Spier RE (ed) Encyclopedia of Cell Technology, vol 1. Wiley VCH, New York, pp 70–76

Groot CAMV (1995) Microcarrier technology, present status and perspective. In: Beuvery EC, Griffiths JB, Zeijlemaker WP (eds) Animal Cell Technology: Developments Toward the 21st Century. Kluwer, Dordrecht, pp 899–905

Guardia MJ Hu WS (1999) Mammalian cell bioreactors. In: Flickinger MC, Drew SW (eds) Encyclopedia of Bioprocess Technology: Fermentation, Biocatalysis and Bioseparation, vol 3. Wiley VCH, New York, pp 1587–1594

Guo XM, Zhao YS, Chang HX, Wang CY, E LL, Zhang XA, Duan CM, Dong LZ, Jiang H, Li J, Song Y, Yang XJ (2006) Creation of engineered cardiac tissue in vitro from mouse embryonic stem cells. Circulation 113:2229–2237

Haughney H, Aranha H (2003) A novel aseptic connection device: Considerations for use in aseptic processing of pharmaceuticals. Pharm Technol: Reprint

Heidemann R, Riese U, Lutkemeyer D, Buntemeyer H, Lehmann J (1994) The Super-Spinner: A low cost animal cell culture bioreactor for the CO_2 incubator. Cytotechnol 14:1–9

Hoerstrup SP, Sodian R, Sperling JS, Vacanti JP, Mayer JE (2000) New pulsatile bioreactor for in vitro formation of tissue engineered heart valves. Tissue Eng 6:75–79

Houtzager E, van der Linden R, de Roo G, Huurman S, Priem P, Sijmons PC (2005) Linear scale-up of cell cultures. The next level in disposable bioreactor design. BioProcess Int 6:60–66

Howaldt M, Walz F, Kempken R (2006) Kultur von Tierzellen. In: Chmiel H (ed) Bioprozesstechnik. Elsevier, München, pp 323–359

Huang W (2005) The shrinking biomanufacturing facility. GEN 25:42–46

Hundt B, Best C, Schlawin N, Kassner H, Genzel Y, Reichl U (2007) Establishment of a mink enteritis vaccine production process in stirred-tank reactor and Wave® Bioreactor microcarrier culture in 1–10L scale. Vaccine 25:3987–3995

Jackson LR, Trudel MJ, Lipman NS (1999) Small-scale monoclonal antibody production in vitro: Methods and resources. Lab Anim 28:38–50

Jasmund I, Bader A (2002) Bioreactor developments for tissue engineering applications by the example of the bioartificial liver. Adv Biochem Eng Biotechnol 74:99–109

Kang SH, Lee GM, Kim BG (2000) Justification of continuous packed-bed reactor for retroviral production from amphotropic ΨCRIP murine producer cell. Cytotechnol 34:151–158

Kieran PM, Malone DM, MacLoughlin PF (2000) Effects of hydrodynamic and interfacial forces on plant cell suspension systems. In: Schügerl K, Kretzmer G (eds) Influence of Stress on Cell Growth and Product Formation, vol 67. Springer, Berlin Heidelberg New York, pp 139–177

Kim Y, Wyslouzil BE, Weathers PJ (2002) Secondary metabolism of hairy root cultures in bioreactors. In Vitro Cell Dev Biol Plant 38:1–10

Kompier R, Kislev N, Segal I, Kadouri A (1991) Use of a stationary bed reactor and serum free medium for the production of recombinant proteins in insect cells. Enzyme Microb Technol 13:822–827

Krahe M (2000) Biochemical Engineering. Reprint from Ullmann's Encyclopedia of Industrial Chemistry

Kranjac D (2004) Validation of bioreactors: Disposable vs. reusable. BioProcess International. Industry Yearbook:2

Kretzmer G (2002) Industrial processes with animal cells. Appl Microbiol Biotechnol 59:135–142

Kullig KM, Vacanti JP (2004) Hepatitic tissue engineering. Transpl Immunol 12:303–310

Kumar S, Wittmann C, Heinzle E (2004) Minibioreactors. Biotechnol Lett 26:1–10

Kunas K (2005) Stirred tank single-use bioreactor: Comparison to traditional stirred bioreactor. Vortrag Biotechnika 2005, Hannover, Germany, October 18–20

Kunz B, Marx B, Glover K (2005) Klassifikationssystem für Fermentationsprozesse. Chem Ing Tech 77:711–721

Labecki M, Bowden B, Piret J (1996) Two dimensional analysis of protein transport in the extracapillary space of hollow-fibre bioreactors. Chem Eng Sci 51:4197–4213

Langhammer S, Brecht R, Marx U (2007) Novel bioreactors for fragile proteins. GEN 27:34

Larcher Y, Oram G, Misener L, Tommasini R, Hagg R (2006) Advanced systems for automated tissue engineering. htpp://www.millenium-biologix.com/Html/scientificpapers/ACTES_C.html. Cited 14 January 2006

Legallais C, Dore E, Paullier P (2000) Design of a fluidized bed bioartificial liver. Int J Artif Organs 24:519–525

Lehmann J, Vorlop J, Büntemeyer H (1998) Bubble-free reactors and their development for continuous culture with cell recycle. In: Spier RE, Griffiths JB (eds) Animal Cell Biotechnology, vol 3. Academic Press, New York, pp 221–237

Lindl T (2002) Zell- und Gewebekultur. Spektrum Akademischer Verlag, Heidelberg

Liszewsky K (2006) Biomanufacturing for regenerative medicine. GEN 26:38–41.

Looby D, Racher A, Griffiths JB, Dowsett AB (1990) The immobilisation of animal cells in fixed and fluidized bed porous sphere bioreactors. In: De Bont JAM, Visser J, Mattiasson B, Tramper J (eds) Physiology of Immobilized Cells. Elsevier, Amsterdam, pp 255–246

Lundgren B, Blüml G (1998) Microcarriers in cell culture production. In: Subramanian G (ed) Bioseparation and Bioprocessing, vol 2. Wiley-CH, Weinheim, pp 165–222

Lyons E, Pandit A (2004) Design of bioreactors for cardiovascular applications. In: Ashammakhi N, Reis RL (eds) Topics in Tissue Engineering, vol 2. EXPERTISSUES e-book, pp 2–32

Marolt D, Augst A, Freed LE, Vepari C, Fajardo R, Patel N, Gray M, Farley M, Kaplan D, Vunjak-Novakovic G (2006) Bone and cartilage tissue constructs grown using human bone marrow stromal cells, silk scaffolds and rotating bioreactors. Biomaterials 27:6138–6149

Martin I, Wendt D, Heberer M (2004) The role of bioreactors in tissue engineering. Tibtech 22:80–86

Marx U (1998) Membrane-based cell culture technologies: A scientifically economically satisfactory alternative malignant ascites production for monoclonal antibodies. Res Immunol 6:557–559

McArdle J (2004) Report of the workshop on monoclonal antibodies. ATLA 32, Supplement 1:119–122

Meier SJ, Hatton TA, Wang DIC (1999) Cell death from bursting bubbles: Role of cell attachment to rising bubbles in sparged reactors. Biotechnol Bioeng 62:468–478

Meis P, Peltier J, Blumenthals I, Amanullah A (2005) Evaluation of the CellFerm-Pro STBR system. GEN 25:50–52

Menkel F (1992) Einführung in die Technik von Bioreaktoren. Oldenbourg Verlag, München, pp 13–15

Merchuk JC, Gluz M (1999) Bioreactors: Air-lift reactors. In: Flickinger MC, Drew SW (eds) Encyclopedia of Bioprocess Technology: Fermentation, Biocatalyis and Bioseparation, vol 1. Wiley VCH, New York, pp 320–353

Meuwly F, Loviat F, Ruffieux PA, Bernard AR, Kadouri A, von Stockar U (2005) Oxygen supply for CHO cells immobilized on a packed-bed of Fibra-Cel disks. Biotechnol Bioeng 93:791–800

Minuth WW, Strehl R, Schumacher K (2003) Zukunftstechnologie Tissue Engineering. Wiley VCH, Weinheim

Morrow KJ (2006) Disposable bioreactors gaining favor. GEN 26:42–45

Morrow KJ (2007) Improving protein production strategies. GEN 28:37–39

Moser A (1981) Bioprozesstechnik. Springer Verlag, Wien, pp 57–58

Muller M, Derouazi M, Van Tilborgh F, Wulhfard S, Hacker DL, Jordan M, Wurm FM (2006) Scalable transient gene expresssion in Chinese hamster ovary cells in instrumented and non-instrumented cultivation systems. Biotechnol Lett (online) doi:10.1007/s10529-006-9298-x

Müller S (2001) Kultivierung von HEK-U293 Zellen als Suspensionskultur im miniPerm Bioreaktor. In Vitro News 1:4

Negrete A, Kotin RM (2007) Production of recombinant adeno-associated vectors using two bioreactor configurations at different scales. J Virol 145:155–161

Nielsen LK (1999) Bioreactor for hematopoietic cell culture. Annu Rev Biomed Eng 1:129–152

Nienow AW (2006) Reactor engineering in large scale animal cell culture. Cytotechnol 50:9–33

Nilsson K, Buzsaky F, Mosbach K (1986) Growth of anchorage dependent cells on macroporous microcarrier. Biotechnol 4:989–990

Odum J (2005) Trends in Biopharmaceutical Manufacturing Facility Design: What's Hot! Pharmaceutical Engineering 25: Reprint

Ohyabu Y, Kida N, Kojima H, Taguchi T, Tanaka J, Uemura T (2006) Cartilaginous tissue formation from bone marrow using rotating wall vessel (RWV) bioreactor. Biotechnol. Bioeng. 95:1003–1008

Okamura A, Zheng YW, Hirochika R, Tanaka J, Taniguchi H (2007) In-vitro reconstitution of hepatic tissue architecture with neonatat mouse liver cells using three-dimensional culture. J Nanosci Nanotechnol 7:721–725

Ozturk SS (2006) Cell culture technology – An Overview. In: Ozturk SS, Hu WS (eds) Cell Culture Technology for Pharmaceutical and Cell-Based Therapies. CRC Press, New York, pp 1–14

Ozturk SS (2007) Comparison of product quality: Disposable and stainless steel bioreactor. Presented at the BioProduction 2007, Berlin, Germany, October 29–31

Pàca J (1987) Bioreaktoren. In: Weide H, Pàca J (eds) Biotechnologie. Gustav Fischer Verlag, Jena, pp 125–175

Petrossian A, Cortessis GP (1990) Large-scale production of monoclonal antibodies in defined serum-free media in airlift bioreactors. Biotechniques 8:414–422

Pierce LN, Shabram PW (2004) Scalability of a disposable bioreactor from 12–500 L run in perfusion mode with a CHO-based cell line: A tech review. BioProcessing J 3:1–6

Prenosil JE, Villeneuve PE (1998) Automated production of cultured epidermal autografts and sub-confluent epidermal autografts in a computer controlled bioreactor. Biotechnol Bioeng 59:679–683

Purdue GF, Hunt JL, Still JM, Law EJ, Herndon DN, Goldfarb W, Schiller WR, Hansbrough JF, Hickerson WL, Himel HN, Kealey GP, Twomey J, Missavage AE, Solem LD, Davis M, Totoritis M, Gentzkow GD (1997) A multicenter clinical trial of a biosynthetic skin replacement, Dermagraft-TC*, compared with cryopreserved human cadaver skin for temporary coverage of excised burn wound. J Burn Care Rehab 18:52–57

Puskeiler R, Kaufmann K, Weuster-Botz (2005) Development, parallelization, and automation of a gas-inducing milliliter-scale bioreactor for high-throughput bioprocess design (HTBD). Biotechnol Bioeng 89:512–513

Puskeiler R, Weuster-Botz D (2004) Rührkesselreaktoren im mL-Massstab: Kultivierung von *Escherichia coli*. Chem Ing Tech 76:1865–1869

Pörtner R (2006) Bioreaktoren und Bioprozesstechnik. Chem Ing Tech 78:1129–1131

Pörtner R, Giesse G (2007) An overview on bioreactor design, prototyping and process control for reproducible three-dimensional tissue culture. In: Marx U, Sandig V (eds) Drug Testing In Vitro: Breakthrough and Trends in Cell Culture Technology, Wiley-VCH, Weinheim, pp 53–78

Ratcliffe A, Niklason LE (2002) Bioreactors and bioprocessing for tissue engineering. Annals New York Academy of Sciences 961:210–215

Safinia L, Panoskaltsis N, Mantalaris A (2005) Haematopoietic culture systems. In: Chaudhuri JB, Al-Rubeai (eds) Bioreactors for tissue engineering, Kluwer, Dordrecht, pp 309–334

Schneider F (2004) Disposable Spinner und Kultivierungssysteme für die Zellkultur. Workshop,, Disposable Kultivierungstechnologie, Wädenswil, Switzerland, October 5–6

Scott LE, Aggett H, Glencross DK (2001) Manufacture of pure monoclonal antibodies by heterogeneous culture without downstream purification. Biotechniques 31:666–668

Schügerl K (1980) Neue Bioreaktoren für aerobe Prozesse. Chem Ing Tech 52:951–965

Schwander E, Rasmusen H (2005) Scaleable, controlled growth of adherent cells in a disposable, multilayer format. GEN 25:29

Senker J, Mahdi S (2003) Human tissue-engineered products: Research activities and future developments of human tissue engineering in Europe and the US. http://lifesciences.jrc.es/docs/TE_WP3_FinalReport.pdf. Cited 14 January 2006

Sinclair A, Monge M (2004) Biomanufacturing for the 21st century: Designing a concept facility based on single-use systems. BioProcess International 10(Supplement):26–31

Singh V (1999) Disposable bioreactor for cell culture using wave-induced motion. Cytotechnol 30(1/3):149–158

Song K, Yang Z, Liu T, Zhi W, Li X, Deng L, Cui Z, Ma X (2006) Fabrication and detection of tissue-engineered bones with bio-derived scaffolds in a rotating bioreactor. Biotechnol Appl Biochem 45:65–74

Stadler EL (1998) Dual purpose fermentor and bioreactor? A capital quandary! Pharmaceutical Engineering 18: Reprint

Suck K, Behr L, Fischer M, Hoffmeister H, van Griensven M, Stahl F, Scheper T, Kasper C (2006) Cultivation of MC3T3-E1 cells on a newly developed material (Sponceram®) using a rotating bed system bioreactor. J Biomed Mater Res A 80(2):268–275

Tang YJ, Ohashi R, Hamel JF (2007) Perfusion culture of hybridoma cells for hyperproduction of IgG$_{2a}$ monoclonal antibody in a wave bioreactor-perfusion culture system. Biotechnol Prog 23:255–264

Thermo Fisher Scientific (2007) Scale comparison of the Thermo Scientific Hyclone 50 L Single-Use Bioreactor for SP2/0 cell culture. Application note:AN003 Rev 1

Trebak M, Chong JM, Herlyn D, Speicher DW (1999) Efficient laboratory-scale production of monoclonal antibodies using membrane-based high density cell culture technology. J Immun Meth 230:59–70

Tzanakakis ES, Verfaillie CM (2006) Advances in adult stem cell culture. In: Ozturk SS, Hu WS (eds) Cell culture technology for pharmaceutical and cell-based therapies. CRC Press, New York, pp 693–721

Ulloa-Montoya F, Verfaille CM, Hu WS (2006) Review: Culture systems for pluripotent stem cells. J Bioeng Biosci 100(1):12–27

Varley J, Birch J (1999) Reactor design for large scale suspension animal cell culture. Cytotechnol 29:177–205

Voss H (1991) Fermentationstechnik und Aufarbeitung. In: Ruttloff H (ed) Lebensmittelbiotechnologie: Entwicklung und Aufarbeitung. Akademie Verlag, Berlin, pp 28–64

Walsh G (2005) Current status of biopharmaceuticals: Approved products and trends in approvals. In: Knäblein J (ed) Modern biopharmaceuticals, vol 1. Wiley-VCH, Weinheim, pp 1–34

Walsh G (2003) Biopharmaceuticals: Biochemistry and biotechnology. Wiley, The Atrium, Southern Gate, Chichester, West Sussex

Wang D, Liu W, Han B, Xu R (2005) The bioreactor: A powerful tool for large-scale culture of animal cells. Curr Pharmaceut Biotechnol 6:397–403

Wang G, Zhang W, Freedmann D, Eppstein L, Kadouri A (1992) Continuous production of monoclonal antibodies in Celligen packed bed bioreactor using Fibra-Cell carier. In Spier RE, Griffiths JB, Mac Donald C (eds) Animal Cell Technology: Development, Process and Products. Butterworth-Heinemann, Oxford, pp 460–464

Wang MD, Yang M, Huzel N, Butler M (2002) Erythropoietin production from CHO cells grown by continuous culture in a fluidized-bed bioreactor. Biotech Bioeng 77:194–203

Weber W, Weber E, Geisse S, Memmert K (2001) Optimization of protein expression and establishment of the Wave Bioreactor for baculovirus/insect cell culture. Cytotechnol 38:77–85

Whiteside JP, Spier RE (2004) The scale-up from 0.1 to 100 liter of a unit process system based on 3-mm-diameter glass spheres for the production of four strains of FMD from BHK monolayer cells. Biotechnol Bioeng 23:551–565

Wurm FM (2005) Manufacture of recombinant biopharmaceutical proteins by cultivated mammalian cells in bioreactors. In: Knäblein J (ed) Modern Biopharmaceuticals, vol 3. Wiley VCH, Weinheim, pp 723–759

Wurm FM (2004) Production of recombinant protein therapeutics in cultivated mammalian cells. Nat Biotechnol 22:1393–1397

Wurm FM (2007) Novel technologies for rapid and low cost provisioning of antibodies and process details in mammalian cell culture based biomanufacturing. Presented at the BioProduction 2007, Berlin, Germany, October 29–31

Zhong JJ (2001) Biochemical engineering of the production of plant-specific secondary metabolites by cell suspension cultures. In: Scheper T (ed) Advances in Biochemical Engineering/ Biotechnology, vol 72. Springer, Berlin Heidelberg New York, pp 1–26

Complementary Reading

BioProcess International (2004) The Disposable Option: Supplement Series

BioProcess International (2005) Disposables: Supplement Series

Chauduri J, Al-Rubeai M (2006) Bioreactors for Tissue Engineering. Kluwer, Dordrecht

Freshney RI, Stacey GN, Auerbach Jm (2007) Culture of Human Stem Cells. Wiley, Hoboken, New Jersey

Vunjak-Novakovic G, Freshney RI (2006) Culture of Cells for Tissue Engineering. Wiley, Hoboken, New Jersey

Chapter 4
Special Engineering Aspects

P. Czermak, R. Pörtner, and A. Brix

Abstract The specific characteristics of mammalian cells discussed in Chap. 2 require adapted solutions for bioreactor design and operation. Especially, cell damage by shear stress and aeration has to be considered. Therefore this chapter starts with a detailed discussion of shear stress effects on mammalian cells (anchorage-dependent and suspendable cells) in model systems and bioreactors, respectively, and consequences for reactor design. Appropriate oxygen supply is another critical issue, as adapted oxygen supply systems are required. Techniques for immobilization of cells, either grown on microcarriers in suspension culture or within macroporous carriers in fixed bed or fluidized bed reactors, are discussed as well. With respect to the operation of bioreactors, the characteristics of different culture modes (batch, fed-batch, chemostat, perfusion) are introduced and practical examples are given. Finally, concepts for monitoring of bioreactors, including offline and online methods as well as control loops (e.g. O_2, pH), are considered.

4.1 Cell Damage by Shear and Aeration

4.1.1 General Aspects

Unlike other microorganisms (e.g. bacteria), mammalian cells are not surrounded by a cell wall adapted to support survival as an individual within the natural environment, but only by plasmatic membrane, which can only withstand weak mechanical

P. Czermak
Institute of Biopharmaceutical Technology, University of Applied Sciences
Giessen-Friedberg, Giessen, Germany and Department of Chemical Engineering,
Kansas State University, Durland Hall 105, KS 66506-5102, Manhattan, USA

A. Brix
Department of Chemical Engineering, Kansas State University, Durland Hall 105,
KS 66506-5102, Manhattan, USA

R. Pörtner
Hamburg University of Technology (TUHH), Institute of Bioprocess and Biosystems
Engineering, Denickestr. 15, D-21073 Hamburg, Germany
poertner@tuhh.de

R. Eibl et al., *Cell and Tissue Reaction Engineering: Principles and Practice,*
© Springer-Verlag Berlin Heidelberg 2009

forces. The cell membrane is basically composed of phospholipids that build a double layer. In conventional bioreactor systems, aerated stirred tanks and bubble columns, adherent cells grown on microcarriers or suspended cells are exposed to variously intense hydrodynamic forces. A sufficiently intense force will destroy the cell wall. Forces of lesser magnitude may induce various physical responses, including reduced growth, cell death (apoptosis) or other metabolic reactions. A number of reviews have summarized the main fluid-mechanical and biological aspects of cell damage (Cherry 1993; Papoutsakis 1991; Chisti 2000, 2001). According to Papoutsakis (1991) fluid mechanical questions include the following:

- What forces affect cells in a flow environment, and how?
- Are the flow effects due to the intensity and/or the frequency of the forces?
- What types of interactions are most damaging to the cells in various reactors and/or processing devices?

Biological questions include the following:

- Does fluid-mechanical stress cause cell death, or simply reduce cell growth?
- Do fluid mechanical forces affect the physiology (e.g. the cell cycle), product expression, molecular processes, receptor-mediated processes and the cytoskeleton of the cells? How?
- Do cells react to and adapt in response to fluid-mechanical forces?

In addition it is relevant to ask:

- Do fluid-mechanical forces affect adherent and suspendable cells differently?
- Is it possible to protect the cells against this kind of damage?
- Does the sensitivity of cells to fluid-mechanical stress depend on the scale?

Finally, the objective should be to formulate the extent and the mechanism of the damages as well as allow calculation and lay-out of bioreactors with respect to parameters such as reactor size, stirrer rotation speed or aeration rate for bubble aeration (Tramper et al. 1993).

Generally, the effects caused by mechanical forces in a mixed culture and damaging effects caused by gas bubbles are discussed separately. The many kinds of mechanical forces (e.g. the motion of microcarriers or free suspended cells relative to the surrounding fluid, to each other and to moving or stationary solid surfaces) are collectively referred to as "shear forces", as the underlying damaging phenomena cannot be ascribed to gas bubbles (Chisti 2001).

By definition, shear in a fluid system has two components, shear stress τ and shear rate γ. A shear stress is a force per unit area acting on and parallel to a surface. Shear rate is a measure of a velocity gradient (velocity/length). The two quantities are related; thus, in laminar Newtonian fluid

$$\tau = \eta_{fl}\gamma, \tag{4.1}$$

where η_{fl} is the viscosity of the fluid. In model systems such as laminar flow between two parallel plates, in a cone-and-plate viscometer or a coaxial cylinder Searle viscometer the shear rate and the corresponding shear stress can be calculated by simple mathematical equations (see below) and the damaging effects can be related to exact values. In a complex, mostly turbulent environment within a bioreactor, the local shear rate varies within the vessel and it is more difficult to associate cell damage with the magnitude of the prevailing shear rate or the associated shear stress. Even more complex are the mechanisms of cell damage caused by gas bubbles in sparged bioreactors.

In general, cell damage typically follows first-order kinetics. Thus, the rate of cell loss (dX_v/dt) depends on the concentration of viable cells X_v and can be described by

$$\frac{dX_v}{dt} = -k_d X_v,\qquad(4.2)$$

where k_d is the cell specific death rate. The k_d value is influenced by the rate of shear stress and the operating conditions within the cultivation system (stirring, bubble aeration, and medium composition) as well as the properties of the cells. In the following sections, available data for the different culture conditions will be discussed.

In the next sections, discussions about model systems will be presented in order to describe the main effects. Later, the effect of shear stress and bubble aeration in real bioreactors will be shown. Finally some recommendations will be given for the design and operation of bioreactors. Note that a distinction between adherent and suspendable cells will be made.

4.1.2   Model Analysis

The intention of these most studies was mainly to understand the basic mechanism of cell death due to flow shear forces or bubbles. In most cases the cells were exposed to defined levels of shear stress in defined flow geometries in order to correlate the effects with known values for the level of shear stress. From these findings, conclusions for design and operation of bioreactors were drawn. Furthermore it was to ask whether critical shear stress levels found in model systems can be used for the estimation of design criteria for large scale bioreactors (e.g. power input, stirrer speed or aeration rate).

4.1.2.1   Shear Stress on Anchorage-Dependent Cells

The damaging effect of shear on anchorage-dependent cells adhered to a solid surface is frequently studied in a flow chamber (Fig. 4.1) (Diamond et al. 1990; Frangos et al. 1985; 1988; Ludwig et al. 1992; Shiragami and Unno 1994; Spraque et al. 1987; Stathopoulos and Hellums 1985). In this apparatus, a laminar flow causes a defined shear stress at the wall τ_w on the bottom plate where the cells are adherently growing. The flow velocity $U(y)$ can be calculated for given values of the flow rate F, the viscosity η_{fl}, the height h and the width b of the flow chamber (Fig. 4.2):

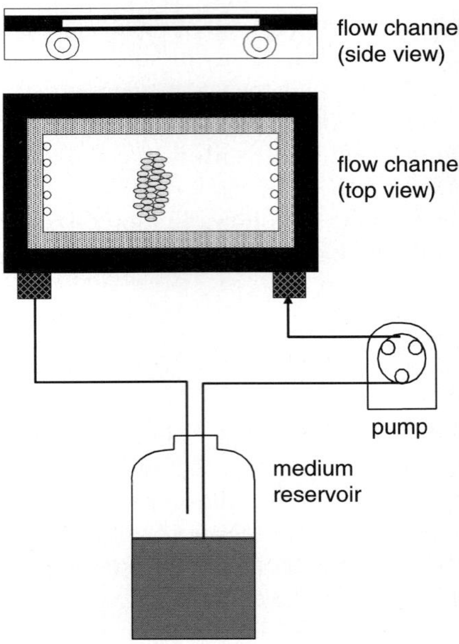

Fig. 4.1 Flow chamber for investigation of shear effects on adherent cells (reproduced from Shiragami and Unno 1994, modified, with kind permission of Springer)

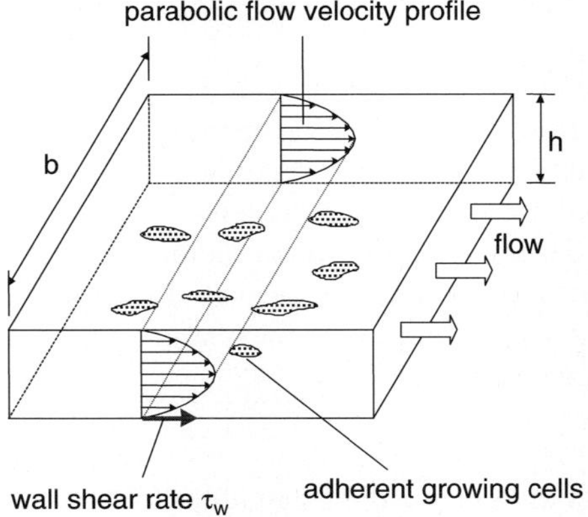

Fig. 4.2 Determination of wall shear rate in a flow chamber for investigation of shear effects on adherent cells

$$U(y) = \frac{6F}{bh^3}(hy - y^2).$$

(4.3)

The wall shear stress τ_w acting on the cells is given by:

$$\tau_w = \eta_{fl}\left(\frac{\mathrm{d}U}{\mathrm{d}y}\right)_{y=0} = \frac{6\eta_{fl}\dot{F}}{bh^2}. \tag{4.4}$$

In experiments performed with the aim to determine a critical shear stress, the cells are allowed to grow on the bottom plate of a flow chamber until they have reached confluence (Fig. 4.3). Afterwards, the medium flows with a defined velocity through the chamber for a certain time and the cell morphology, the number of viable cells and the concentration of substances that are released by the damaged cells [i.e. lactate dehydrogenase (LDH)] are evaluated (Chisti 2001). The following conclusions can be drawn from the available data:

- Cell attachment to a surface is affected by shear stress levels between 0.25 and 0.6 Nm^{-2} (Olivier and Truskey 1993)
- Under stress the cells orientate themselves within the liquid flow lines (as shown in Fig. 4.3)
- Laminar shear stress in the order of 0.5–10 Nm^{-2} may remove adherent cells from the surface (Aunins and Henzler 1993). Kretzmer (1994) as well as Shiragami and Uno (1994) gave some numbers for CHO and BHK cells for the influence of shear stress on the living cells ratio (50% viability): 3 h $> \tau_w = 6$ Nm^{-2}; 24 h $> \tau_w = 0{,}75$–1 Nm^{-2}
- Laminar shear stress in the order of 0.1–1.0 Nm^{-2} may affect cellular morphology, permeability, and gene expression (Aunins and Henzler 1993).
- The combined effect of shear stress and the stress duration can be expressed by the energy ε dissipated during a certain time (Shiragami and Unno 1994)

$$\varepsilon = t \int \eta_{fl}\left(\frac{\mathrm{d}U}{\mathrm{d}y}\right)^2 \mathrm{d}y = \frac{h\tau_w^2 t}{3\,\eta_{fl}} \tag{4.5}$$

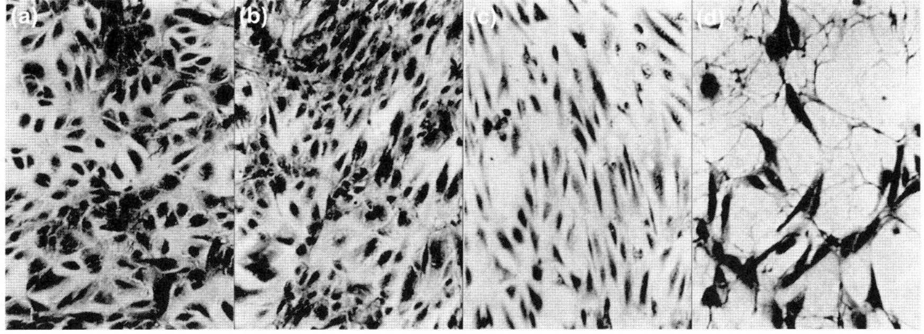

Fig. 4.3 Effect of shear stress on adherent growing primary epithelial cells in a flow chamber after (**a**) 1 week of culture without shear stress (controls), (**b**) 4 h with a shear stress of 1.3 Nm^{-2} (**c**) 24 h with a shear stress of 1.3 Nm^{-2}, (**d**) 24 h with a shear stress of 5.4 Nm^{-2} (adapted from Stathopoulos and Hellums 1985, with kind permission of John Wiley & Sons)

4.1.2.2 Shear Stress on Suspended Cells

The reaction of freely suspended cells on shear forces in a bubble-free environment can be regarded as similar to the behaviour of drops or suspended particles in a shear field. They experience extensional or elongational forces depending on the flow pattern, e.g. whenever the cross-sectional area of a flow channel reduces, as at the entrance of capillaries or within bioreactors in the direction of flow. The cells subjected to extension or extensional flow can rupture similar to drops. From this point of view the mechanical strength of the cell, expressed by the bursting membrane tension, could be an appropriate parameter to describe the allowed level of stress. But as the cells can rotate or tumble in an extensional flow field to relax the imposed stress, in addition to hydrodynamic stress, the possibility of rotation-associated stress relaxation also needs to be considered. The effect of shear forces on suspendable cells was investigated as follows:

- Determination of the surface tension of the cell membrane
- Correlation of the cell damage with the shear stress in a surrounding fluid.

The basic theory for the behaviour of emulsified, viscous liquid drops was developed by Taylor (1934). According to this model, the resistance and stability of a cell exposed to shear forces are basically influenced by the surface tension and the size of the cell. Cell deformation is the result of the action of external hydrodynamic forces and the cell would burst if the following condition is fulfilled:

$$\tau_{crit} r_Z^2 = 2 r_Z \sigma_Z \tag{4.6}$$

with the the critical shear stress τ_{crit}, the cell radius r_Z and the membrane tension at cell burst σ_Z. In order to determine the membrane tension at cell burst, Zhang et al. (1992b) applied an experimental set-up as shown in Fig. 4.4. Initially, the single cells were fixed between two glass fibres (diameter approx. 50 μm) that can slide against each other. After the immobilization, the cells were compressed in the gap between the fibres until they burst. The burst strength and cell diameter were recorded to calculate the membrane tension. By this, values of $\sigma_Z = 1.7$–2.5×10^{-3} Nm^{-1} for the membrane tension were determined leading to critical shear rates of 500–700 Nm^{-2}.

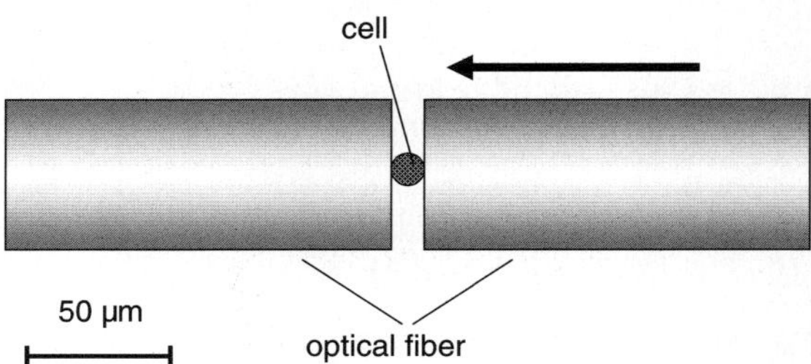

Fig. 4.4 Experiment for determination of the force required to burst the cell. Cell entrapped between two optical fibers (adapted from Zhang et al. 1992b)

Based on the hypothesis that a cell should burst whenever its bursting membrane tension is exceeded in a laminar flow, Born et al. (1992) developed a model predicting cell damage in laminar flow. For a critical discussion see Chisti (2001). Even if the model seems to be supported by experimental data from cells sheared in a cone-and-plate viscometer, the practical utility of this approach is regarded as limited.

Cell damage due to shear stress was investigated in certain devices, where the cells could be exposed to defined high-shear levels in laminar or turbulent flow (e.g. cone-and-plate-viscometer, coaxial cylinder Searle viscometer, capillary flow (Abu-Reesh and Kargi 1989; Born et al. 1992; Kramer 1988; Smith et al. 1987; Tramper et al. 1986) (reviewed extensively by Chisti 2001). In most cases the cells were cultivated under unharmful conditions e.g. in a small spinner reactor and then transferred to one of the devices mentioned above. Here the cells were exposed to a certain shear level for a certain time.

For a coaxial cylinder Searle viscometer (couette viscometer), within a narrow gap between the two cylinders, the shear rate and shear stress are constant within the gap and proportional to the rotation speed. From the measured torque M_d, height h and inner radius $r_{i,cou}$ the shear stress τ can be calculated.

$$\tau = \frac{M_d}{2\Pi h r_{i,cou}^2}. \tag{4.7}$$

The laminar-turbulent flow transition is defined by the Taylor number Ta

$$Ta = \left(\frac{\rho_{fl} U_T w}{\eta_{fl}}\right)\left(\frac{2w}{d_{i,cou}}\right)^{0.5}. \tag{4.8}$$

with gap width w, diameter of inner cylinder $d_{i,cou}$ rotating at peripheral speed U_T and fluid density ρ_{fl}. The flow is laminar when Ta < 41.3. Laminar flow with Taylor vortices occurs for 41.3 < Ta < 400. When Taylor number exceeds 400, the turbulent flow is fully developed. An example is shown in Fig. 4.5.

As can be seen from Table 4.1, critical values between 1 and 2×10^3 Nm^{-2} have been reported. Reasons for this large range of values can be seen in (i) different criteria for the threshold shear stress (e.g. total cell disruption, certain level of cell death, appearance of cell lyses, and induction of cell inactivation among others), (ii) different cell lines, (iii) different devices, and (iv) different culture conditions (medium with or without serum, etc.). Nevertheless some general conclusions can be drawn from these data:

- Turbulent flow is more damaging than laminar
- Duration of shear stress is relevant
- The mechanism of cell damage is very complex and depends on the device used for the studies
- The shear rate seems not to be the appropriate parameter to describe the effect.

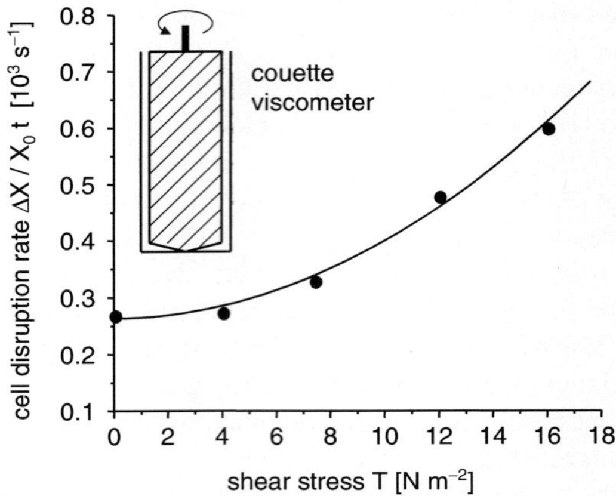

Fig. 4.5 Influence of shear stress on the cell disruption rate of hybriodm cells in a couette viscosimeter (adapted from Kramer 1988)

Table 4.1 Impact of shear stress on suspendable cells

Reference	Cell type	System	Type of flow	Threshold shear stress [N m^{-2}]	Exposure time
Kramer (1988)	Hybridom	Coaxial cylinder	n.d.	4	
Tramper et al. (1986)	Insect cells	Coaxial cylinder	n.d.	1–4	1–3 h
Born et al. (1992)	Hybridom	Cone-and-plate	Laminar	208	20 min
				350	180 s
Abu-Reesh and Kargi (1989)	Hybridom	Couette	Turbulent	5	
McQueen et al. (1987)	Myeloma	Capillary	Turbulent	180	
Cited by McQueen et al. (1987)	Hybridom	Capillary	n.d.	0.87	
Cited by McQueen et al. (1987)	Insect	Capillary	n.d.	1.5	
Augenstein et al. (1971)	HeLa/L293	Stainless steal capillaries	n.d.	0.1–2×10^3	

According to data from Abu-Reesh and Kargi (1989), cell damage followed first-order kinetics both under laminar and turbulent conditions. For turbulent shear stress levels of 5–30 Nm^2, the death rate constant k_d varied between 0.1 and 1 h^{-1} and increased exponentially with increasing stress level.

4.1.2.3 Cell Damage Caused by Bubbles

The specific events responsible for the bubble-associated cell damage include

- Bubble formation at the sparger
- Bubble detachment
- Rise through the fluid
- Break-up of bubbles at the surface and foam formation (Chisti 2000).

Among others, Jordan (1993) studied the damaging effect of bubbles to understand the main mechanisms. The interaction between cells and bubbles was strongly influenced by the concentration of protecting agents (e.g. Pluronic F68, see below) and the retention time of the bubbles in the medium. Based on his phenomenological observations he suggested subdividing the way of a bubble from the sparger to the surface into four zones (Fig. 4.6). Cells getting in contact with a bubble directly after bubble formation are damaged instantly, mainly due to a difference in surface tension (zone 1). After longer retention of the bubble in the medium, several medium compounds, especially high molecular weight compounds or proteins, are adsorbed at the surface of the bubble, and the surface tension is reduced. Thus the cells are not damaged any more, but are still attached at the surface and therefore float with the rising bubble (zone 2). In zone 3, attachments of cells at the surface is further reduced. In zone 4 the gas bubbles burst at the surface of the fluid and form a foam layer. Cells can be damaged either during bubble collapse or when trapped in the foam (Fig. 4.7). When a bubble collapses at the surface, cells adsorbed to the bubble surface or trapped in the wake of the bubble are exposed to relatively high

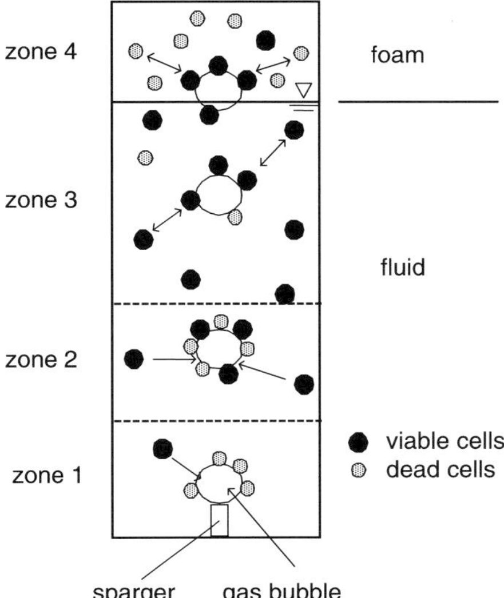

Fig. 4.6 Mechanisms of cell damage in a bubble column (adapted from Jordan 1993)

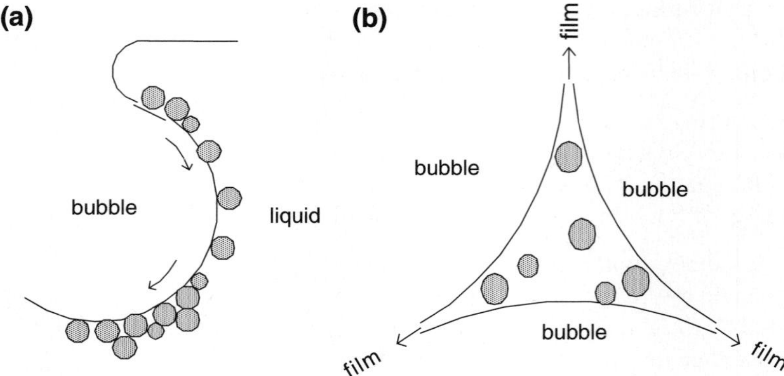

Fig. 4.7 Cell damage in a foam layer (adapted from Papoutsakis 1991, with kind permission of Elsevier). (a) cells near the bubble interface, large shear stress due to bubble break up, (b) cells are sheared in the thinning films either between bubbles or around bubbles

shear forces. These forces are strong enough to rupture the cell's membrane. The shear forces can originate from the velocity gradient arising from the upward movement of the bubbles and the downward jet resulting from the bursting bubbles. Cells entrapped in the foam layer are exposed to shear by the movement of the bubbles surrounding a cell in different directions and nutritional effects (e.g. oxygen limitation). Meanwhile it is generally accepted that these effects are mainly responsible for cell damage in aerated systems.

Some general statements can be made regarding bubble-associated damage (Chisti 2000):

- Cell lines differ tremendously in their sensitivity to aeration (e.g. in sparged bioreactors, mouse hybridom are generally more robust than insect cells)
- Small bubbles (e.g. <2 mm diameter) are more damaging than large bubbles (e.g. ~10 mm diameter). Large bubbles have mobile interfaces (wobbling) and, because they rise faster, they carry fewer attached cells to the surface. Furthermore, larger bubbles (unlike smaller ones) do not remain on the surface as stable foam.
- Furthermore cell damage is directly affected by the aeration rate.
- Sparging-associated damage may be enhanced by impeller agitation.
- Cells can be protected by certain media additives.

Implications on design and operation of sparged bioreactors will be discussed in Sect.. 4.1.3.3.

4.1.3 Cell Damage in Bioreactors

The above sections described the main mechanisms responsible for cell damage due to defined shear forces or bubbles. In a complex, mostly turbulent environment within a bioreactor, the local shear rate varies within the vessel and it is more difficult to associate cell damage with the magnitude of the prevailing shear rate or the associated shear stress. Even more complex are the mechanisms of cell damage caused by gas bubbles in sparged bioreactors, bubble columns or air-lift-bioreactors. In the following sections, some basic studies and approaches to describe cell damage mathematically are discussed, focussing mainly on those strategies and equations suitable for bioreactor design and scale-up. Comprehensive reviews can be found in the literature (Chisti 2000, 2001).

4.1.3.1 Anchorage Dependent Cells Grown on Suspended Microcarriers

The impact of shear forces on anchorage-dependent cells in stirred bioreactors was largely studied for microcarrier cultures as anchorage-dependent cells can be grown on the surface of suspendable particles (Fig. 4.8) (Sect. 4.3.1.2). Microcarrier culture is still applied widely for large-scale production of viral vaccines among others. Cells grown on microcarriers experience more severe hydrodynamic forces than do freely suspended cells (for review see Chisti 2001). This is because in the mostly turbulent environment within highly agitated or aerated systems, the length scale of fluid eddies can easily approach the dimensions of microcarriers, resulting in high local relative velocities between the solid and the liquid phases. Additionally, collision among microcarriers and between the impeller and microcarriers may damage attached cells.

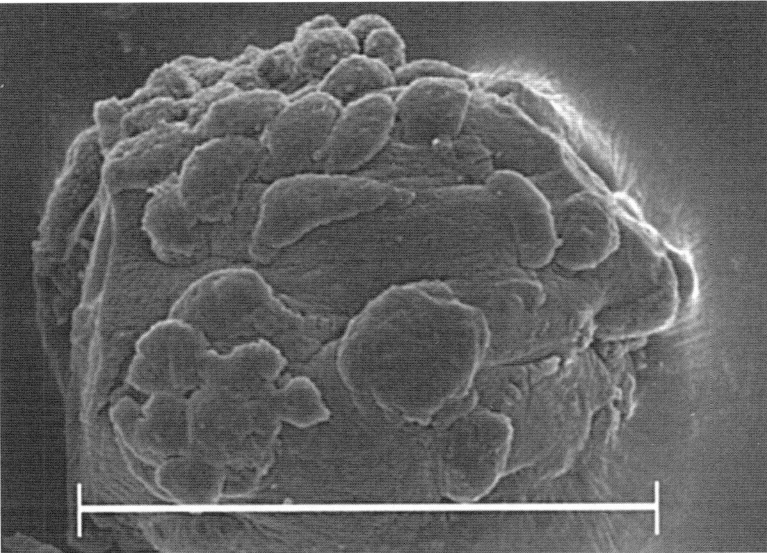

Fig. 4.8 Microcarrier for suspension culture of adherent cells, pig chondrocytes grown on Cytodex 3 (GE Healthcare, bar approx. 150 µm)

In a turbulent flow caused by a stirrer, relatively large eddies are created at first. These eddies break up into smaller ones stepwise, until the energy is completely dissipated. Cherry and Papoutsakis (1986, 1988) distinguished between three different microcarrier-eddy-situations depending on the ratio between eddy size and microcarrier diameter (Fig. 4.9).

Several concepts to describe the complex shear effects on cells in a turbulent flow were suggested and judged by available data on the impact of shear forces on microcarrier cultures in small scale bioreactor systems (compare Fig. 4.10) Among others the concept of an "Integrated Shear Factor" ISF – a measure of the strength of the shear field between the impeller and the spinner vessel walls – was developed to describe shear damage to mammalian cells (Croughan et al. 1987; Sinskey et al. 1981). The ISF is defined as

$$ISF = \frac{2\Pi n_R d_R}{D_R - d_R} \tag{4.9}$$

with rotation speed n_R, vessel diameter D_R and stirrer diameter d_R. Data from Hu et al. (1985) can be described well. For a range of stirrer vessels (volume 0.25–2.0 L, d_R 3.2–8.5 cm) the viable cell density dropped sharply when the ISF value exceeded 18–20 s^{-1}. Furthermore Croughan et al. (1988) and Croughan and Wang (1989) discussed other concepts such as the correlation between cell damage and impeller tip speed or a time averaged shear rate. Even if in all cases a satisfying

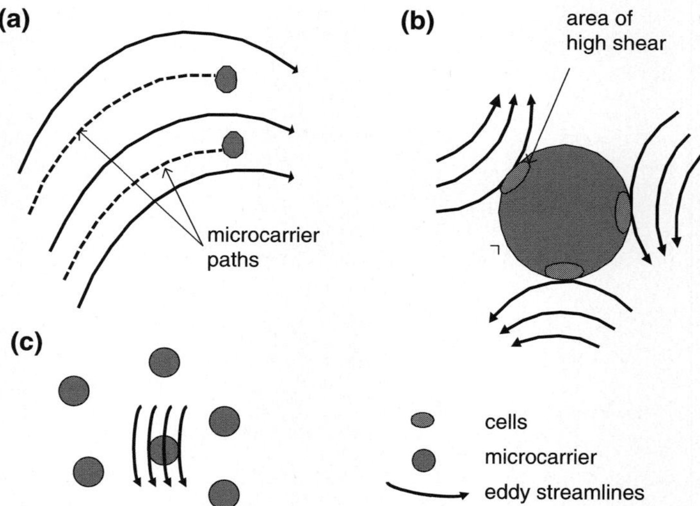

Fig. 4.9 Shear forces on microcarriers in a turbulent flow. Microcarrier-eddy interactions: (a) eddies much larger than beads, (b) multiple eddies same size as bead, (c) eddy size same as interbead spacing (adapted from Cherry and Papoutsakis 1986, modified, with kind permission of Springer)

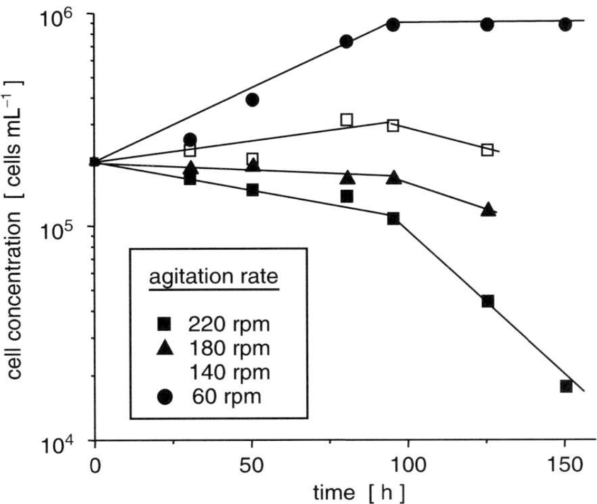

Fig. 4.10 Growth of human FS 4-cells on microcarriers Cytodex 3 in a 250 mL spinner reactor (adapted from Croughan et al. 1987, with kind permission of John Wiley & Sons)

success in correlating cell damage could be obtained, these concepts are regarded as not a relationship for scale-up. For example, if standard stirred tank reactors ($D_R = 3d_R$) are scaled up to a geometrically similar larger vessel, the ISF depends solely on the rotational speed of the impeller (ISF = πn_R). Therefore the rotational speed would be independent of scale and the impeller tip speed $d_R n_R$ would therefore increase with scale. In general it is doubtful whether concepts based on the stirrer speed are appropriate for scale-up of cell damage. Instead, between cell damage and volume specific power input was proposed. This will be discussed in detail in the following sections.

According to Cherry and Papoutsakis (1986, 1988) cells grown on microcarriers within turbulent eddies of the same size as the microcarriers are exposed to the largest shear stress. The energy of the eddies is transferred to the surface of the microcarriers, resulting in high local velocities between the microcarriers and the fluid, and the highest shear rates on the cells. The microcarriers are caused to rotate within these eddies. From Kolmogorov´s theory the length scale l_{Kol} of those smallest eddies are in the order of

$$l_{Kol} = \left(\frac{v^3}{\varepsilon} \right)^{1/4} \tag{4.10}$$

with the kinematic viscosity v and ε the energy dissipation rate per unit mass. The Kolmogorov eddy length scale corresponds to the diameter of the smallest eddy generated in the reactor. In a turbulent environment, eddies break down to form smaller eddies. As eddies break down, the energy of the larger eddy is passed down. At the smallest eddy diameters the flow actually becomes laminar, rotation forces and friction forces are in equilibrium, rotation of eddies stops and fluid flow is characterized by the formation of flow lines. When the Kolmogorov eddy length scale becomes equivalent to the cell diameter, the movement of the flow lines can shear the cells. The Kolmogorov eddy length scale is affected by stirrer speed, liquid properties and impeller design. Croughan et al. (1987) plotted data from Hu et al. (1985) by using the mean energy dissipation rate per unit mass

$$\varepsilon = \frac{P}{\rho_{fl} V}$$
(4.11)

with power input P and volume V. As can be seen from Fig. 4.11, for human fibroblasts the number of viable cells decreased rapidly when the length scale declined to approximately below 125 µm. The mean diameter of microcarriers was about 185 µm, or roughly similar to the length scale of the microeddies. Even if this correlation seems to support the underlying theory, the Kolmogorov eddy length scale allows only for a rough estimation of the expected cell damage and is regarded as not suitable for scale-up. For example, if the mean energy dissipation rate per unit mass is increased ten times, the Kolmogorov length scale decreases only by a factor of 0.56. Furthermore this approach cannot be transferred to suspendable cells.

Croughan et al. (1987) further showed that the specific death rate k_d can be linked to the mean energy dissipation rate per unit mass

$$k_d \approx \varepsilon^{3/4}.$$
(4.12)

Hülscher (1990) confirmed this relationship for microcarrier cultures as well as for suspendable cells (see below).

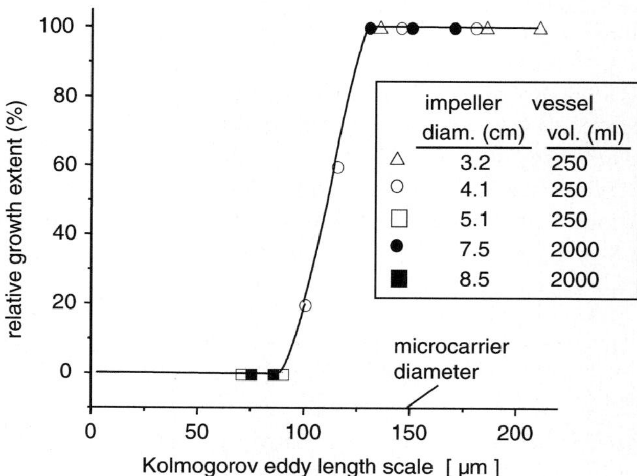

Fig. 4.11 Growth of human FS 4-cells on microcarriers Cytodex 3 in a 250 mL spinner reactor – correlation of growth with Kolmogorov length of eddies (adapted from Croughan et al. 1987, with kind permission of John Wiley & Sons)

4.1.3.2 Non-Anchorage Dependent Cells Grown in Suspension

Similar to shear effects on cells grown on microcarriers, several studies have addressed damage to suspendable cells (mostly hybridom or myeloma, insect cells, or anchorage dependent cells adapted to growth in suspension, e.g. CHO, BHK) in stirred tanks. Again, factors such as cell type, medium, amount of serum, type of stirrer, and type of vessel, among others, have to be considered. A comprehensive overview is given by Chisti (2001). As an example, data from Kramer (1988) are summarized in Table 4.2. In this case, cell damage on suspended hybridom cells grown in a small stirred vessel was first observed at a stirrer speed of 300 rpm; at 580 rpm cell growth stopped completely. Similar results were reported by Shiragami (1997), who observed a maximum specific antibody production rate at ~ 180 rpm in a 250 mL reactor.

The Reynolds number Re is expressed by

$$\mathrm{Re} = \frac{n_R d_R^2 \rho_{fl}}{\eta_{fl}} \qquad (4.13)$$

In view of the available studies, damaging effects caused by intense mechanical forces and those associated with aeration or bubble entrainment have to be distinguished (compare Chisti 2001):

- Cell damage in stirred tanks operated at high stirrer speeds can be reduced by applying baffles to prevent vortexing and gas entrainment.
- Cell damage increases significantly in serum- or protein-free medium.
- Intense mechanical forces may cause physical damage to cells, leading to necrosis or lysis, or may induce apoptosis (a genetically controlled cell death) at sublethal stress levels.

Stevens (1994) compared data from different sources on damage to suspended hybridom cells. The mean shear stress and the mean energy dissipation rate per unit mass were calculated for those conditions, where growth rate and death rate were equal. Critical values for the mean shear stress in stirred vessels were between 0.39 and 1.5 Nm^{-2}, significantly lower than those reported for coaxial cylinder Searle viscometer (compare Table 4.1). A good correlation of these data can be obtained

Table 4.2 Cell damage of suspended hybridom cells (data from Kramer 1988) Culture conditions: stirred tank reaktor (volume 750 ml, vessel diameter 15 cm, rushton turbine (four blades), stirrer diameter 5 cm, medium RPMI 1640 + 10% Serum

Stirrer speed [rpm]	200	300	400	500
Re [–]	12,000	18,000	24,000	30,000
ε [W m^3]	26	88	210	410
μ [h^{-1}]	0.047	0.059	0.049	0.034
k_d [h^{-1}]	0	0	0.01	0.034

by plotting the critical mean energy dissipation rate per unit mass vs. the serum content in the medium (Fig. 4.12). The data for different hybridom cell lines agree quite well with those cases with surface aeration and without bubble entrainment. In the case of bubble aeration resulting in foam formation, significantly lower critical values were observed. For bubble aeration, but without free surface (no foam), the critical values were much higher. This proves that effects caused by bubble and foam formation have to be considered separately (see below).

Even if some data from the literature could be used to describe the approach used in Fig. 4.12, this does not provide a concept to be used for the design and scale-up of bioreactors with respect to shear stress. But it attempts to describe the impact of culture conditions on the damage of suspended cells by the mean power input. A similar approach was suggested by Märkl et al. (1987) for microbial cultures in stirred vessels. Strategies for design and scale-up of large-scale reactors will be discussed in Sect. 4.1.3.4.

4.1.3.3 Sparged Bioreactors

Gas sparging or bubbling air or another gas in the culture broth through a sparger located relatively deep within the bioreactor may cause damage to suspended mammalian cells or cells grown on microcarriers. As sparged bioreactors are still the preferred means of cell culture aeration, much effort has been expended on understanding the underlying mechanisms involved in bubble-associated cell damage and the methods available to control such damage (reviewed by Chisti 2000). The basic

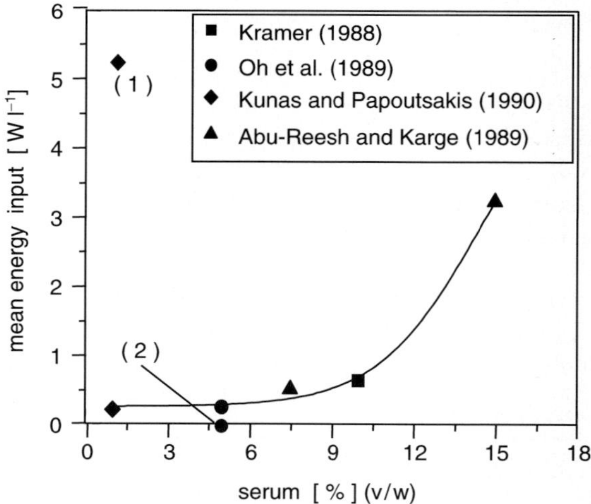

Fig. 4.12 Volume specific energy required for cell death. Mean energy input at 50% of maximal growth versus concentration of serum in the culture medium, (1) bubbles, but without free surface (no foam), (2) surface- and bubble aeration (foam), else: surface aeration

mechanisms have been outlined in Sect. 4.2.1.3. Here aspects relevant for the design and operation of sparged bioreactors (bubble aerated stirred vessels, air-lift bioreactors, bubble column bioreactors) are discussed.

The main mechanisms for cell damage in sparged bioreactors are discussed with reference to a mechanistic analysis suggested by Tramper et al. (1987) and Tramper and Vlak (1988). Assuming a first-order kinetic for cell damage (4.2), the first-order rate constant k_d for cell death was related to all causes of cell death in a sparged bioreactor by

$$k_d = \frac{24 Q V_k}{\pi^2 d_B^3 d_T^2 h_L}$$ (4.14)

with the volumetric aeration rate Q, the height of fluid h_L, the vessel diameter d_T, the bubble diameter d_B and a hypothetical killing volume V_k. This approach is based on the idea that cells trapped in the circulation wake of a bubble are carried with the bubble to the surface, where most of the damage takes place. The volume of the bubble wakes is addressed as "killing volume" surrounding the bubbles. From (4.14) some general conclusions can be drawn:

- The specific cell-death rate constant (first-order rate constant) should be proportional to the frequency of bubble generation ($6Q/\pi d_B^3$), the frequency of bubble rupture and the killing volume associated with the bubbles.
- Bubble break-up and foam formation at the culture surface are the principal cause of damage to suspended cells in sparged bioreactors.
- The specific cell-death rate constant should decline with the increasing height of the fluid.
- Events at the sparger and rise stages can be regarded as of minor importance, in particular, bubble coalescence and break-up within the bulk culture.
- Cell damage will increase in direct proportion to the superficial gas velocity in the column or the specific power input.
- Large bubbles should be less damaging than small bubbles.
- Serum or other proteins should have a protective effect.

Most of these statements could be confirmed experimentally (Chisti 2000). Nevertheless the concept of the killing volume itself has some week points. The killing volume V_K seems to be independent of the aeration rate and the height of the fluid, provided that the bubble diameter is not affected. But V_K should increase with bubble diameter indicating increasing cell damage with increasing bubble diameter. This seems to be incorrect, as there are numerous contrary experimental observations in the literature showing reduced cell damage with increasing bubble diameter. [For further discussion of this as well as other concepts of cell damage in sparged bioreactors, see Chisti (2000)].

Cell damage caused by bubbles can be significantly reduced by adding shear-protective substances to the culture medium, including serum and serum proteins (e.g. serum albumin), pluronics or polyethylene glycols among others (Chisti 2000). The protective effect can be physical and/or physiological (biochemical) depending on the specific agent. The protective effect of serum increases with increasing serum concentration up to 10% v/v serum and seems to be physical and physiological. The physical protective effect can be attributed to reduced plasma-membrane fluidity, increased medium viscosity or a turbulence-dampening effect. The precise nature is not yet clear. Physiological effects seem to take longer to become effective, meaning that a prolonged exposure of cells to serum will reduce their shear sensitivity. In addition to whole serum, serum compounds such as serum albumin also protect cells (e.g. bovine serum albumin (BSA) at >0.4 g L^{-1} (Hülscher et al. 1990)). But despite these shear protecting effects, serum or serum components are mostly avoided in industrial production processes, as discussed earlier (Sect. 2.3).

Alternatively, chemical agents can be used as shear protective agents. The most prominent one is the non-ionic surfactant polyol Pluronic®–F68, a block polymer of poly-oxyethylene and poly-oxypropylene (molar mass 8,358 Da). Typically Pluronic®–F68 is added at 0.5–3 g L^{-1} to the cell culture medium. The protective effect increases with concentration, whereas 0.5 g L^{-1} seems to be a minimum level, and is associated with several, mostly physical mechanisms:

- The attachment of cells to bubbles is reduced; therefore fewer cells are carried to the surface.
- The plasma-membrane fluidity of the cells is reduced.
- The gas–liquid interface is stabilized and therefore the film drainage during bubble rupture is slower.
- The nutritional transport is improved due to a reduced cell-fluid interfacial tension.
- The membrane of the cells is stabilized by incorporating Pluronic®–F68 into the membrane, thus strengthening the cells.

Furthermore, physiological factors, e.g. an effect on the permeability of some cells, have been observed.

Poly-ethyleneglycol (PEG) has been shown to protect some cells but not others. A protecting effect has been found for molecular weights exceeding 1,000 Da and concentrations above 0.25 g L^{-1}. The protective mechanism of PEG seems to be due to an effect on the surface tension and/or the presence of PEG at the gas–liquid interface (Chisti 2000).

Despite the many different protective mechanisms discussed above, the main effect can be attributed to a reduced number of cells adhering to the bubble surface just before its rupture. Additives that rapidly lower the gas-liquid interfacial tension will obviously prevent cell adhesion to bubbles and so reduce cell damage. For a certain culture process, a feasible additive (combination of additives) and an appropriate concentration, as well as the best time for addition, has to be found empirically.

4.1.3.4   Consequences for Reactor Design

For design and operation of sparged and agitated bioreactors some general conclusions can be drawn (Chisti 2000; Fenge and Lüllau 2006; Ma et al. 2006, modified)

- The aeration rate should be kept as low as possible (below 0.1 vvm).
- The sparger should be located such that the rising bubbles do not interact with any impellers.
- The sole purpose of the impeller should be to suspend cells and to mix the fluid gently. Therefore the impeller should be of a type that does not produce excessively high local rates of energy dissipation.
- For air-lift and bubble column bioreactors the aspect ratio (ratio between height and diameter) should be ~14 for small scale and 6–7 for large scale.
- In stirred tanks the mean energy dissipation rate should be below ~ $1{,}000\,W\,m^{-3}$.
- Impeller tip speed should be between 1 and $2\,ms^{-1}$.
- A suitable additive such as Pluronic®–F68 should be used whenever feasible.

Despite a number of concepts having been suggested in the literature to describe shear effects caused by fluid-mechanical forces or bubbles, none of these concepts allow for a precise layout of large-scale bioreactors (design, operating parameters such as aeration rate, power input or stirrer speed) without additional experiments. Furthermore, experimental findings on a small scale have to be validated on a pilot scale before being transferred to the final industrial scale. A simple strategy for design and scale-up of large scale-reactors based on the concept of a constant power-to-volume-ratio could be as follows:

- Determine the critical mean energy dissipation rate (power input per volume) in a laboratory scale bioreactor geometrically similar to the large scale reactor.
- Calculate the stirrer speed in the large scale bioreactor by considering the mean energy dissipation rate to be constant (e.g. by applying a correlation between the power-number or Newton-number and the Reynolds-number).

In applying this strategy, some restriction should be considered. First an increase in the scale of more than 10:1 might lead to unreasonable values for the stirrer speed. Therefore several scale-up steps from the laboratory to the industrial scale are required and the design criteria have to be evaluated at each step. Furthermore it is not taken into account, that due to an uneven distribution of energy dissipation within a stirred vessel suspended cells or cells grown on microcarriers are exposed to varying levels of shear stress on their way through the vessel. The maximal energy dissipation can differ significantly from the mean energy dissipation, depending on the geometry of the vessel and the stirrer. Additional frequency of exposure to the maximal energy dissipation has to be considered. Henzler and Kauling (1993) have shown some important scale-up relationships (Fig. 4.13). The mechanical stress produced by the liquid increases with the size of the reactor, even if the mean energy dissipation (power input per volume) remains constant. In contrast, the shear produced by the bubbles decreases simultaneously. If the size of the reactor is increased by a factor of 1,000 (from 1 to 1,000 L) the main shear due to aeration decreases by a factor of 10. As cell damage seems to be caused mainly by sparging, this is a very important indication. It suggests that cell damage due to shear stress observed on the laboratory scale is of minor importance in larger reactors. This seems to be one reason why sparging of bioreactors is often regarded as

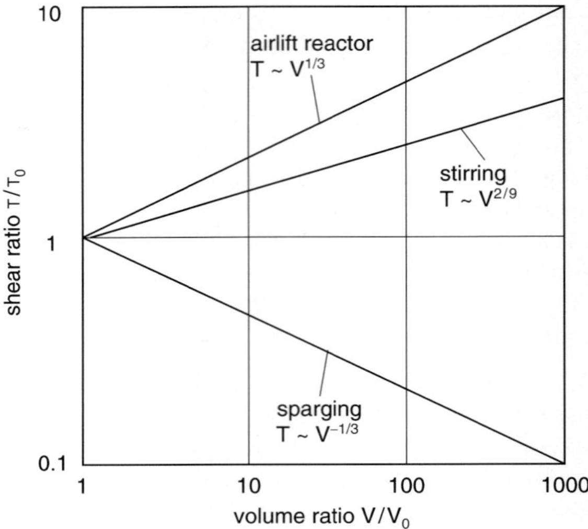

Fig. 4.13 Variation of mechanical stress with increasing reactor volume for P/V = constant (adapted from Henzler and Kauling 1993, with kind permission of Springer)

problematic on a laboratory scale, whereas on a large scale it continues to be the preferred and robust method for supplying oxygen to the cell culture.

Varley and Birch (1999) draw some conclusions from research concerned with shear stress on mammalian cells:

- Many suspension cells, especially hybridoma, are not as shear sensitive as thought at first.
- Most cell death is caused by bursting bubbles, even if other reasons for cell death have to be considered and cell death increases when the Kolmogoroff eddy length approaches the size of the cells.
- The range of tolerable aeration rates and stirrer speed for altering mixing is limited.

Implications on design and operation of bioreactors from mammalian cells, especially stirred tanks and air-lift reactors, will be discussed in Sect. 5.1

4.2 Oxygen Supply

4.2.1 Introduction

Mass transfer, including oxygen supply and carbon dioxide removal, is one of the most important factors in operating cell culture bioreactors. Because oxygen is only slightly soluble in medium (approximately $7\,mg\ L^{-1}$) and the oxygen consumption rate (or oxygen uptake rate OUR) is between 3×10^{-10} and 2×10^{-8} mg cell^{-1} h^{-1} (Aunins and Henzler 1993), a typical culture of 2×10^{6} cells mL^{-1} would deplete the oxygen dissolved in the medium in under one hour and run out of oxygen rapidly. Therefore it is clearly necessary to supply oxygen during the culture period. Some methods used for this purpose on a small scale are surface aeration, membrane aeration, gas sparging and increase of the partial pressure of oxygen in the headspace or the supplied air (Aunins and Henzler 1993). The scale-up of animal cell cultures is greatly dependent upon the ability to supply sufficient oxygen without causing cell damage. Because of high mass transfer rates and operational simplicity, gas sparging is the preferred method in large scale cell culture. However, cultured animal cells may be damaged by gas sparging (Handa et al. 1987; Kunas and Papoutsakis 1990; Michaels et al. 1996; Zhang et al. 1992a). Cell bubble interactions are generally regarded as the likely cause of this cell damage but the mechanism is not fully understood (Sect. 4.1.3.3) (Chisti 2000; Meier et al. 1999; Wu 1995). It is hypothesized that shear associated with bubble rupture on the bubble surface is the main cause.

The supply of oxygen in the cultivation of animal cells in bioreactors still poses a problem. Even though oxygen consumption rates of animal cell cultures are low compared to cultures of other microorganisms, they must be supplied with a sufficient amount of oxygen. But animal cells are more sensitive to shear stress caused by vigorous mixing and gas sparging. This necessitates a maximum possible oxygen transfer into the liquid phase to keep the mechanical force low (Kunas and Papoutsakis 1990; Fenge et al. 1993; Papoutsakis 1991). In bubble aerated reactor systems, damage to animal cells often occurs due to a wide range of bubble size distribution (Kunas and Papoutsakis 1990; Fenge et al. 1993). It occurs when ascending bubbles force the cells in the medium to the surface (Grima et al. 1997). The stress to the cells is caused by the displosion of bubbles breaking through the surface, resulting in permanent damage (Kunas and Papoutsakis 1990; Papoutsakis 1991).

The size of the bubble appears to be an influential parameter for the tendency of the cells to rise to the surface. Another disadvantage of bubble aeration is that foam could be generated on the surface of the medium, which then could lead to a push-out of the medium out of the bioreactor. But there are several methods available to fight foam production within the vessel (Tan et al. 1994; Zhang et al. 1993). Widely used are silicone based antifoaming agents. The most significant advantage of bubble aeration is its high oxygen mass transfer into water. This high rate is due to a high volume-specific phase surface area and a lower transport resistance compared to membrane procedures.

Because of the relatively high transport resistance, bubble-free aeration systems such as membranes require high membrane surface areas to gain a sufficient oxygen supply into the medium. Also greater effort in installation and maintenance is required in comparison to bubble aeration. Also, the need for sterile operation makes the use of membrane procedures more problematic. In consequence, membrane aeration systems are mostly used in the laboratory and pilot scales (Marks 2003; Henzler and Kauling 1993). Other indirect aeration systems used are spin filters and vibro mixers. In addition to this, systems with external or perfluoro-carbon mediated aeration can be used Gotoh et al. (2001).

But these procedures involve high apparatus costs, which make them unsuitable for large scale production. Also a scale up can be prone to errors. With increase of the inner volume, oxygen transfer progressively becomes more difficult, because the mass transfer efficiency of stirred tank reactors generally decreases as the scale is increased (Marks 2003; Henzler and Kauling 1993). Even though all other aeration methods are technically more complex and lead to less reliability due to the greater number of connection tubes, direct sparging is still the method of choice for the pilot and production scale (Henzler and Kauling 1993; Krahe 2003) Three different shear hypotheses were considered by Henzler and Kauling (1993), who also showed a relation for scale up (Sect. 4.1.3.4). The shear ratio related to increasing volume of the reactor from $1\,L$ to $1\,m^3$ was calculated and a tenfold decrease of the main shear was found. This finding indicated that shear problems due to bubble aeration observed in the laboratory scale are of minor importance in larger bioreactors (Wu et al. 1995). In recent studies it has been shown that the bubble size has a significant influence on cell damage (Meier et al. 1999). The bigger the cell size the higher the relative speed between the gas bubble and the cell. Three possible causes of cell death are associated with bursting bubbles: bursting bubbles may destroy nearby cells within the suspension due to a shear associated with the burst; cells which are trapped inside a bubble may be damaged by the burst, and a bursting bubble may also damage cells which attached to the bubble during its ascent. The relative effect of each of these is a result of the bubble size and medium properties (Meier et al. 1999; Marks 2003).

Due to the contact of a cell with a bubble, shear stress on the cell membrane is created, which damages the cell. Floating of a cell in foam can also cause cell damage (Wu et al. 1995). Once a foam bubble collapses, the high surface tension results in a destruction of the cell membrane (Ghebeh et al. 2002). Experiments performed by Kunas and Papoutsakis in completely filled bioreactors with micro-bubbles of sizes between 50 and $300\,\mu m$ in diameter showed that the cultured hybridomas sustained only very little damage, even when small bubbles were present (Kunas and Papoutsakis 1990). Microbubbles have less relative speed in the fluid, less collision chance and damage ability, due to their higher mean residence time. It was shown by Michaels et al. that it is possible to minimize the frequency of bubble coalescence and breakup events by employing microbubbles for oxygenation at low agitation rates for mixing (Michaels et al. 1996).

As a result of smaller bubble size, the relative surface of the gas in the liquid increases. This leads to a higher mass transfer of oxygen into the liquid. Less gas flow is required to meet the oxygen demand of the living cells. If the reactor is aerated with high oxygen partial pressures and with pulsed aeration profile, most of the bubbles will be completely dissolved before reaching the surface, resulting in less foam. Nowadays nearly every industrial cell culture reactor is equipped with standard ring spargers, spargers or micro spargers. In industrial production with cell cultures the strategy is to use cell lines, which are not so sensitive to cell damage by direct sparging.

4.2.2 Limitations for Oxygen Transfer

Oxygen concentration in media is indicated usually as percentage air saturation, percentage dissolved oxygen, concentration in weight per volume (μg mL^{-1} or mg L^{-1}), p_{O_2} or molar units (mmol L^{-1}). Percentage air saturation (identical to percentage dissolved oxygen, % DO) is given by:

$$\% \text{ DO} = (c_{L}/c^*)\ 100\%,$$

with c_{L} as the actual oxygen concentration in the medium and c^* as the oxygen concentration in the equilibrium with air (21% oxygen).

Oxygen tension or p_{O_2} is the partial pressure of oxygen in the liquid phase in equilibrium with the partial pressure in the gas phase.

The interrelationship between the equilibrium oxygen concentration in media and the partial pressure in the gas phase is given by Henry's law:

$$P_{O2} = c^*\text{He} \tag{4.15}$$

He is Henry's constant which is a function of temperature. Oxygen solubility in pure water between 0 and 36 °C is given by the following equation (Truesdale et al. 1955):

$$c^* = 14.16 - 0.3943T + 0.007714T^2 - 0.0000646T^3. \tag{4.16}$$

With solubility of oxygen c^* in units of mg L^{-1} and temperature T in °C.

The solubility of oxygen in media is also influenced by the presence of electrolytes and organic components so the equilibrium oxygen concentration in medium has to be corrected (Schumpe et al. 1978). Quicker and Schumpe have given an empirical correlation to correct values of oxygen solubility in water for the effects of cations, anions and sugars (Quicker et al. 1981):

$$\text{Log}_{10}(c_0^*/c^*) = 0.5 \sum_i H_i z_i^2 c_{iL} + \sum_j K_j c_{jL}. \tag{4.17}$$

With:
c_0^*: oxygen solubility at zero solute concentration (mol m^{-3})
c^*: oxygen solubility (mol m^{-3})
H_i: constant for ionic component i (m^3 mol^{-1})
Z_i: valency of ionic component i
c_{iL}: concentration of ionic component i in the liquid (mol m^{-3})
K_j: constant for non-ionic component j in the liquid (m^3 mol^{-1})
c_{jL}: concentration of non-ionic component j in the liquid (mol m^{-3})

Values of H_i and K_j for use in the above equation are listed in literature (Schumpe et al. 1978; Quicker et al. 1981). In a typical fermentation medium oxygen solubility is between 5 and 25% lower than in water as a result of solute effects (Doran 2006).

For RPMI medium the oxygen concentration has been reported to be 0.2 mmol L^{-1} (or 7.1 mg L^{-1} – in equilibrium with air $p_{O_2} = 0.21$, $T = 37$ °C) (Oller et al. 1989).

4.2.2.1 Evaluation of Oxygen Consumption and Oxygen Transport Efficiency

High concentrations of oxygen can be toxic for the cells. Effects of different oxygen concentrations on cells are reported in the literature. A good summary is given by Doyle and Griffith (1998) and Ruffieux et al. (1998). Established cell lines function well over a wide range of oxygen concentrations (between 15 and 90% of air saturation). Primary cells are cultivated best at lower oxygen concentrations to simulate the in vivo environment.

The oxygen uptake rate (OUR) or the specific oxygen consumption rate (q) is influenced by the cell type, cell density, proliferative state of the culture, and glucose and glutamine concentration and is a good indicator of cellular activity; under some conditions, it is even a good indicator of the number of viable cells (Doyle and Griffith 1998; Ruffieux et al. 1998; Riley 2006).

For the quantification of the oxygen uptake rate (OUR) and the oxygen transfer rate (OTR), it is necessary to give an unsteady-state mass balance of oxygen within the bioreactor. The mass balance can be written as follows: the change in oxygen concentration over the time in the reactor is equal to the rate of oxygen transfer into the culture (OTR) minus the oxygen consumed by the cells:

$$dc/dt = k_L a(c^* - c_L) - \text{OUR X} \tag{4.18}$$

with:

$k_L a$: mass transfer coefficient, which is the product of k_L, the overall mass transfer coefficient from the gas to the liquid phase (two film model), and a, the gas–liquid interfacial area per unit of the reactor liquid volume (s^{-1})
c_L: concentration of oxygen in solution (kg m^{-3})
c^*: equilibrium solubility of oxygen – oxygen saturation (kg m^{-3})
X: cell density (cells L^{-1})
OUR: oxygen uptake rate (kg O$_2$ 10^{-6} cells s^{-1})
t: time (s)
The oxygen transfer rate (OTR) is given by:

$$\text{OTR} = k_L a(c^* - c_L) \tag{4.19}$$

A very common method for measuring the $k_L a$ and the OUR is the "dynamic method". The main advantage of the dynamic method is the low cost of equipment needed. There are several different versions of the dynamic method described in the literature. Here only the method with cells in a batch culture is described.

As shown in Fig. 4.14 at some time t_0 the culture is de-oxygenated by stopping the aeration if the culture is oxygen consuming. Dissolved oxygen concentration c_L drops during this period. Before c_L remains c_{crit} the aeration is started and the increase in c_L is monitored as a function of time. Assuming re-oxygenation is fast relative to cell

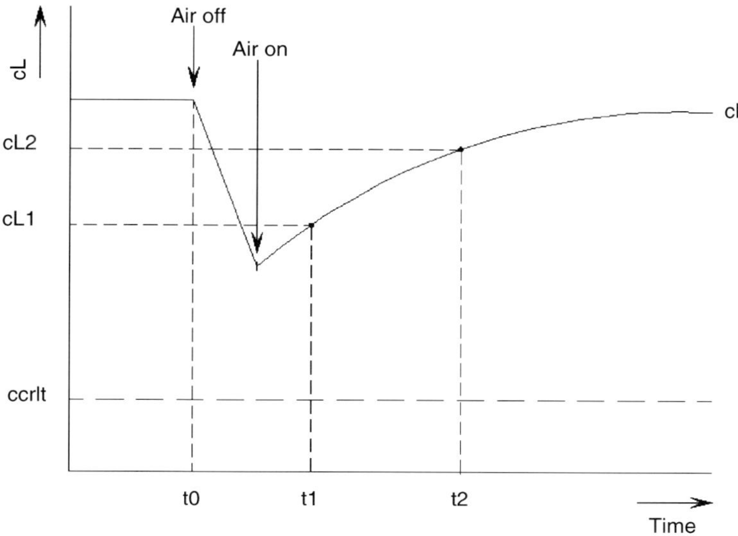

Fig. 4.14 Variation of oxygen tension for dynamic measurement of $k_L a$

growth, the dissolved oxygen level will soon reach a steady state value c_{AL} which reflects a balance between oxygen supply and oxygen consumption in the system.

During the re-oxygenation step the system is not in steady state and we can use the above unsteady-state mass balance (4.19). When $c_L = c_{AL}$, $dc/dt = 0$ because there is no change in c_L with time. Therefore:

$$\text{OUR } X = k_L a(c^* - c_{AL}).\tag{4.20}$$

Substituting the result into (4.19) and cancelling the term $k_L a\ c^*$ gives:

$$dc/dt = k_L a(c_{AL} - c_L).\tag{4.21}$$

Assuming $k_L a$ is constant with time, we can integrate (4.21) between t_1 and t_2. The resulting term for $k_L a$ is:

$$k_L a = \ln[(c_{AL} - c_{L1})/(c_{AL} - c_{L2})]/(t_2 - t_1).\tag{4.22}$$

When $\ln[(c_{AL} - c_{L1})/(c_{AL} - c_{L2})]$ is plotted against $(t_2 - t_1)$ as shown in Fig. 4.15 the slope is $k_L a$.

Equation (4.22) can be applied to actively oxygen consuming cultures or to systems without oxygen uptake.

In the de-oxygenation period the slope of the dissolved oxygen concentration profile is the oxygen uptake rate (OUR) because (4.18) reduces to

$$dc_L/dt = -\text{QUR } X\tag{4.23}$$

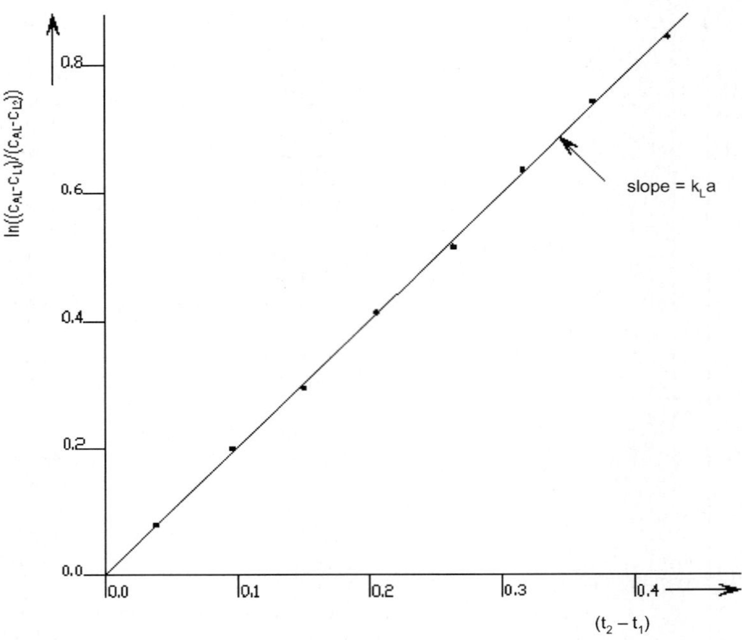

Fig. 4.15 Evaluating k_La using the dynamic method

for a reactor where the oxygen transfer term can be neglected. The most critical step is the removal of all air bubbles from the reactor.

4.2.3 Oxygen Supply Systems (Aeration Systems)

A culture can be aerated by one, or a combination, of the following methods: surface aeration, direct sparging, indirect and/or membrane aeration (diffusion), medium perfusion, increasing the partial pressure of oxygen, and increasing the atmospheric pressure (Griffith 2000; Varley and Birch 1999). Methods of oxygen supply for submerse fermentation of animal cells can be seen in Fig. 4.16.

4.2.3.1 Surface Aeration

In surface aeration, mass transfer occurs only through the surface of the liquid. To increase the mass transfer by increasing the effective phase interface, a second impeller should be located at the interface or near the liquid interface (Fig. 4.17).

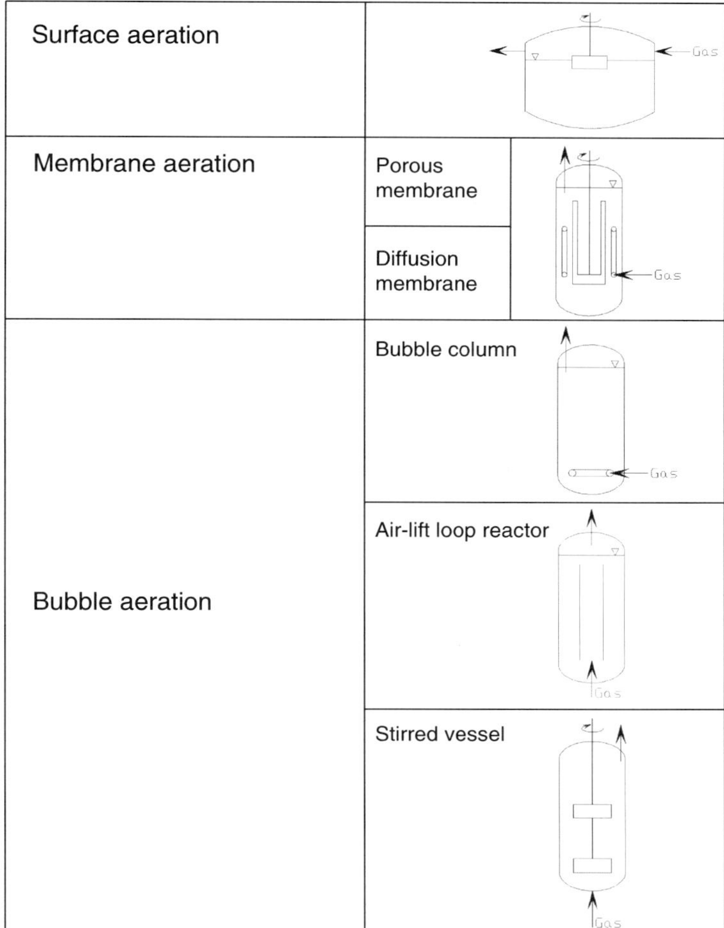

Fig. 4.16 Methods of oxygen supply for submers fermentation of animal cells [adapted from Henzler and Kauling (1993) with kind permission of Springer Science and Business Media]

Alternatively an intensive wave motion could be inserted (wave bioreactor) (Singh 1999; Sect. 5.1.4). The partially submerged second impeller (surface aerator) was seen to increase the mass transfer coefficient four times for laboratory scale stirred reactors (Hu et al. 1986). Because bubble introduction should be avoided, oxygen input cannot be increased indefinitely by surface aeration. Henzler and Kauling (1993) gave correlations in the low power input range based on experimental data for the volumetric mass transfer coefficient $k_L a$, the limiting speed of the impeller located at the interface and for the scale-up.

$$k_L a(H_R/D_R) = f(d_R/D_R)(n_R^2 d/g) \tag{4.24}$$

                                                                                     P. Czermak et al.

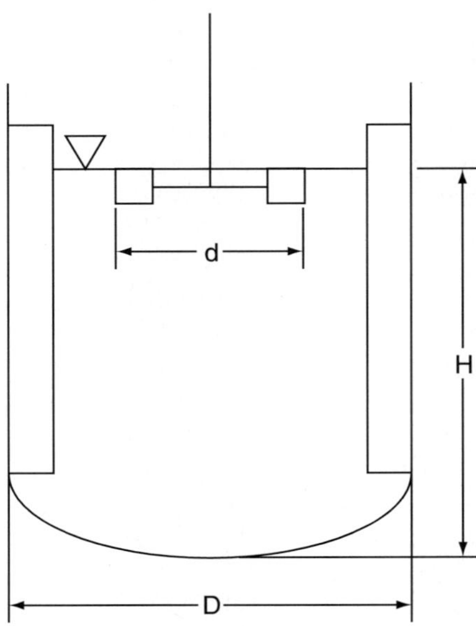

Fig. 4.17 Schematic of a reactor with surface aeration [stirrer at interface – adapted from Henzler and Kauling (1993) with kind permission of Springer Science and Business Media]

with

H_R: filling height of the stirred reactor – impeller located at the interface (m)
D_R: reactor diameter (m)
d_R: impeller diameter (m)
n_R: impeller speed (s^{-1})
g: acceleration due to gravity (m^2 s^{-1})
and

$$f(d_R/D_R) = [(d_R / D_R) - 0.13]400 \qquad (4.25)$$

for $d_R/D_R > 0.15$

Henzler and Kauling (1993) show that even for semi-industrial scale, sufficient oxygenation of cell cultures cannot be provided by the surface of the liquid. Therefore surface aeration is sufficient only for low density cultures in small vessels like T-flasks, roller bottles and bench scale vessels, where a relatively large ratio of air-medium interface to volume exists.

4.2.3.2 Direct Sparging

Direct gas sparging into the medium is the most common and simple way to supply oxygen into the bioreactor (Marks 2003; Varley and Birch 1999; Chalmers 1994). Aeration with spargers gives high oxygen transfer rates caused by large interfacial area, a, for bubbles (Fenge et al. 1993; Henzler and Kauling 1993; Varley and Birch 1999; Moreira et al. 1995; Nehring et al. 2004). Disadvantages of direct sparging are possible cell damage and foaming of the culture (Kunas and Papoutsakis 1990; Michaels et al. 1996; Chisti 2000; Meier et al. 1999; Wu 1995; Henzler and Kauling 1993; Zhang et al. 1992a,b). Anyway many different cell types have been grown successfully in air-sparged bioreactors such as bubble columns, airlift reactors or agitated/stirred vessels (Doyle and Griffith 1998; Ma et al. 2006).

Interfacial area (a) will also obviously affect $k_L a$. Oxygen transfer rates increase as the size of bubbles decreases. Bubble size will generally be in the region of 4–6 mm for sparging with standard spargers (metal sparger, ring injector) or 0.1– 1 mm for sparging with micro spargers (e.g. ceramic) (Varley and Birch 1999; Nehring et al. 2004). The effects of additives including new born calf serum and Pluronic F-68 were found to depend on the size of the bubbles (Henzler and Kauling 1993; Varley and Birch 1999; Moreira et al. 1995). Although it is known that bubble coalescence is important, as it affects bubble size and hence interfacial area and $k_L a$, the effect of fluid properties and surface properties is still not well understood (Varley and Birch 1999).

Correlations based on the most relevant bubbling regimes for animal cell culture are available for predicting the bubble size at the sparger: separate bubble formation and chain bubbling (Varley and Birch 1999).

For separate bubbling:

$$d_s = 1.7 [\sigma d_o / (\Delta \rho g)]^{1/3} \tag{4.26}$$

where
d_s: Sauter mean diameter (m)
σ: surface tension (kg s^{-2})
d_o: orifice diameter (m)
$\Delta \rho$: density difference (kg m^{-3})
g: acceleration due to gravity (m^2 s^{-1})
For the chain bubbling regime:

$$d_s = 1.17 v_0^{0.4} d_0^{0.8} g^{-0.2} \tag{4.27}$$

with
v_0: gas velocity at the sparger (m s^{-1})
Equations have been proposed for bubble size for turbulent flow (generally stirred sparged vessels) far from the sparger, both for non-coalescing and coalescing solutions. However, at the low gas flow rates and low agitation rates used in animal cell culture, the bubble size far from the sparger is likely to be equal to the bubble size at the sparger (Varley and Birch 1999).

There are several correlations for the oxygen transfer coefficient in the literature (Henzler and Kauling 1993; Moreira et al. 1995; Ma et al. 2006; Moo-Young and Blanch 1981).

For specified medium, specified reactor dimensions and type of sparger, the mass transfer coefficient $k_L a$ in bubble columns and loop reactors (e.g. airlift) depends only on the superficial gas velocity (Henzler and Kauling 1993).

$$k_L a = C v^\alpha \qquad (4.28)$$

with

C: constant

v: superficial gas velocity (m s^{-1})

α: exponent (further function of superficial velocity).

The influence of sparger holes/pores is noticeable at low gas velocity in the range desired for the oxygenation of cell cultures (Henzler and Kauling 1993; Nehring et al. 2004). With increasing gas velocity the aeration efficiency ($k_L a/v$) decreases as a result of increasing coalescence, so it makes sense to use oxygen-enriched air to minimize shear due to gas bubbles (Henzler and Kauling 1993). Using special ceramic micro-spargers or increasing gas velocities gives a uniform distribution of the gas over all the sparger holes/pores and a more homogenous bubble size distribution (Figs. 4.18 and 4.19). This leads to reduced coalescence because of the decrease in the local gas hold-up (Henzler and Kauling 1993). Henzler and Kauling (1993) showed that the mass transfer is hindered in protein-containing media by the presence of protein at the interface, especially for smaller bubbles, because of the increased foam formation tendency of protein-containing media and a resulting increase in coalescence of bubbles.

The mass transfer can be enhanced by stirring. Therefore lower gas velocities are necessary in sparged bioreactors with agitation than in bubble column reactors. For industrial animal cell reactors, gas flow rates of 1 vvh (volume gas per volume reactor per hour) for bubble columns and 0.1–1 vvh for aerated and stirred reactors have been reported (Bliem et al. 1991). These lower gas flow rates may reduce the level of required antifoam agents and shear protectants (Czermak and Nehring 1999). For cell culture, the region of low impeller power is of interest. The $k_L a$ is independent of the type and geometry of the impeller if sufficient mixing takes place in the bioreactor. As the viscosity of cell culture media is low, this assumption is usually satisfied (Henzler and Kauling 1993).

For stirred reactors (4.27) becomes the form

$$k_L a = C v^\alpha (P/V)^\beta \qquad (4.29)$$

with

P/V: power input per unit volume of gas dispersion (kg m s^{-3})

β: exponent independent of scale and impeller type.

Table 4.3 gives $k_L a$ values, determined using the dynamic method (Nehring et al. 2004), for standard stirred bioreactors (Braun Biotech/Sartorius) on different scales, equipped with propeller or three-blade segment impeller (both are axial mixing stirrers) and different sparger systems (microsparger – stainless steel, Braun Biotech/Sartorius; microsparger – ceramic, BIM/Applikon Biotechnology; ring sparger – stainless steel, Braun Biotech/Sartorius).

Hydrophilic materials such as porous ceramics are suited to the production of microbubbles (Figs. 4.18 – 4.20) in water-based fluids. The formation of microbubbles (bubbles sizes differ in diameter from 100 to 500 µm (Nehring et al. 2004;

Table 4.3 k_La values for standard stirred bioreactors (Braun Biotech/Sartorius) in different scale equipped with propeller or three blade segment impeller (both are axial mixing stirrers) and different sparger systems (microsparger – stainless steel, Braun Biotech/Sartorius; microsparger – ceramic, BIM/Applikon Biotechnology; ring sparger – stainless steel, Braun Biotech/Sartorius)

Reactor type	rpm	Aeration System	Sparging rate (vvm)	Medium	k_La (h⁻¹)	References
Biostat ECD 10 L	50	Microsparger 100 µm pores (stainless steel)	0.1	Tap-water	9.3	(Fenge et al. 1993)
Biostat UCD 40 L	20	Microsparger 100 µm pores (stainless steel)	0.1	Tap-water	7.1	(Fenge et al. 1993)
Biostat UCD 35 L	20	Ring sparger 1 mm pores	0.1	Tap-water	4.9	(Fenge et al. 1993)
Biostat B 5 L	50	Microsparger 2–5 µm pores (ceramic – TiO_2)	0.1	RPMI 1640	33.1	(Nehring et al. 2004; Bliem et al. 1991)
Biostat B 5 L	50	Microsparger 100 µm pores (stainless steel)	0.1	RPMI 1640	10.1	(Nehring et al. 2004; Bliem et al. 1991)
Biostat B 5 L	50	Microsparger 2–5 µm pores (ceramic – TiO_2)	0.05	RPMI 1640	30.6	(Nehring et al. 2004; Bliem et al. 1991)
Biostat B 5 L	50	Microsparger 100 µm pores (stainless steel)	0.05	RPMI 1640	9.4	(Nehring et al. 2004; Bliem et al. 1991)
Biostat B 5 L	100	Microsparger 2–5 µm pores (ceramic – TiO_2)	0.1	RPMI 1640	50.4	(Nehring et al. 2004; Bliem et al. 1991)
Biostat B 5 L	100	Microsparger 100 µm pores (stainless steel)	0.1	RPMI 1640	10.4	(Nehring et al. 2004; Bliem et al. 1991)
Biostat B 5 L	100	Microsparger 2–5 µm pores (ceramic – TiO_2)	0.05	RPMI 1640	32.8	(Nehring et al. 2004; Bliem et al. 1991)
Biostat B 5 L	100	Microsparger 100 µm pores (stainless steel)	0.05	RPMI 1640	9.7	(Nehring et al. 2004; Bliem et al. 1991)

Fig. 4.18 The ceramic sparger tip in buffer and DMEM medium

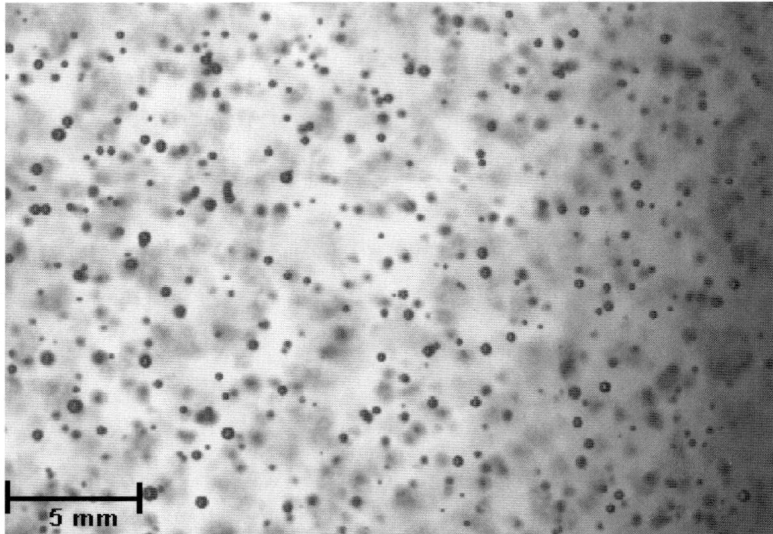

Fig. 4.19 Bubbles in phosphate buffer aerated by a ceramic sparger. Bubbles sizes differ in diameter from 100 to 500 μm (from Nehring et al. (2004) with kind permission of American Chemical Society)

Bliem et al. 1991)) is due to the small contact angle between the water and ceramic. The high k_La-values of the ceramic microsparger aeration system clearly show that a significantly better oxygen transfer into the medium was achieved by the larger phase surface area and in addition by the fact that the gas bubbles remained longer in the bioreactor (Nehring et al. 2004).

Fig. 4.20 SEM images of the porous structure of a zirconium dioxide microsparger cross section (The image was taken after breaking the ceramic tube across the diameter)

The culture of cells on microcarriers in sparged reactors requires low gas flow rates, small bubble diameter for high oxygen transfer rates, low mechanical agitation and the use of antifoams to avoid shear stress to the anchorage dependent cells and to avoid flotation of the carriers (Doyle and Griffith 1998). By using a ceramic microsparger system, five times higher $k_L a$ values (two microspargers integrated into the baffles) in comparison to stainless steel microspargers (Applikon Biotechnology) were shown for a microcarrier culture (3 g L^{-1} Cytodex 3 microcarrier; DMEM medium) in a 75 L stirred reactor (H/D = 1.5; 2 propeller stirrers with different diameter; 33 rpm) at a very low aeration rate of 0.001 vvm (Czermak and Nehring 2000).

4.2.3.3 Membrane Aeration

Bubble free oxygenation of cell culture media by membrane aeration with open-pore membranes or diffusion membranes can be seen in (Figs. 4.21–4.23) (Aunins and Henzler 1993; Henzler and Kauling 1993; Qi et al. 2003; Schneider et al 1995).

For microporous membranes, the medium is in direct contact with air in the micropores of the membrane. The air–liquid interface in the pores is controlled by pressure and hydrophobic force. For diffusion membranes oxygen diffuses first from the gas phase into the oxygen soluble membrane (very common: silicone rubber) and then into the culture medium.

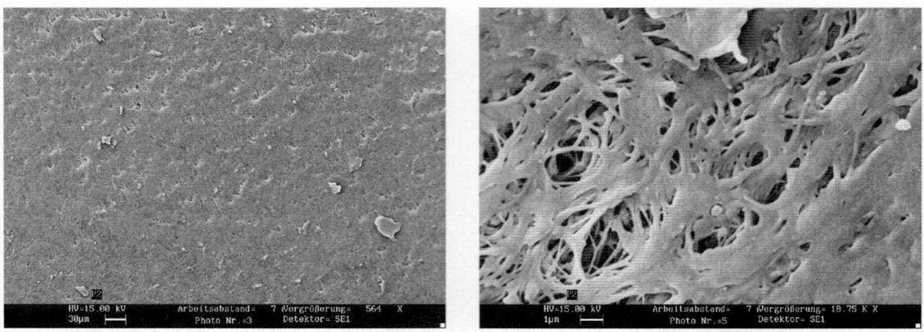

Fig. 4.21 REM cross section of a porous oxygenation membrane (PP)

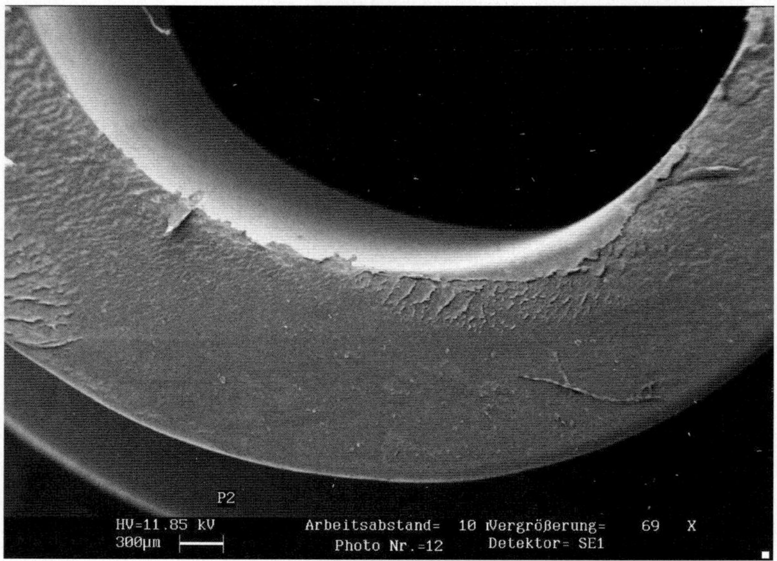

Fig. 4.22 REM cross section of diffusion membrane (silicon tube)

In the case of oxygenation with microporous membranes there is the mass transfer coefficient at the gas–liquid interface k equal to k_L (4.29) (Aunins and Henzler 1993; Henzler and Kauling 1993). In aeration with diffusion membranes, k is an overall mass transfer coefficient at the liquid side coefficient k_L and additional membrane diffusion (tubular membrane):

$$k=1/[(1/k_L+[d_2\ln(d_2/d_1)]/(2D_sHe_w/H_s)]\tag{4.30}$$

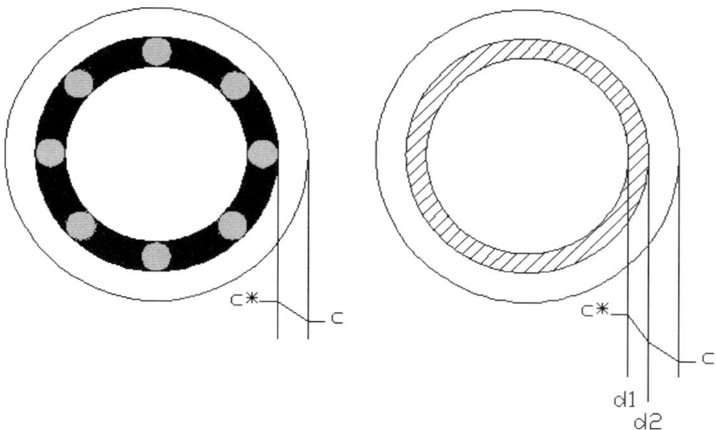

Fig. 4.23 Concentration profiles for oxygen supply via membranes [adapted from Henzler and Kauling (1993)]

with:

k: mass transfer coefficient at the gas–liquid interface (m s^{-1})

k_L: mass transfer coefficient at the liquid-side interface (m s^{-1})

d_2: outer diameter of the membrane tube (m)

d_1: inner diameter of the membrane tube (m)

D_s: diffusion coefficient of oxygen in the membrane (m^2 s^{-1})

He_w: Henry constant of oxygen in water (m^2 s^{-2})

He_s: Henry constant of oxygen in membrane material (m^2 s^{-2})

For the material constant $D_s He_w / He_s$ of silicone the temperature dependence in the range of 10–40 °C is as follows (Henzler and Kauling 1993; Bräutigam 1985):

$$D_s He_w / He_s \sim 7.3 \times 10^{-9} \exp(0.024T)(\text{m}^2 \text{ s}^{-1}) \qquad (4.31)$$

For the description of the mass transfer coefficient k_L it is suitable to use dimensionless relations in the form of Sh = f (Re, Sc) with the Sherwood number Sh = $k_L d_2 / D$. Correlations are given by Aunins and Henzler (1993) and Henzler and Kauling (1993).

The main factors which affect the oxygen transfer are the oxygen concentration in the supply gas, the membrane porosity and the surface area of the membrane, as well as the value of k_L influenced by agitation of the medium by a stirrer or by movement of the membranes (dynamic membrane aeration).

The advantages of membrane aeration for cells sensitive to sparging, microsparging and foam formation are superposed by limitations such as the complexity of the process, limited design data, large membrane area, difficulties in maintenance and cleaning. In addition, proteins can foul the membrane potentially altering the hydrophobic properties of the membranes, especially at porous membranes (Aunins and Henzler 1993; Henzler and Kauling 1993; Doyle and Griffith 1998; Ma et al. 2006;

Qi et al. 2003; Schneider et al. 1995). With membrane aeration, oxygen transfer rates comparable to direct sparging are possible, but high gas pressures and flow rates are necessary for it (Aunins and Henzler 1993; Henzler and Kauling 1993; Moreira et al. 1995). Because of the large membrane area required (2–3 m of silicone tubing per litre of medium) membrane aeration is used to small and intermediate scale bioreactors (5–500 L) only (Doyle and Griffith 1998; Ma et al. 2006).

To overcome the limitations of standard membrane aeration (Fig. 4.24), especially with increasing scale, the newly developed dynamic membrane aeration system is of interest (Fig. 4.25) (Frahm et al. 2007; Rampe 2006). The silicone tubes (membranes) are wrapped on a rotor. This membrane carrier is placed in an oscillating circular motion. Therefore the flow of the medium at the membranes and the mass transfer is enhanced. In addition there are no moving gaskets for the gas supply, with the resulting advantages for cleaning, design and scale-up (Frahm et al. 2007).

In Table 4.4 the mass transfer coefficient (for the membrane and the liquid film for oxygen) is listed for different power inputs to show the efficiency increase by oscillating the silicone membrane tubing through the medium (dynamic membrane aeration) compared to a standard membrane aeration system, where the silicone membrane tubing is wrapped in a basket (stator) and fluid flow around the tubing is generated by an anchor stirrer (Frahm et al. 2007). Figure 4.25 shows a dynamic aeration rotor at a 200 L scale. It is easy to scale-up the membrane area by adjusting the number of rotor arms and/or by branching out the rotor arms and therefore the volume specific mass transfer coefficient k_a of the dynamic membrane aeration system is much higher than a comparable standard membrane aeration system with silicone tubing.

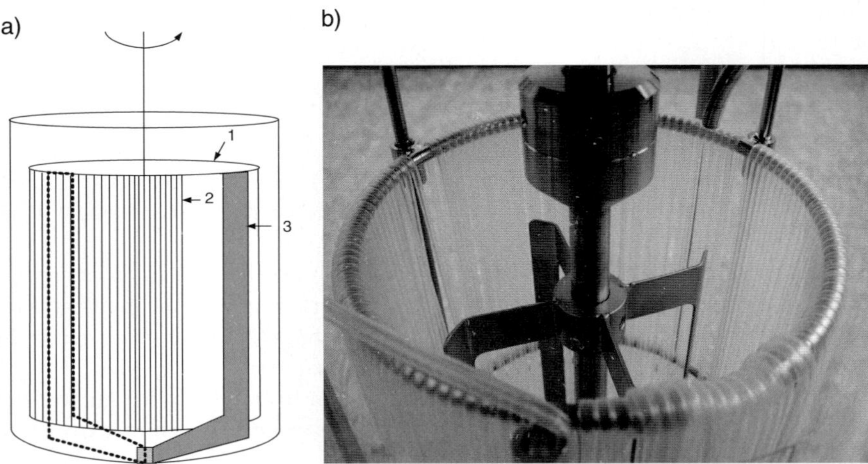

Fig. 4.24 (**a**) Schematic of a standard membrane aeration system (1) membrane basket (2) silicone tubing (3) anchor stirrer (**b**) photograph of membrane basket and anchor stirrer (with courtesy from Bayer Technology Services GmbH, Germany)

Fig. 4.25 Photograph of the dynamic membrane aeration system (DMA) for a 200 L cell culture reactor (with courtesy from Bayer Technology Services GmbH, Germany)

Table 4.4 Mass transfer coefficient for oxygen (membrane and liquid film) in dependence of power input per liquid volume (100 L reactor volume) adapted from (Frahm et al. 2007)

Aeration system	P/V (W m^{-3})	k (m h^{-1})
Dynamic membrane aeration	10	0.075
Membrane stator	10	0.055
Dynamic membrane aeration	100	0.09
Membrane stator	100	0.07

4.2.4   Consequences for Reactor Design and Operation

Ozturk calculated the limit in the cell densities for several aeration systems based on the mass transfer coefficients reported by Henzler and Kauling see Fig. 4.26 [adapted from Ozturk (1996)].

Surface aeration is not sufficient for high density cultures. Membrane aeration as a better method for oxygen delivery at high density cultures can be used inside or outside of the bioreactor, in the form of a basket or gas exchanger, respectively. The mass transfer in membrane aeration depends on the length of tubing per reactor volume. Handling membrane aeration systems increases by reactor size and is not manageable for high density systems larger than to 200 L. Membrane aeration using silicone tubing can provide enough oxygen to support 5×10^7 cells mL^{-1} (Fig. 4.26). In cultures aerated by spargers, higher mass transfer rates and thus higher cell densities can be achieved. Cell lysis and foaming limit the sparging rates to about 0.1 vvm (Ozturk 1996). Selection of a good sparger is important for minimizing the foaming and maximizing the mass transfer rates (Fenge et al. 1993; Nehring et al. 2004; Zhang et al. 1992a,b). The use of micro spargers (such as sintered metal or ceramic), provides more mass transfer area and therefore higher k_La. Micro bubbles can increase the foaming problems by producing more dense foam (Fig. 4.27).

Macro spargers create bubbles greater than 2 mm and are more appropriate for minimizing the foaming (Ma et al. 2006; Ozturk 1996). Though the mass transfer coefficients are lower for macro spargers, and higher gas flow rates should be used to achieve the same cell density as micro spargers. However, Henzler and Kauling

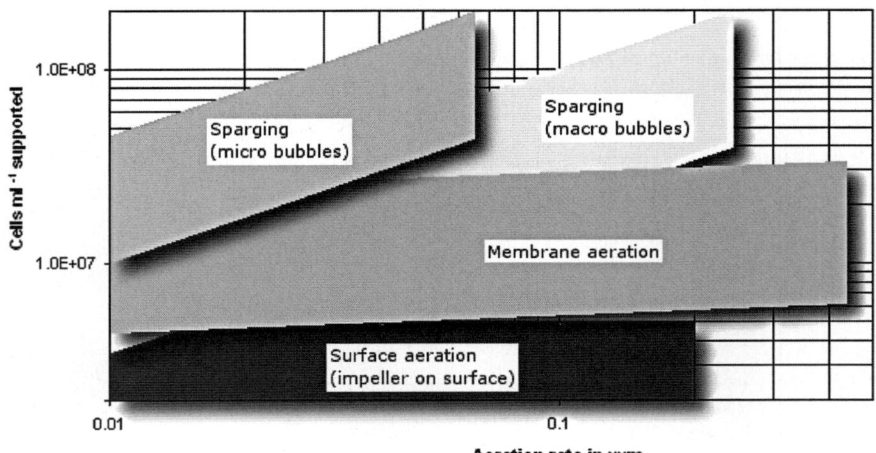

Fig. 4.26 Limitations in cell density based on oxygen delivery in different aeration systems (adapted from Ozturk (1996) with kind permission of Springer Science and Business Media)

Fig. 4.27 5–10 mm foam layer at the medium surface (cell line: CHO-easyC – CHO-K1 derived line, Cell Culture Technologies GmbH, Zürich, Switzerland; CHO-master HP-1 medium; aerated with a ceramic micro sparger Fig. 4.18

(1993) showed that if the size of the reactor is increased, e.g., from V = 1 L to V = 1,000 L, the main shear due to the aeration decreases by a factor of 10. They suggest that problems observed on lab scale with sparger aeration are of minor importance in larger reactors; thus in many cases sparger aeration can be used for oxygenation even in production processes with animal cells. Nehring et al. (2004) compared a microsparging aeration system made of porous ceramic with bubble-free membrane aeration in the cultivation of MDCK cells and in the cultivation of virus infected MDCK cells (2, 5, 30 and 100 L reactor scale). They achieved similar results for cell densities. Antifoaming agents such as Pluronic F-68 and lowering gassing rates and agitation speed (possible when using micro sparger systems) can decrease the problems related to sparging, such as foaming and cell damage (Aunins and Henzler 1993; Nehring et al. 2004; Ozturk 1996; Ma et al. 2004; Dey et al. 1997). However, the introduction of such additives decreases mass transfer and results in problems in the downstream processing (Aunins and Henzler 1993).

4.2.4.1 CO_2 Accumulation

CO_2 plays an important role in cell culture. The solute species of CO_2 in the form of CO_2, HCO_3^-, H_2CO_3 or CO_3^{2-}, and therefore the variation of CO_2 concentrations, influences the pH of the medium when using sodium bicarbonate as the pH buffer. In addition CO_2 is a potentially inhibitory byproduct in cell cultures. So CO_2 accumulation and excessive removal should be avoided (Ma et al. 2006; Ozturk 1996). Aeration removes CO_2 from the system and leads to a more complex gas exchange with respect to the balance between O_2 addition and CO_2 removal. The partial pressure of metabolically produced carbon dioxide (pCO_2) and hence mass transfer of carbon dioxide from solution to the gas bubbles is important in animal cell culture. A rise in pCO_2 can cause a rise in H_2CO_3 and hence a drop in pH. pCO_2 must therefore be carefully monitored and controlled (Marks 2003; Varley and Birch 1999; Ma et al. 2006). The accumulation of CO_2 depends on the mass transfer characteristics of the aeration system. The mass transfer coefficients for CO_2 and O_2 are not very different when using sparger for aeration and CO_2 accumulation is a minor problem. If a micro sparger system is used for aeration, CO_2 stripping can still be a problem. Small bubbles get saturated by the CO_2 more easily than large bubbles and the efficiency of CO_2 removal decreases. While using membrane aeration, CO_2 accumulation is of greater importance. Depending on the material, the mass transfer coefficients for O_2 and CO_2 can be very different. For silicone tubing the CO_2 mass transfer coefficient is measured to be about four times lower than that of O_2 (Ozturk 1996). However, a proper pH control is necessary.

4.3 Immobilization of Cells

Biocatalysts (e.g. cells) are regarded as immobilized when they are restricted in their motility while their metabolic or catalytic activity is maintained (Lundgren and Blüml 1998). *In vivo,* mammalian cells are immobilized in tissues and organs and perfused by lymph, blood, etc. *In vitro,* immobilization is primarily used to increase the stability and the process intensity of the culture. Immobilization of mammalian cells was first described as early as 1923 by Carrel (1923). Since then, different techniques for immobilizing mammalian cells have been proposed, including entrapping cells on a particle surface or in the interstices of a porous particle, encapsulation of cells within gels, or growth of cells within compartments formed by membranes (Lundgren and Blüml 1998; Butler 2004; Davis and Hanak 1997; Davis 2007). These techniques are widely applied, as they offer some solutions to problems inherent in mammalian cell culture:

- Attachment of anchorage-dependent cells is the only way these cells can grow.
- Immobilization in macroporous carriers can protect cells against shear stress, promoting the use of serum- or protein-free medium (Lüdemann et al. 1996, Sect. 5.1.2).

- Cell densities in immobilized systems are considerably higher compared to those without immobilization (e.g., suspension culture approx. 10^6–10^7 cells mL^{-1}, immobilized culture approx. 10^7–10^8 cells mL^{-1}, tissue approx. 10^9 cells mL^{-1}) and allow for smaller reactor volumes.
- Immobilization techniques enable preliminary separation of (extracellular) products and cells, easing requirements for downstream.
- Immobilization systems are preferably run in perfusion mode (Sect. 4.4), where the medium feed rate is not dependent on the growth rate of the cells and higher volume-specific productivities can be obtained.

Besides these advantages of immobilization techniques, several problems have to be solved depending on the specific system:

- Materials have to be selected according to the requirements of the cells and the culture system. Especially in the case of adherent cells, they have to promote cell attachment and spreading, a complex mechanism as shown in Fig. 2.1 (Sect. 2.2.1).
- All techniques involving the growth of cells in a three-dimensional structure, mass transfer effects, especially nutrient and oxygen supply, as well as removal of toxic metabolites and carbon dioxide, have to be considered.

In the following sections, the different techniques for cell immobilisation are briefly discussed, as this has been done extensively by a number of reviews and text books (Lundgren and Blüml 1998; Butler 2004; Davis and Hanak 1997; Kelly et al. 1993; Howaldt et al. 2005; Ozturk and Hu 2006; Hübner 2007). Special attention is paid to mass transfer aspects to give a deeper understanding of the underlying mechanisms and resulting consequences. The main focus is on solid and macroporous carriers for immobilization of mammalian cells as well as on encapsulation techniques. Membrane-based bioreactor concepts are discussed in Sect. 5.1.3 for production of biopharmaceuticals and in Part II, Chap. 2 for tissue engineering.

4.3.1 Carriers for Cell Immobilization

4.3.1.1 General Aspects

The development of carriers to support cell growth was initially driven by the idea to provide a large-scale production system for adherent cells, e.g., for vaccine production. In the early sixties facilities for vaccine production used batteries of roller bottles. Scale-up was possible by increasing the number of roller bottles used. The high labour intensity of this technology is quite obvious. An important breakthrough was the development of the so called "microcarrier" by van Wezel (1967) in the beginning of the seventies, a solid particle with a 100–200 µm structure, on which the cells could grow. As the microcarriers are suspendable in stirred tanks, it became possible to work with large scale suspension reactors (nowadays up to 4,000 L (Lundgren and Blüml 1998). The great success of this technology led to the development of a large number of different microcarriers for suspension culture and furthermore for different reactor systems, especially for fixed-bed and fluidized-bed reactors (Lundgren and Blüml 1998). The desired features of a carrier can be defined according to Lundgren and Blüml (1998) as well as Pörtner and Märkl (1995):

- Autoclavable
- Available in large batches for industrial customers
- Available with documents required for approval (Drug master or regulatory support files)
- High batch-to-batch consistency
- Suitable for a large number of cell lines (adherent and non-adherent cells)
- Good long-term stability
- High surface-to-volume ratio (large multiplication steps)
- Material of non-biological origin (minimal viral risks)
- Macroporous for high cell density and shear force protection
- Efficient diffusion from the medium into the center of the carriers
- Non-toxic, non-immunogenic matrix
- Size appropriate to reactor system
- Simple immobilization and harvesting of cells
- Mechanical stability
- Uniform size distribution
- Reusable
- Possibility to count cells
- Transferable between vessels (ease of scale-up)
- Transparent

Even if an "ideal" carrier does not exist, the above list gives some arguments for selecting a carrier or evaluating different suppliers. Although for research purposes aspects such as good growth conditions might be mainly important, for production processes, other aspects related to the approval of the process and maintaining the process for the lifespan of the product have to be considered.

Depending on the intended culture system, the properties of the carriers, especially the size, vary significantly (from $10\,\mu m$ to $5\,mm$). In Fig. 4.28 different carriers are grouped according to size. The term "microcarrier" is traditionally used basically for those carriers used in suspension culture with an ideal size of 150–$500\,\mu m$. Larger carriers are used for fluidized and fixed bed reactors due to the higher sedimentation rate. For fluidized bed reactors, carriers having a size of 0.6 to $1\,mm$ are used. For fixed-bed cultures, larger carriers (3–$5\,mm$) are better, to prevent blocking of free channels between the carriers. Obviously larger carriers will have limitations in terms of mass-transfer-effects, especially with respect to oxygen supply, because of the poor solubility of oxygen in the medium and the high uptake rate of the cells. This will be discussed later in this chapter.

A wide variety of different materials has been suggested for carriers (Lundgren and Blüml 1998; Pörtner and Platas Barradas 2007; Pörtner et al. 2007) including dextran, collagen, polymers, glass, and ceramic (among others), as they have a great impact on parameters such as toxicity, hydrophobicity, micro- and macroporosity,

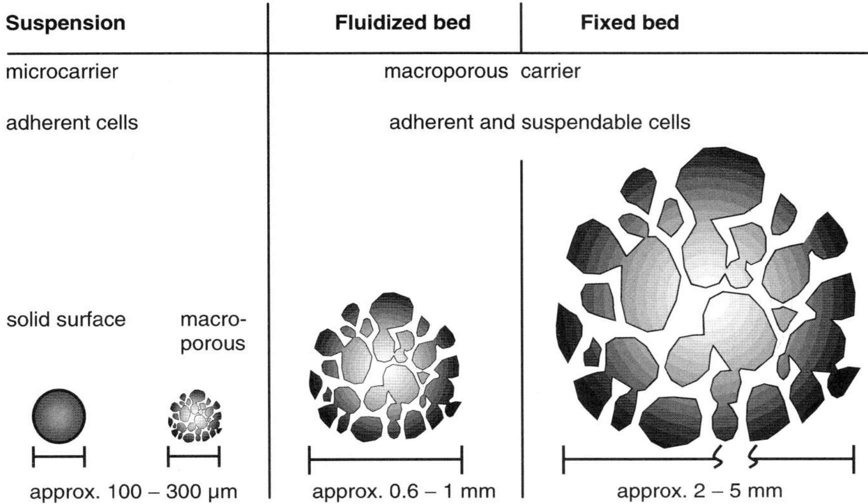

Fig. 4.28 Carrier for immobilization of mammalian cells

mechanical stability, mass-transfer, specific gravity, and shape. Different shapes are available, such as fibers, flat discs, woven discs, and cubes, though the sphere is the most common. The surface is usually positively or negatively charged to ensure cell attachment. The density of the electrostatic charge on the surface is critical to allow cell attachment and growth. Non-charged carriers are also available, which are normally coated with collagen or gelatine (denatured collagen) or have fibronectin or fibronectin peptides coupled on the surface. While carriers consisting of gelatine may be autoclaved, this is not the case for protein-coated carriers.

The density of the carriers is determined by the intended application. Smooth microcarriers for suspension culture have a density just above the medium ($1.02–1.04\,g\ cm^{-3}$), while materials used for macroporous carriers have densities between 1.04 and $2.5\,g\ cm^{-3}$. Lundgren and Blüml (1998) suggest the use of sedimentation velocity as a better parameter to judge the suitability of a carrier for a specific reactor type, especially for fluidized bed reactors, as this includes not only the specific density but also the size and shape of the carrier. Sedimentation velocities lower than $30\,cm\ min^{-1}$ do not create sufficient mixing for efficient nutrient supply throughout the carrier. For adherent cells tending to form bridges between the carriers higher sedimentation rates ($150–250\,cm\ min^{-1}$) are recommended.

The surface of most carriers can be either solid or microporous. In the case of microporosity, the pore size is not sufficient to allow cells to grow in the inner parts of the carrier. Micropores are intended to take up larger molecules, up to a molecular weight of approximately 100 kda, to allow the cells to create a micro-environment on the beads and to support cell attachment and growth.

A carrier is considered macroporous if the average pore size is between 30 and 400 μm, allowing for cell growth within the carriers. The porosity of these carriers, defined as the ratio between the volume of the pores and the total carrier volume (as percentage) is normally between 60 and 99%. Macroporous carriers are suitable for immobilizing adherent as well as non-adherent cell lines. Furthermore they allow cells to grow in a three-dimensional, almost tissue-like structure at higher densities compared to solid or microporous carriers. The tissue-like growth stabilizes the cell population, protects against shear stress and decreases the need for external growth factors, so that the use of serum-free or protein-free media becomes easier. As the stability of the culture is improved at high cell density (Sect. 4.4), macroporous carriers are preferably used for long-term continuous culture. Due to the high cell density, the different culture technologies applied for macroporous carriers (stirred tanks for macroporous microcarriers, fixed bed and fluidized bed reactors) are generally run at high perfusion rates. This provides a sufficient nutrient supply as well as removal of toxic metabolites. Furthermore, the residence time of an extracellular product at 37 °C is reduced and it can quickly be separated from the cells and cooled.

Besides these advantages, some disadvantages of carriers have also to be considered (Lundgren and Blüml 1998):

- Some carriers have to be prepared before use.
- Scale-up of a culture using cells harvested from microcarriers is more complex than expansion of a suspension culture.
- Cell harvest from macroporous carriers can be difficult.
- Cell count, especially in the case of macroporous carriers, requires special techniques.
- For larger macroporous carriers, mass transfer limitations have to be considered.

Despite the large number of carriers and support materials suggested in the literature, only very few are on the market and even fewer fulfill industrial standards for large-scale manufacturing (Lundgren and Blüml 1998). In the following sections, some typical examples are discussed and recommendations for use in culture given.

4.3.1.2 Microcarrier for Suspension Culture

The term "microcarriers" denotes small beads, either solid or macroporous (Fig. 4.29), having a diameter of approximately 100–300 μm and a density of 1.02–1.04 g cm^{-3}, slightly higher than that of the growth medium. When first developed in the late 1960s by van Wezel (1967), microcarrier culture introduced new possibilities and suspension culture of anchorage-dependent cells in high density became possible (Lundgren and Blüml 1998; Nilsson et al. 1986; van der Velden-de Groot 1995). Nowadays beads made of DEAE-Sephadex, DEAE-polyacrylamide, polyacrylamide, polystyrene, cellulose fibers, hollow glass, gelatin or gelatin-coated dextran beads are in use (Lundgren and Blüml 1998; Shuler and Kargi 2002). In microcarrier

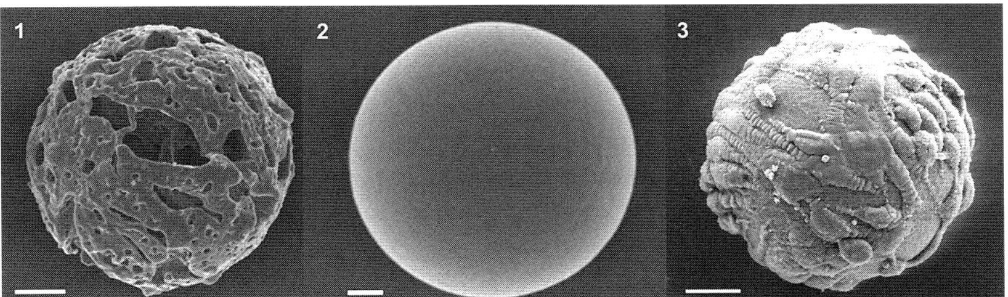

Fig. 4.29 Microcarrier for suspension culture of anchorage dependent cells. (1) Cultisphere (Percell Biolytica, Gelatin), bar 20 μm; (2) Cytodex 3 (GE Healthcare, Gelatin-Dextran), bar 10 μm; (3) Cartilage cells grown on Cytodex 3, bar 20 μm

culture, cells grow as monolayers on the surface of small spheres (Fig. 4.29) or three-dimensional within the pores of macroporous structures. The carriers are usually suspended in culture medium by gentle stirring.

On smooth solid microcarriers, cells grow on the outer surface until a monolayer is formed. Each microcarrier can accommodate approximately 100–200 cells. To reach an optimal growth on all individual microcarriers, an even distribution of cells is required. For most cell lines, more than 7 cells per carrier are required during inoculation to ensure that the population of unoccupied microcarriers is <5% and the use of the available surface area is maximized (Butler 2004). For laboratory scale, it is often recommended to establish a microcarrier culture with 1–3 g (dry weight) of microcarriers per liter with initial cell densities of $1–2 \times 10^5$ cells mL^{-1} (might be lower in serum containing medium) and by using a certain start-up strategy (e.g., intermittent stirring to allow for attachment of cells on carriers). Final cell densities are in the range of 2×10^6 cells mL^{-1} with a multiplication factor of approximately 10–20 (Handbook GE Healthcare). In microcarrier culture usually a serum supplement of 5–10% is added to the medium for general purpose culture, as anchorage-dependent cells require certain adhesion factors contained in serum. For certain types of cell, the concentration of serum supplement can often be reduced. Furthermore, serum- or protein-free media supplemented with certain factors including fibronectin, transferring, insulin or epidermal growth factors, as well as shear protecting agents such as serum albumin, are probably suitable for microcarrier culture (Handbook GE Healthcare).

On the industrial scale, higher multiplication factors (up to 100) can be achieved by optimizing the cell-carrier-system for ease of scale-up (Lundgren and Blüml 1998). The larger the multiplication factor in each step of scale-up to final production volume, the fewer the scale-up steps required. For a multiplication factor of 100 a 1-L suspension would be appropriate to inoculate a 100 L-reactor. A further increase of the scale might be possible by a technique named "bead-to-bead transfer" or "colonization". Basically the microcarrier culture is scaled-up just by adding more microcarriers and more media (Butler 2004). Lundgren and Blüml (1998) reported on a 1:1,000 dilution with Chinese Hamster Ovary Cells (CHO) that do not attach well to the carriers or that detach easily during mitosis. In this case a 1 L-suspension could inoculate a 1,000 L-reactor. If this technique cannot be applied and harvesting of cells from the beads is required, specific techniques for breaking the cell-surface and cell-cell interactions and separation of cells and microcarriers have to be applied.

4.3.1.3 Macroporous Carrier for Fluidized Bed and Fixed Bed Culture

Suitable carriers for the immobilization of mammalian cells in fixed bed and fluidized bed must fulfil certain requirements such as high surface to volume ratio, simple and non-toxic immobilization, optimal diffusion from the bulk phase to the centre of the carrier, mechanical stability, and pH stability. They should be steam sterilisable, reusable and suitable for adherent and non-adherent cells (reviewed by Pörtner and Märkl 1995; Pörtner and Platas 2007; Pörtner et al. 2007). A large number of macroporous carriers and support materials for immobilization have been suggested, including glass (natron, borosilicate), cellulose, collagen, synthetic materials, e.g., polypropylene and polyurethane, and ceramic. Most of these carriers proved to be suitable for a large number of cell types (e.g., hybridom, CHO, hepathocytes etc., see Sect. 5.1.2). In some cases (e.g., carriers made of glass or cellulose) coating with gelatine was recommended to support the growth of strictly adherent cells (unpublished data, Lüllau et al. 1994). The modification of the surface is a critical parameter. Anchorage-dependent cells, especially primary or non-established cell lines, have proved to be much more sensitive to the carrier surface than established cell lines (e.g. hybridom, CHO, L293, NIH3 T3). This has to be taken into account during the selection of a suitable carrier for each specific cell line. Examples are shown in Figs. 4.30 and 4.31. As already mentioned, only very few carriers are on the market and even fewer fulfill industrial standards for large-scale manufacturing.

The cell density within a carrier depends on several parameters such as diameter of the carrier, porosity, cell type, etc. In general, the number of immobilized cells in the carrier increases with decreasing carrier diameter (Fig. 4.32). Therefore smaller carriers are preferable (e.g., 0.6–1 mm for fluidized bed). But in the case of fixed reactors, smaller carriers create narrow free channels, which can be blocked by cells growing in these channels. Therefore, diameters of 3–5 mm are an appropriate choice for fixed-bed reactors (Pörtner and Platas, 2007). Further recommendations regarding design and operation of fixed bed and fluidized bed reactors are given in Sect. 5.1.2.

Fig. 4.30 Macroporous carriers for fixed bed and fluidized bed. I: Fibra-Cel™ (New Brunswick); II: glass carrier SIRAN® (QVF); III: ceramic carrier SPONCERAM® (Zellwerk)

Fig. 4.31 SEM of macroporous carriers for fixed bed and fluidized bed. I: Fibra-Cel™ (New Brunswick), Ia: native carrier, Ib: CHO-K1 cells growing on the carrier. II: glass carrier SIRAN® (QVF), IIa: native carrier, IIb: hybridom cells growing within the carrier. III: ceramic carrier SPONCERAM® (Zellwerk). IIIa, native carrier surface, IIIb: hepatoblastoma cells growing within the carrier

4.3.1.4 Mass Transfer Effects in Macroporous Carriers

Within macroporous carriers a sufficient supply of immobilised cells with nutrients and removal of inhibiting metabolites are essential. On one hand, oxygen supply has to be addressed, as the solubility of oxygen in culture medium (approx. $0.2\,mmol\,L^{-1}$, if the medium is in equilibrium with air) is far lower than the concentration of other nutrients (e.g. glucose $18\text{–}25\,mmol\,L^{-1}$). Therefore, the supply of necessary oxygen to immobilised cells is an important criterion for selection of a macroporous carrier. On the other hand, removal of carbon dioxide is also an important aspect, as increased levels of carbon dioxide might reduce the pH to unfavourable conditions.

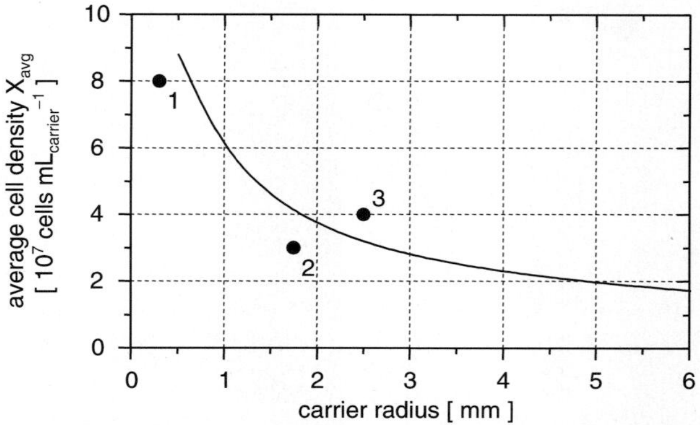

Fig. 4.32 Average cell density X_{avg} as a function of the carrier size (radius) for SIRAN carrier. The symbols indicate experimental data by (1) Yoshida et al. (1993), (2) Pörtner et al. (1995) and (3) Looby et al. (1990). The line shows a model calculation suggested by Fassnacht (2001) using a cell density of $1.75{\times}10^{8}$ cells mL^{-1} for those areas of the carrier loaded with cells

In the following sections, oxygen supply within a macorporous carrier will be addressed to provide a deeper understanding of the mechanisms involved in mass-transfer limitation. Here, only internal effects are considered, and effects due to a boundary layer around the carriers are ignored (Fig. 4.33). It can be assumed that mass transfer within the carriers is mainly diffusive. Some carriers, especially those having a very high porosity (>90%) can be perfused at an early stage of cultivation. But during the progress of the culture the free pores will be closed by growing cells and extracellular matrix. For a spherical particle the oxygen concentration c_{O2} within the particle can be described by the following differential equation:

$$\frac{\mathrm{d}^{2}c_{o_{2}}}{\mathrm{d}r^{2}}+\frac{2}{r}\frac{\mathrm{d}c_{o_{2}}}{\mathrm{d}r}-\frac{Xq_{o_{2},\max}}{D_{e}}\frac{c_{o_{2}}}{k_{o_{2}}+c_{o_{2}}}=0 \tag{4.32}$$

with radius r, diffusiv coefficient D_{e}, cell density X, maximal cell specific oxygen uptake rate $q_{o2,\max}$, Monod-constant k_{o2}, if the cells are distributed homogeneously within the carrier, oxygen can be considered as the limiting substrate, and diffusion can be described by Fick's law. This equation can be solved with the following boundary conditions: for $r = 0 \rightarrow \mathrm{d}c_{o2} / \mathrm{d}t = 0$, and $r = r_{p}$ (radius of the particle) $\rightarrow$ $c_{o2}(r_{p}) = c_{o2,Fl}$ (oxygen concentration in the bulk fluid). The diffusive coefficient is assumed to be $D_{e} = 2.56{\times}10^{-9}$ m^{2} s^{-1}, as suggested by Kurosawa et al. (1989) for oxygen limitation within alginate beads. The cell specific oxygen uptake rate can be described by a Monod-equation. From Miller et al. (1987) and own unpublished data, the Monod-constant k_{o2} is seen to be approximately 0.001 mmol L^{-1}. As this value is very low compared to the oxygen concentration in the medium, it can be ignored in the first place and oxygen uptake can be regarded as following a zero order reaction. Furthermore the term Xq_{o2},max can be combined and replaced by

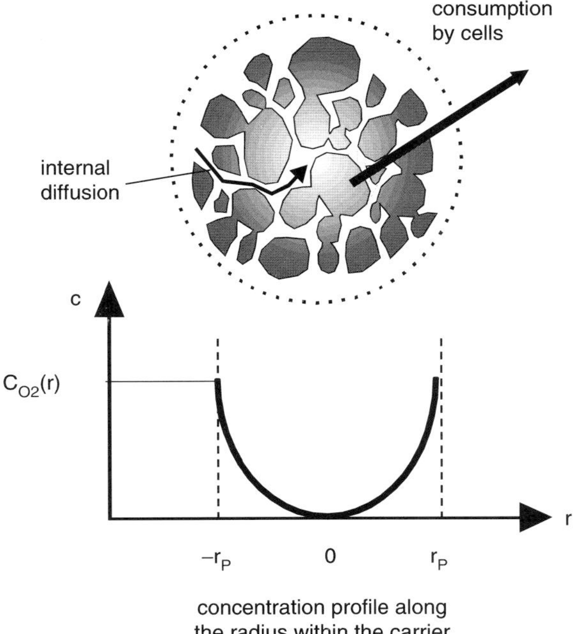

Fig. 4.33 Modeling of oxygen profil in a carrier loaded with cells neglecting the boundary layer effects

the term OUR, the volume specific oxygen uptake rate of a carrier loaded with cells, leading to

$$\frac{d^2 c_{o_2}}{dr^2} + \frac{2}{r}\frac{dc_{o_2}}{dr} - \frac{OUR}{D_e} = 0 \tag{4.33}$$

This equation was solved numerically by a fifth-order Runge–Kutta-method for different values of the oxygen concentration in the fluid and different volume-specific uptake rates. As can be seen from Fig. 4.34, severe oxygen limitation has to be expected, especially for larger particles ($r_p = 2\,\text{mm}$) and higher volume-specific uptake rates. These theoretical considerations are supported by NMR studies described by Fassnacht (2001) (Fig. 4.35) and underline the statements made earlier, regarding an appropriate carrier size in fixed bed and fluidized bed reactors.

Besides solving the mass balance equation for the biocatalyst, a good estimation of the degree of mass transfer limitation within a biocatalyst can be obtained from the Thiele-Modulus Φ and the effectiveness factor η_i for internal mass-transfer-resistance. The Thiele-Modulus (Thiele 1939) describes the ratio between the maximum conversion within a biocatalyst and the maximum mass transfer to the biocatalyst. There are several definitions of the Thiele-Modulus in the literature.

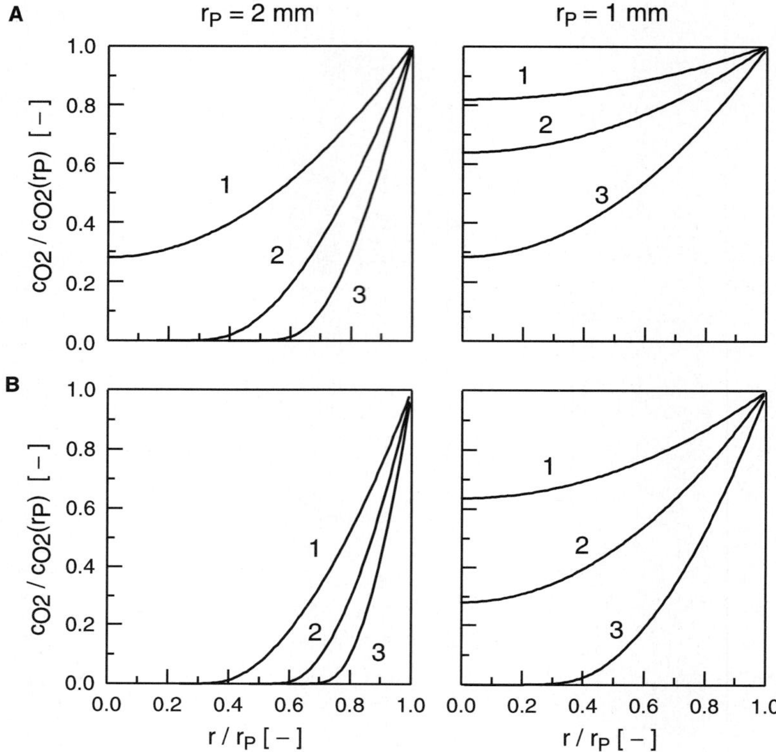

Fig. 4.34 Oxygen profile in a spherical carrier loaded with cells. Oxygen concentration $C_{O2}(r_P)$ in the bulk medium: 1–200 µmol L^{-1}, 2–100 µmol L^{-1}, 3–50 µmol L^{-1}. Volumetric oxygen uptake rate OUR: A – 2 mmol L^{-1} h^{-1}, B – 4 mmol L^{-1} h^{-1}

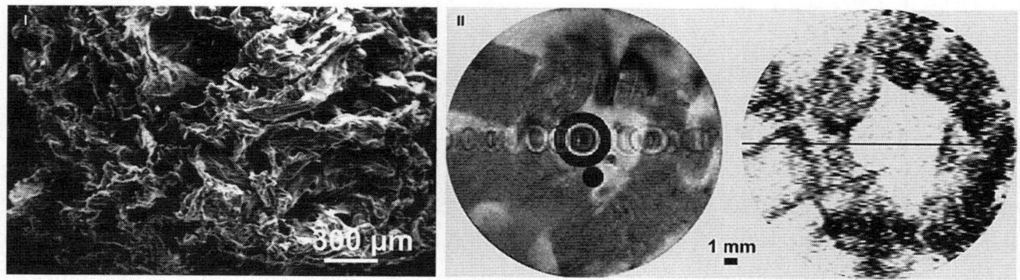

Fig. 4.35 I – SEM of CellSnow (Biomaterials), a macorporous cellulose carrier (approx. 5×5 mm). II – Cross-section of an NMR fixed-bed (2 mm slice thickness) reactor with CHO-K1 cells immobilized on CellSnow. On the left a spin echo image is shown. The *darker* parts show the cubic-shaped carriers, and the lighter parts indicate the interstitial space. The center shows the cross-section of the tube leading to the bottom of the bed. The small black circle located below the center is the glass capillary that contains a standard for 31P spectra. On the right a STEAM image of the same slice. This image shows viable cells at a dark color because of the low apparent diffusion coefficient of the intracellular water. Extracellular water is not detected and remains white. The black dashed circles indicate the cross-section of a single carrier showing that cell growth is only possible to a critical depth of approx. 0.7 mm (adapted from Fassnacht 2001)

In general, separate equations are required for different reaction kinetics and catalyst geometries. An overview is given by Doran (2006). For the case discussed above, a spherical particle and a zero-order kinetic for oxygen uptake, the Thiele-Modulus Φ_0 can be calculated as follows:

$$\phi_0 = \frac{r_P}{3\sqrt{2}}\sqrt{\frac{OUR}{D_e c_{O_2}}}.$$

(4.34)

The Thiele-Modulus Φ_0 was calculated for different values of the carrier radius, the oxygen uptake rate and the oxygen concentrations c_{fl} in the medium (Fig. 4.36). As expected, the Thiele-Modulus increases with increasing oxygen uptake rate, increasing carrier radius and decreasing oxygen concentration in the bulk medium.

An effectiveness factor of the carrier η_i can be defined as the ratio of the average observed reaction rate throughout the carrier to the rate without mass-transfer resistance:

$$\eta_i = \frac{\text{average reaction rate throughout the carrier}}{\text{rate without mass-transfer resistance}}$$

The effectiveness factor can also be interpreted as the ratio of the average reaction rate to the reaction rate on the surface of the carrier. For a spherical particle and a zero-order kinetic for oxygen uptake, the following equations can be used to calculate $\eta_{i0} = f(\Phi_0)$:

for $0 < 0 < 0.577 \rightarrow \eta_{i0} = 1.$

for $\Phi_0 < 0.577.$

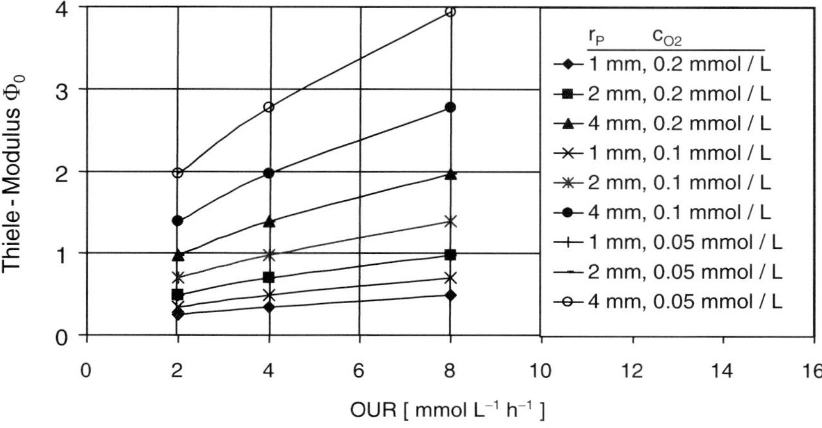

Fig. 4.36 Effectivity of oxygen supply in carriers with immobilized cells. Thiele-Modulus Φ_0 vs. volume specific oxygen uptake rate OUR calculated for different values of carrier radius r_p and different concentrations c_{O2} in the medium

$$\eta_{i0} = 1 - \left[\frac{1}{2} + \cos\left(\frac{\Psi 4\pi}{3} \right) \right]^3, \tag{4.35}$$

where

$$\Psi = \arccos\left(\frac{2}{3\phi_0^2} - 1 \right). \tag{4.36}$$

The effectiveness factor η_{i0} was calculated according to these equations for different values of the carrier radius, the oxygen uptake rate and the oxygen concentrations c_{O2} in the medium (Fig. 4.37). As expected, all data points can be described by one curve. It confirms that calculation of the Thiele-Modulus and the effectiveness factor provides a good estimation of the degree of mass-transfer-limitation based on available process parameters, without solving the exact mass balance equations.

4.3.2 Encapsulation

Microencapsulation is another method for the immobilization of mammalian cells (Butler 2004; Hübner 2007; Huebner and Buchholz 1999). Basically, it can be used to protect cells against hazardous environmental conditions. The capsules can be cultivated in suspension reactors similar to microcarriers. Even direct aeration in combination with higher aeration rates, high agitation rates or stirrer speeds can be used without damaging the cells. The fragile nature of the microcapsule is

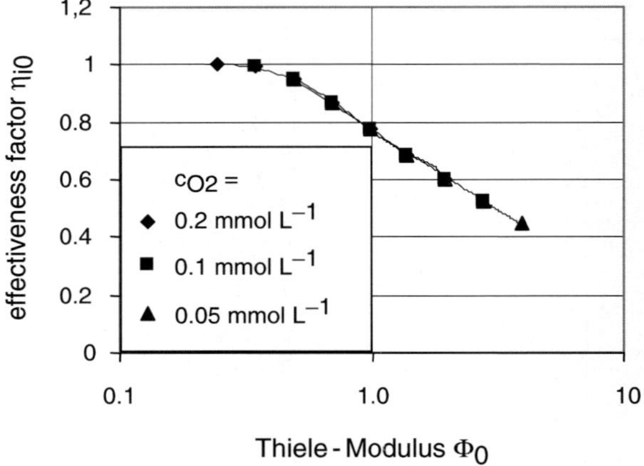

Fig. 4.37 Effectivity of oxygen supply in carriers with immobilized cells. Effectiveness factor η_{i0} vs. Thiele-Modulus Φ_0 calculated for carrier radius 1, 2 and 4 mm, oxygen uptake rate 2, 4 and 8 mmol L^{-1} h^{-1} and different oxygen concentrations c_{O2} in the medium

detrimental to scale-up and long-term operation. Handling of the cells is simplified, as medium exchange can be easily performed without cell loss. Furthermore, within the capsules, a microenvironment favourable for the cells can develop. Autocrine growth factors can accumulate and in an ideal case the cells switch directly to exponential growth after immobilisation, without a significant lag phase. This is probably the most important aspect.

The three basic encapsulation systems available are the bead, the coated bead and the membrane-coated hollow sphere (Hübner 2007; Murtas et al. 2005; Sielaff et al. 1995). Typical size for beads made of polysine-alginate is 300–500 µm, and the molecular weight cut-off of these capsule membranes is 60–70 kDa (Shuler and Kargi 2002). The simplest way to encapsulate cells in solid alginate beads is shown in Fig. 4.38. Immobilization in Ca-alginate beads often leads to problems when the cells release organic acids such as lactate into the medium or when phosphate ions exist as buffering agents in the medium and compete for the binding with the calcium ions. This difficulty led to the development of the coated beads, where additional polyelectrolyte layers, which are also stable in the presence of other ions, are applied to the beads. So the dissolving of the bulb core later on could be done without any problem, and the production of pure polyelectrolyte immobilisates without a solid bead core was a consequent improvement. Common polymers and appropriate

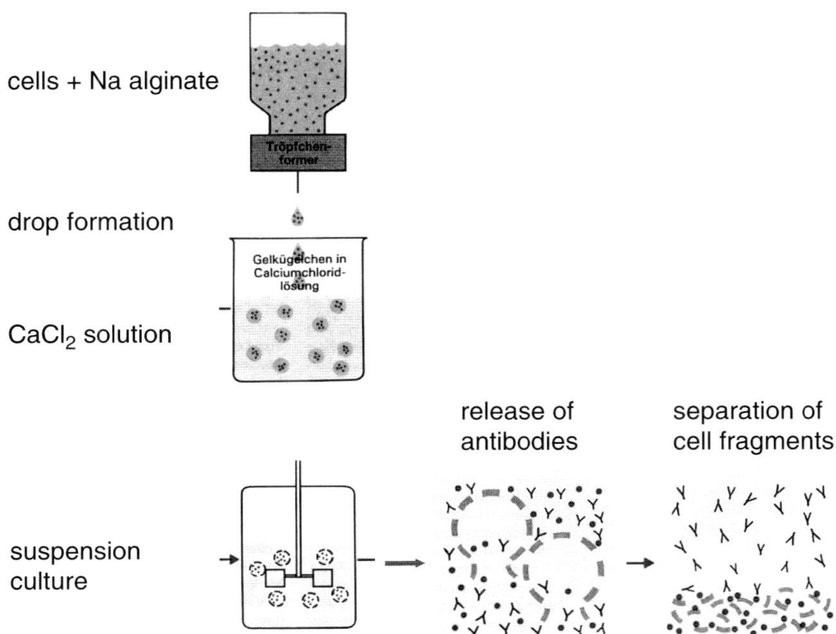

Fig. 4.38 Encapsulation of non-anchorage dependent, suspendable hybridom cells for production of monoclonal antibodies in calcium alginate

encapsulation devices as well as their application have already been described (Huebner and Buchholz 1999).

Cells grow within capsules upto tissue-like cell density before their growth seems to be stopped by space limitations. Cell densities of 10^8 cells per millilitre are standard. These values can only be reached, though, when the nutrient supply of the cells is adequate and the diameter of the capsules is not too large. Again, the sufficient oxygen supply is usually most critical (see above). Cells with a high consumption rate can only be cultivated in relatively small capsules (<1 mm) whereas cells with a lower consumption rate can be cultivated in larger capsules.

4.4 Culture Modes

4.4.1 Principles of Culture Modes

For all biotechnological processes, process strategies for the operation of bioreactors can be classified as discontinuous mode (batch, repeated batch or fed-batch), continuous mode (chemostat, perfusion) or dialysis mode. The basic principles of these modes and some general aspects of their application are discussed here briefly, as this is basic knowledge in biochemical engineering [Doran (2006), among others]. In Sect. 4.4.2 examples of cultivations performed with a specific cell line to provide a deeper understanding of the characteristics of different culture modes are given.

4.4.1.1 Batch and Fed-Batch

A batch culture as shown in Fig. 4.39a is characterized by the growth of cells without further addition of substrate (balance equations in Table 4.5). Before inoculation an appropriate amount of substrates (e.g., a culture medium as mentioned in Sect. 2.3) is provided in the culture system. After inoculation the cells start to grow by converting substrates and oxygen into cell mass and metabolites (including the desired product such as monoclonal antibodies or a recombinant protein). At first the cells go through a lag phase with reduced cell growth. Then exponential growth follows until either substrates are consumed or inhibiting metabolites have accumulated (decline phase and death phase). Cell death due to substrate limitation or metabolite inhibition is mainly induced by apoptosis, a genetically controlled cell death program (Sect. 2.5.2.1). During the cultivation additional compounds such as air or oxygen for aeration, acid or base for pH-control or anti foam agents to prevent foaming have to be added. Therefore in a strict sense a batch culture cannot be regarded as a "closed system".

Batch cultures are regarded as simple and reliable and are often applied in the lab as well as on an industrial scale. Nevertheless the productivity of batch cultures

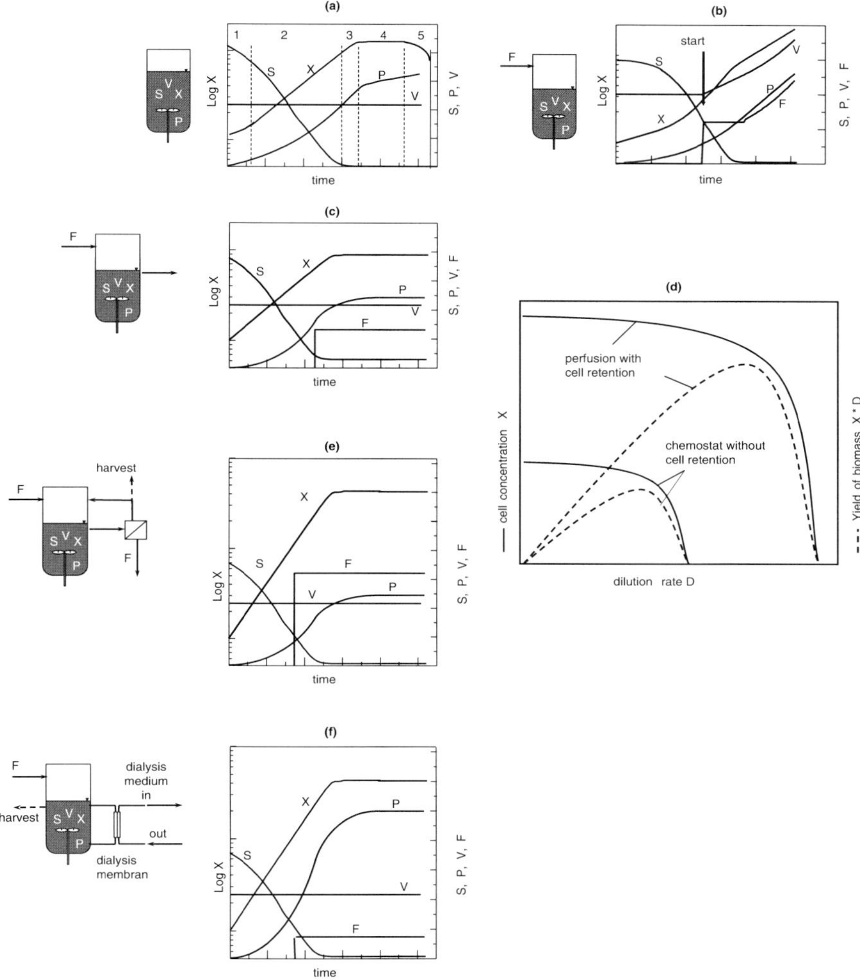

Fig. 4.39 Principles of culture modes (**a**) batch, (**b**) fed-batch, (**c**) chemostat, (**d**) cell concentration and yield vs. dilution rate for chemostat and perfusion, (**e**) perfusion with cell retention, (**f**) dialysis (F: feed rate, P: product, S: substrate, V: volume, X: cell conc.)

is limited by the initial concentration of substrates. On one hand, the solubility of certain medium compounds is limited. On the other hand, too high substrate concentrations my induce substrate inhibition or a high cell specific substrate uptake rate leading to increased production of inhibiting metabolites. Furthermore a batch culture runs at a high volume specific productivity only for a short period of time when a high cell concentration is reached and substrates are not yet consumed completely. After harvest the bioreactor has to be cleaned and refurbished before the cycle is repeated (Krahe 2003).

Table 4.5 Balance equations for batch and fed-batch mode

Concentration of total cells X_t	$\dfrac{dX_t}{dt} = \left(\mu - \dfrac{F}{V}\right)X_V$	(4.37)
Concentration of viable cells X_v	$\dfrac{dX_v}{dt} = \left(\mu - k_d - \dfrac{F}{V}\right)X_V$	(4.38)
Concentration of dead cells X_d	$\dfrac{dX_d}{dt} = \left(k_d \dfrac{F}{V}\right)X_V$	(4.39)
Substrate concentration c_S	$\dfrac{dS}{dt} = \left(\dfrac{F}{V}\right)(c_{s0} - c_s) - q_s X_V$	(4.40)
Product concentration c_P	$\dfrac{dc_p}{dt} = \left(\dfrac{F}{V}\right)c_p + q_p X_V$	(4.41)
Volume V	$\dfrac{dV}{dt} = F$	(4.42)

Concentration of total cells X_t, concentration of viable cells X_v, concentration of dead cells X_d, substrate concentration c_S, product concentration c_P, volume V, flow rate F, cell specific rates: growth rate μ, death rate k_d, substrate uptake rate q_S, metabolite production rate q_P. For batch mode the flow rate $F = 0$

A first step to improve the efficiency of a batch process can be seen in a so-called "repeated batch". Here the main part of the cultivation broth is harvested at the end of the exponential growth phase. The remaining part is used as inoculum for the next run. The culture system is filled with fresh medium and the process is newly started. This procedure can be repeated several times.

The limited productivity of a batch culture can be further overcome by means of a fed-batch (Fig. 4.39b, balance equations Table 4.5). Here the culture system is initially filled to one-third or half only and started as a batch. As soon as substrates are consumed or have reached growth limiting values, fresh medium is added, mostly in concentrated form. Cell and product yield can be significantly improved by applying a fed-batch strategy, where nutrients are added after depletion according to an appropriate feeding protocol (Pörtner et al. 2004). The main advantages of a fed-batch compared to a batch are a longer growth phase, higher cell and product concentrations and a higher product yield. On the other hand a larger effort for equipment and control is required. Several control strategies are available for fed-batch culture, which are discussed in detail in Sect. 4.4.4.

4.4.1.2 Chemostat without Cell Retention

In chemostat culture, also named "Continuous Stirred Tank Reactor" (CSTR), fresh medium is continuously pumped to the culture system and spent medium is removed. The volume of the culture broth remains constant. The spent medium (harvest) contains all medium compounds including the cells. An alternative to the chemostat is the "turbidostat", where the cell concentration is kept constant by controlling the dilution rate. The time course of a chemostat culture is shown in Fig. 4.39c, the balance equations are given in Table 4.6. Again cultivation is started as batch. When substrate concentrations are below critical values, feeding of fresh medium and harvest of spent medium are started. After some time all process parameters become constant (average values – steady state), if homogeneous mixing within the culture system can be assumed.

During steady state the values for cell concentration, as well as substrate and product (metabolite) concentration, depend on the flow rates for medium supply and harvest, expressed by the dilution rate D (4.42). As shown in Fig. 4.39d, a constant or increasing cell concentration (depending on the death rate of the cells) can be observed with increasing dilution rate. But when the dilution rate approaches the maximal growth rate of the cells, the cells cannot compensate for wash-out and the cell concentration decreases. At $D > D_{crit}$ (4.47) the cells are washed out of the culture system completely. Substrate concentration is very low at low dilution rates and increases close to the critical dilution rate. The course of the product concentration looks similar to the cell concentration. Chemostat cultures without cell retention are a valuable tool for research (e.g., kinetic studies) (Pörtner and Schäfer 1996; Sect. 2.5), but not for production scale due to low cell and product concentration. Multi-stage chemostat cultures have been suggested to improve the yield of this culture mode (van Lier et al. 1994).

Table 4.6 Balance equations for chemostat culture

Dilution rate	$D = \dfrac{F}{V}$	(4.43)
Concentration of total cells X_t	$\dfrac{dX_t}{dt} = (\mu - D)\,X_V$	(4.44)
Substrate concentration c_S	$\dfrac{dc_S}{dt} = (c_{S0} - c_S)\,D - q_S X_V$	(4.45)
Product concentration c_P	$\dfrac{dc_P}{dt} = -c_P D + q_P X_V$	(4.46)
Steady state ($dX/dt = 0$; $dc_S/dt = 0$)	$\mu = D$	(4.47)
Max. dilution rate D_{crit} (for monod type kinetic)	$D_{crit.} = \mu_{max}\,\dfrac{c_{S0}}{k_S + c_{S0}}$	(4.48)

4.4.1.3 Perfusion with Cell Retention

The drawbacks of a continuous chemostat culture can be overcome to some extent by retaining the cells in the culture system and perfusing the system continuously with medium. This can be done by connecting cell retention devices (e.g., settlers, spin filters, and centrifuges among others) to a suspension bioreactor or by immobilizing the cells (e.g., fixed bed or fluidized bed bioreactor). The specific techniques are discussed in detail in Chap. 5. Similar to a chemostat, here fresh medium is pumped continuously to the culture system and spent medium (harvest) is removed with the same flow rate. But as the cells are retained in the bioreactor, significantly higher flow rates can be applied leading to higher cell concentrations within the bioreactor, as shown in Fig. 4.39d and e.. Therefore higher volume specific productivities can be expected. The critical perfusion rate $D_{crit,perf}$ can be calculated from

$$D_{crit,perf} = D_{crit,chemostat} \frac{1}{1 + \alpha_{perf} - \alpha_{perf}\beta_{perf}} \tag{4.49}$$

with the critical dilution rate in the chemostat $D_{crit,chemostat}$, recirculation rate α_{perf} (ratio of feed and recirculated medium) and rate of concentration β_{perf} (ratio of cell concentration within the bioreactor and the recirculated medium).

Continuous perfusion presents several advantages (Voisard et al. 2003; Geserick et al. 2000; Lüllau et al. 2003; Bödeker et al. 1994). Among these are the ability to grow cells to a very high density, the ease of handling media exchanges for the purpose of fresh feed and product harvest, the easy removal of metabolites and other inhibitors, and the prospect of easy scale up. When dealing with adherent cells, perfusion reactor design becomes slightly simpler (e.g. fixed bed and fluidized bed) due to immobilization of cells within macroporous carriers compared to microcarrier suspension culture (Pörtner and Platas Barradas 2007).

4.4.1.4 Dialysis Culture

Non-porous dialysis membranes allow for simultaneous enrichment of cells and high molecular weight compounds. According to Strathmann (1979) dialysis in liquid media is defined as transport of a solved compound from one homogeneous phase, through a membrane, into a second homogeneous phase, driven by a concentration gradient between the two phases. Mass transfer through the membrane can be described in terms of Fick's law:

$$S = P_{mem}A(c_1 - c_2) \tag{4.50}$$

The molecular flow S is caused by the concentration gradient of a specific compound $c_1 - c_2$ across the membrane and is proportional to the membrane area A and the permeability coefficient P_{mem}. The permeability coefficient depends on the specific compound, the type of membrane and other process parameters (boundary layers on both sides of the membrane, fluid velocity, stirrer speed, etc.). If process parameters are maintained constant, a relationship can be found between the permeability coefficient and the molecular weight of the compound, similar to that between the diffusion coefficient and the molecular weight. Therefore the permeability coefficient decreases with increasing molecular weight (Pörtner and Märkl 1998). At high values of the molecular weight, the mass transport over the membrane ceases. Usually membranes are characterised by a specific "cut-off", meaning that above this molecular weight, no mass transport over the membrane occurs. The basic principle of a dialysis process, as well as the time course of important process parameters, is shown in Fig. 4.39f. The dialysis process can be run as batch, fed-batch or continuously. An overview of the numerous process strategies available for dialysis cultures can be found in Pörtner and Märkl (1998) and Pörtner (1998).

4.4.2 Examples of Different Culture Modes

In the following sections, the characteristic features of batch, fed-batch and continuous cultures (chemostat, perfusion and dialysis) are discussed to provide a better understanding of these culture modes with respect to mammalian cells. All experiments were performed with one cell line, a hybridoma cell line producing an IgG-1 monoclonal antibody (data from Pörtner 1998 and Fassnacht 2001, if not mentioned otherwise). The intention is not to benchmark the different culture modes on the basis of, e.g., cell concentration, time–space yield or antibody concentration. Recommendation as to which culture mode should be used for a specific process has to consider the respective process, the desired complexity of the process control and for how long the same cell line will be used.

4.4.2.1 Batch and Fed-Batch

Batch processes (Fig. 4.40) usually start with an initial cell density of approximately $1\text{--}2\times10^5$ cells mL^{-1} and last 1–2 weeks. The required initial cell density depends on the type of cell and can be higher in serum- and protein-free media compared to serum containing medim (Lee et al. 1992; Lüdemann et al. 1996). Final cell densities are in the range of $2\text{--}5\times10^6$ cells mL^{-1}, depending on the medium. Therefore during a batch culture the cell number increases only approximately ten times (Griffiths 1992). Metabolites (lactate, ammonia) do not usually accumulate to inhibiting values in a batch culture, as in most cell culture media, the substrate concentrations (e.g., glucose and glutamine) are quite low and consumed quite rapidly.

A typical example of a fed-batch process is shown in Fig. 4.41 (data from Pörtner et al. 1996). In this case the appropriate feed rate was determined by model simulation before starting the feeding pumps ("a priori"). Data at time 24 h were used as start values for simulation of the expected time course for cells, glutamine

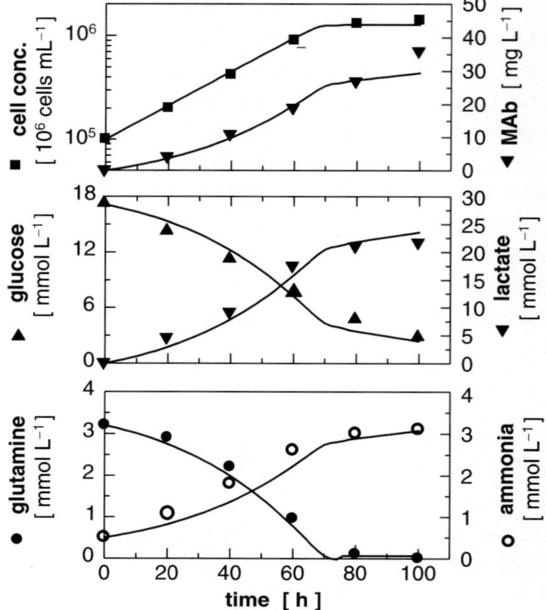

Fig. 4.40 Batch culture of hybridom cell line IV F 19.23 in T-25 flask (data from Pörtner 1998)

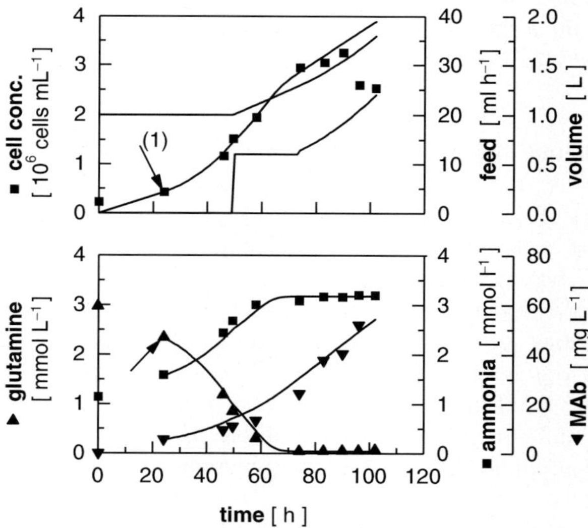

Fig. 4.41 Fed-batch culture with hybridom cell line IV F 19.23 - model based process control (data from Pörtner et al. 1996), suspensions reactor 2 L, 2 rushton turbine (110 rpm), surface aeration; (1) start values for model prediction

ammonia, glucose, lactate and monoclonal antibody (MAb) according to the following strategy: At a calculated glutamine concentration of 1 mmol L^{-1} (t = 50 h) the feed pump was started with a constant pump rate (12 mL h^{-1}). At a calculated glutamine concentration of 0.04 mmol L^{-1} (t = 73 h) the pump rate was increased to maintain this concentration. The "a priori"– determined course of the culture could predict the course of the culture quite well. The maximum cell concentration was 3.2×10^6 cells mL^{-1}, the final MAb-concentration was 54 mg L^{-1}. This example gives a basic idea of the complexity of fed-batch-processes. Other process strategies are discussed in Sect. 4.4.3.

4.4.2.2 Chemostat

As mentioned earlier, in chemostat culture a "steady state" is obtained, if the medium flow rates (feed, harvest) are kept constant. Figure 4.42 shows the relationship between the steady state concentrations for cells, glutamine, glucose, lactate, ammonia and monoclonal antibodies (MAb). Similar data were shown by others for hybridom cells (Miller et al. 1988; Hiller et al. 1991). As expected, the cell

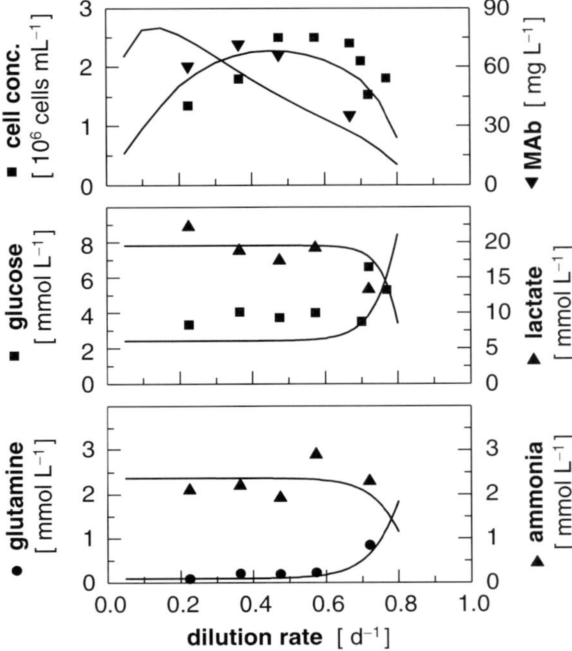

Fig. 4.42 Chemostat culture of hybridom cell line IV F 19.23 (data from Pörtner 1998) suspensions reactor 400 ml, paddle impeller (35 rpm), surface aeration

concentration first increases with the increasing dilution rate. The low cell densities at very low dilution rates are mainly due to cell death caused by apoptosis. Later, for dilution rate D approaching the maximal growth rate, the cell concentration decreases and finally becomes zero for $D > D_{crit}$. A similar course can be found for antibody concentration. Substrate concentrations, especially for glutamine as the limiting substrate in this case, remain at very low values at low dilution rates and increase at dilution rates approaching the maximal growth rate. Ammonia and lactate did not reach limiting concentrations (Pörtner and Schäfer 1996).

In general, the maximal cell and product concentrations obtained in chemostat cultures are quite low and in a similar range as those found in batch culture. Therefore chemostat cultures are a valuable tool for kinetic studies, especially as reliable data at very low substrate concentrations can be obtained. For industrial applications usually cell and product yield are regarded as too low (Griffiths 1992; Werner et al. 1992).

4.4.2.3   Continuous Fixed Bed Culture with Cell Retention

Cultivation of the hybridom cell line in a 100 ml fixed-bed reactor applying oxygen concentration above air saturation (compare Sect. 5.1.2) is shown in Fig. 4.43 (data from Fassnacht 2001). After inoculation, the cells started to proliferate in the fixed-bed, which can be seen in the decreasing glucose and increasing lactate concentrations. A high dilution rate D_{FB} of 10 mL medium per ml fixed-bed per day was used to minimize substrate limitation. A first steady-state was obtained on the fifth day after inoculation at a dissolved oxygen concentration of air saturation at the entry of the fixed-bed. This steady-state was controlled by oxygen limitation,

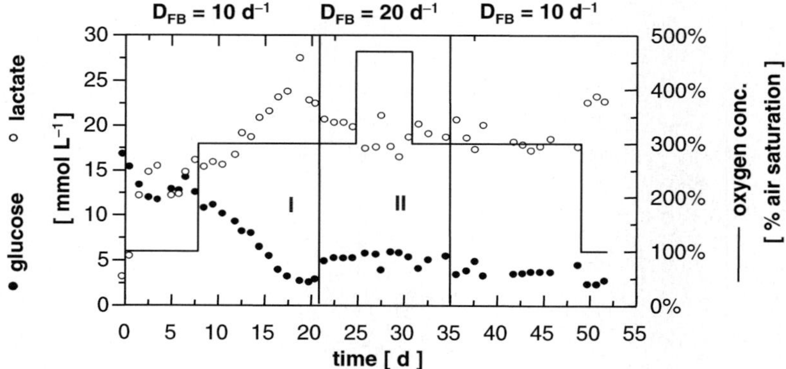

Fig. 4.43 Fixed bed culture of hybridom cell line IV F 19.23 – variation of dissolved oxygen concentration (data from Fassnacht 2001) fixed bed 100 ml, axial flow, conditioning vessel 1 L, macroporous carrier SIRAN®, steady state values for phase I: $c_{Gln} = 0.52$ mmol L⁻¹, $c_{MAb} = 30.7$ mg L⁻¹; phase II: $c_{Gln} = 1.36$ mmol L⁻¹, $c_{MAb} = 20.2$ mg L⁻¹

as indicated by the high steady-state glucose concentration and a steady-state glutamine concentration of 2.46 mmol L^{-1} (data not shown). The dissolved oxygen concentration was raised to 300% air saturation on the eighth day. This resulted in an increase of the glucose consumption rate and a new steady-state was reached on the 21st day. The glutamine concentration dropped to 0.52 mmol L^{-1} (data not shown), and the steady-state antibody concentration was 30.7 mg L^{-1}. The low glucose and glutamine concentrations suggest that the steady-state was now controlled by substrate limitation. Therefore, the dilution rate D_{FB} was increased to 20 d^{-1}. This increased the glucose consumption rate significantly, and a new steady-state was obtained with a corresponding non-limiting glutamine concentration of 1.36 mmol L^{-1} and an antibody concentration of 20.2 mg L^{-1}. However, setting the dissolved oxygen concentration to oxygen saturation (477% air saturation) on the 26th day did not result in a further increase of the glucose uptake rate indicating that the system was also not controlled by oxygen limitation and that the maximal possible performance was reached. The oxygen concentration was again reduced to 300% air saturation and the dilution rate to 10 d^{-1}, and the same steady-state was reached on day 46 as on the 21st day, with a glutamine concentration of 0.49 mmol L^{-1}. The shown examples underline the potential of long- term cultivation, especially long-term stability in bioreactor systems with cell retention.

4.4.2.4 Continuous Dialysis Culture

Dialysis offers an efficient process strategy for the removal of inhibiting metabolites and a simultaneous retention of cells and macromolecular products. It also allows for a number of different strategies (batch, fed-batch, continuous) for decoupling the flows of high molecular and low molecular substrates, low molecular metabolites and high molecular products. Pörtner and coworkers (Lüdemann 1997; Pörtner et al. 1997; Schwabe et al. 1999; Frahm et al. 2003) used a "nutrient-split" feeding strategy based on suggestions made by Ogbonna and Märkl (1993). The medium is divided into a buffer solution containing all mineral salts and a medium concentrate containing all relevant substrates (glucose, glutamine, other amino acids, etc.). The medium concentrate can be fed directly to the cells with significantly reduced flow rates, whereas the buffer solution is used as dialysis medium (compare Fig. 4.39f).

Figure 4.44 shows as continuous dialysis culture run with the above mentioned "nutrient-split" – feeding strategy in a membrane-dialysis-bioreactor (data from Lüdemann (1997)). The bioreactor is discussed more in detail in Sect.. 5.1.1.1. The process was started as batch culture. Continuous perfusion of the cell chamber was switched on at $t = 74$ h with a perfusion rate of $D_i = 0.11\,d^{-1}$. At $t = 163$ h the perfusion rate was set to $D_i = 0.145\,d^{-1}$. For the dialysis chamber perfusion was started at $t = 123$ h with a perfusion rate $D_a = 0.163\,d^{-1}$. The viable cell concentration increased to a maximum value of 1.1×10^7 cells mL^{-1} at $t = 140$ h and levelled of at 7.5×10^6 cells mL^{-1}. Antibody concentration increased up to 650 mg L^{-1}.

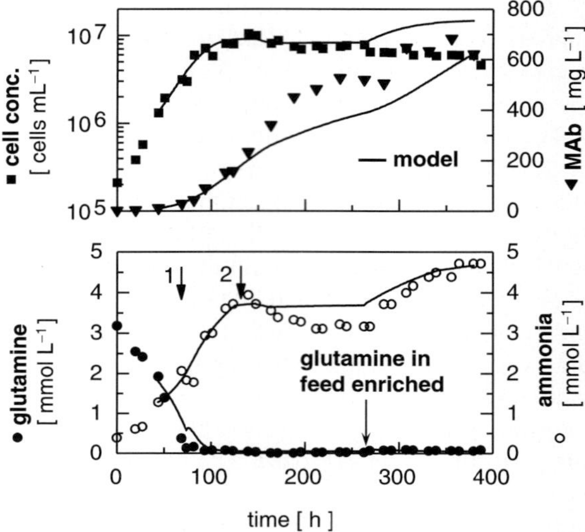

Fig. 4.44 Culture of hybridom cell line IV F 19.23 in Membrane-Dialysis-Reactor – "Nutrient-Split" feeding (data from Lüdemann 1997), suspension culture, dialysis chamber 4.9 L, culture chamber 1.3 L, 2 rushton turbines (110 rpm), membrane and bubble aeration, 1: start feed, 10× medium concentrate, $D_i = 0.11\,\mathrm{d}^{-1}$ until $t = 163\,\mathrm{h}$, then $D_i = 0.145\,\mathrm{d}^{-1}$; 2: start dialysis, buffer, $D_a = 0.163\,\mathrm{d}^{-1}$

4.4.2.5 Comparison of Different Culture Modes

The examples discussed above allow for some general conclusions regarding yield and efficiency of different culture modes, even if the used cell line cannot be regarded as "industrial-like" and optimal process parameters were not chosen in all examples. The main results are summarized in Table 4.7. As expected, the lowest cell and product concentrations were obtained in batch culture. The low efficiency, expressed by a high medium requirement per mg monoclonal antibody, is mainly due to high cell specific substrate uptake rates during exponential growth. Only the utilization level of the substrates is high, as most of the substrates are consumed totally. For the fed-batch culture cell and product concentration are significantly higher. Due to the applied process strategy to control glutamine at a very low level (Kurokawa et al. 1994) the cell specific substrate uptake rates and therefore the medium requirement could be kept low (Sect. 4.4.3). During chemostat the product concentration was quite low and the medium requirement high, as cell and product concentration are determined by the relationship between cell-specific growth rate and perfusion rate. In perfusion mode, shown here by fixed bed cultivation with cell retention, growth rate and perfusion rate are not coupled. Therefore a significantly higher product yield could be obtained, even if product concentration was similar to chemostat culture and lower compared to fed-batch culture. Furthermore Table 4.7 shows a significantly higher product concentration and product yield for the dialysis culture due to simultaneous retention of cells and product. The "nutrient-split"-feeding strategy

Table 4.7 Experimental results for different culture modes with hybridom cell line IV F 19.23 (* estimated from the glucose uptake rate)

Culture system	Batch	Fed-batch	Chemostat	Perfusion (fixed bed)	Membran-dialyse-reactor, continuous, "nutrient-split" feeding
Perfusion rate [d^{-1}]	–	Calculated "a priori"	0.7	10	D_i : 0.145; D_a : 0.163
Cell conc. [cells mL^{-1}]	2×10^6	3.2×10^6	2.4×10^6	$5\times10^{7*}$	6×10^6
MAb-conc. [mg L^{-1}]	35	54	35	30	650
MAb-yield [mg L^{-1} d^{-1}]	8.4	12.7	26.4	300	94.3
Utilisation level glucose [%]	88.6	100	76	80	90
Utilisation level glutamine [%]	100	100	86	83	99.6

further improves the efficiency of this culture mode. Drawbacks of the dialysis technique can be seen in the higher efforts for equipment and process control (Pörtner and Märkl 1998). The above discussion focuses on data obtained in lab-scale culture systems. Implications on the industrial scale will be addressed in Sect. 4.4.4.

4.4.3 Process Strategies for Fed-Batch

The research on fed-batch strategies for suspension cultures of mammalian cells is motivated by the industrial relevance of this process (Glaser 2001; Wrotnowski 2000) (see below). However, despite intensive research carried out in this field during the past few years (reviewed by Pörtner et al. 2004, Wlashin and Hu 2006), it is still difficult to evaluate the performance and suitability of different strategies. In cell culture operations, normally only a few variables are available on-line (e.g., DO, pH, and temperature), and cultivation strategies have to rely on off-line measurements that either have significant time delay or are usually obtained at the end of the cultivation. Conventional control of substrates is therefore rarely applied and alternative procedures are required.

As a simple feeding strategy, pre-defined fixed feed trajectories can be applied. Usually the feed rate has to be increased during the process (e.g., using an exponential or linear profile), as the number of cells and consequently the demand for substrate(s) increases. As there is no feedback or re-adjustment during the process, this control strategy cannot react to changes occurring during the actual process with the risk of suboptimal feeding. However, the fixed feed trajectories can be refined, based on the results of the preceding cultivation. Consequently, it allows improvement on a cultivation-to-cultivation basis. This is subject to the condition that any culture conditions influencing nutrient uptake of the cells remain unchanged.

On-line characterization is a basic approach describing the metabolism and physiological state of cells. The physiological state of the cells determines the uptake of nutrients and the production of metabolites. Metabolic activity is difficult to measure directly, but respiration and energy conversion rates can be monitored by oxygen mass balance (Eyer et al. 1995; Kyung et al. 1994) or spectroscopic analysis of NAD/NADH + H$^+$ (Siano and Muthrasan 1991). A strategy described by Zhou et al. (1995) and Zhou and Hu (1994) determined the physiological state of the cells on-line during cultivation by the oxygen uptake rate (OUR). This strategy showed a high potential for sustaining optimal growth and reducing the formation of toxic metabolites (ammonia and lactate). Mass spectrometry analysis of exhaust gas was further used to determine OUR (Kyung et al. 1994).

Model-based control with model-aided cultivation is one of the most promising approaches in cell culture (Barford et al. 1992; Hansen et al. 1993). In the following, two concepts using process models are illustrated: The first is the *"a priori"-determination* of feed trajectories based on a kinetic model (Fig. 4.41). The model and the calculated feed trajectories remain unchanged once the cultivation has been started. The second is the *adaptive, model-based OLFO-controller*. These concepts react to changes during the cultivation and adapt the model and the feed trajectories accordingly.

The "a priori" determination of feed trajectories, based on a kinetic model, requires intensive kinetic studies of the used cell line. Before starting the feeding phase of the cultivation, feed trajectories are calculated on the basis of the kinetic model. This concept can be considered an extension of the pre-defined feed trajectories concept. In both cases, the trajectories are determined before the cultivation but originate in this case from kinetic studies. Furthermore, the intention of the "a priori" determination of feed trajectories is to predict the death phase and calculate the feed trajectories accordingly. Fixed feed trajectories cause overfeeding during the death phase. Of course, given stable cell lines and culture conditions, the overfeeding can be reduced by adjusting the fixed feed trajectories before the next cultivation. However, the "a priori" determination of feed trajectories allows a more sophisticated control of the death phase in comparison to the fixed feed trajectories. Controlling fed-batch processes with "a priori" simulation of process parameters based on a kinetic model (open loop) demands the ability for accurate prediction, especially at low substrate concentrations when the risk of apoptotic cell death is high. The calculation of optimal feed trajectories (Lee et al. 1999; Ryszcuk and Emborg 1997) using kinetic models shows that small deviations between calculated growth and death rates and actual values have a large effect on the reliability of the trajectories. The parameters of the kinetic models, structured or unstructured, are usually derived from previous experiments and thus the trajectories cannot consider actual culture conditions or changes in metabolism. The feed trajectories obtained from simulation might not result in the expected productivity (de Tremblay et al. 1993).

Alternatively model-based, adaptive control has been suggested, e.g., the open-loop-feedback-optimal (OLFO)-controller (discussed by Pörtner et al. 2004). Major elements of the OLFO-controller are a process model, model parameter identification and an optimization part. They permit adaptation of model parameters and prediction of the future process course of the culture as well as its optimization. Strategies based on the use of expert systems, fuzzy control and neural networks have already been applied in microbial processes but not often investigated for mammalian cell culture.

An overriding goal of the control strategies introduced above was to minimize the production of toxic metabolites such as ammonia and lactate to extend the growth phase. This requires the control of glutamine and glucose at low concentrations so that under these conditions the cell-specific metabolite production is reduced. On the other hand, cell death by apoptosis may be introduced at too low concentrations. Therefore an "intelligent" control strategy has to find an optimum between these two extremes.

Pörtner et al. (2004) compared four different fed-batch strategies including *fixed feed trajectories*, control via *oxygen uptake rate OUR* (stoichiometric feeding), *"a-priori" determination* of feed trajectories based on a kinetic model and the *OLFO-control* strategy. The intention was not to benchmark the different control strategies on the basis of, e.g., time-space-yield or antibody concentration. Recommendation as to which control strategy should be used for a specific process has to consider the respective process, the desired complexity of the process control and for how long the same cell line will be used. However, the illustration of the different strategies points out the different characteristics of each strategy, thus facilitating the choice.

Table 4.8 summarizes the characteristics of the different feeding strategies. A recommendation based on these findings would favour fixed feed trajectories for an established process with a well-characterized and stable production cell line. An adaptive, model-based control strategy could be the method of choice during cell line development or for rapid production of small amounts of the product for clinical trials, because of its universal character and because it does not require intensive process development. From our experience only very few batch experiments are required to adapt the kinetic model. As the OLFO-controller can detect changes in the cell metabolism during a culture, it can adapt the model parameters and the feed trajectories. Our studies illustrate that the adaptive, model-based OLFO-controller is a valuable tool for the fed-batch control of hybridom cell cultures (Frahm et al. 2002). Because of its universal character, it can be transferred to different cell lines or adapted to different boundary conditions (e.g., enhanced fed-batch cultivation using dialysis (Frahm et al. 2003)). More sophisticated optimization criteria, e.g., optimization of final antibody concentration, time-space yield of antibody or another performance measure, can be implemented (Frahm et al. 2005).

4.4.4 Process Strategies Applied in Industrial Processes

Until now, batch and fed-batch strategies are being applied quite frequently on industrial scale (Kelly et al. 1993; Otzturk 2006) due to the several reasons discussed

Table 4.8 Comparision of the characteristics of the feeding strategies (from Pörtner et al. 2004)

	Fixed feed trajectory	Control via oxygen uptake rate (OUR)	"A priori" determination of feed trajectories based on a kinetic model	Model-based adaptive OLFO-control
Possibility of feedback during cultivation	No	Yes, using OUR parameter	No	Yes, using several parameters
Applicability for control at limiting substrate concentrations	No for non-stable cell lines/ varying culture conditions (no feedback)	Yes (online feedback via OUR)	No for non-stable cell lines/varying culture conditions (no feedback)	Moderate (feedback in intervals)
Risk of substrate limitation	High for non-stable cell lines/ varying culture conditions	Low	Moderate/high for non-stable cell lines/ varying culture conditions	Low
Suitable for the the whole process	No, problematic during decline phase/death phase	No, problematic during decline phase/death phase	Yes, problematic during decline phase/ death phase only for not well characterized and unstable production cell line and varying culture conditions,	Yes
Risk of overfeeding	High at the end of the cultivation	High at the end of the cultivation	Moderate at the end of the cultivation	Low
Complexity	Low	Low	Moderate/high	High
Time for implementation (from a first cultivation to satisfactory results)	Moderate	Low	High	Low/moderate
Time expense during cultivation	None	Low	None	High
Operating expense when changing to a different cell line	Moderate	Low/none	High	Low
Possibiltiy of fine-tuning	Yes, from cultivation to cultivation	No/marginal	Yes, from cultivation to cultivation	Yes, during a cultivation and from cultivation to cultivation
Suggested application area	Established process, well characterized and stable production cell line and culture conditions	All areas, method is sub-optimal for optimizing the process performance	Well characterized and stable production cell line and culture conditions, higher effort than fixed feed trajectories	Cell line development, rapid production of small amounts of product (e.g. clinical trials)

already. The drawbacks of discontinuous modes such as large reactor volumes, high maintenance (cleaning, sterilization, etc.) can be overcome to some extent by using a continuous mode, especially perfusion with cell and/or product retention (Table 4.9). There are a growing number of reports on stable perfusion cultures even on an industrial scale lasting several months (Kompala and Ozturk 2006; Bohmann et al. 1995; Fassnacht et al. 1999; Fussenegger et al. 2000). The main advantage of perfusion cultures can be seen in the reduced bioreactor size (approximately one-tenth of a suspension reactor without cell retention). Nevertheless there are some difficulties in the perfusion concept. Further equipment such as the retention device itself, pumps for feeding, harvest and medium circulation, and storage tanks for feed and harvest are required. The amount of media needed to complete a moderate to long-term run can be excessively large. A further possible drawback of continuous cultivation is the possibility of variability over the time of the run. Batch or fed-batch cultivations are much shorter in duration and have fewer chances for random occurrences to happen. This is important in dealing with cGMP (current manufacturing practice) conditions that dictate precise reproducibility or evidence of a minimum of non-effects of minor deviations. In the meantime there is a growing number of pharmaceutical products in the market, produced successfully from perfusion systems. Among these are block busters such as Kogenate-FS® (Bayer Health Care) or Remicade® (Centocor) (Kompala and Ozturk 2006). Previous concerns such as getting the processes approved are not longer an issue and it can be expected that due to an increasing pressure on production costs, more perfusion processes will be installed in the future.

The process that is becoming more and more important for the production of small quantities of recombinant proteins is transient transfection (Meissner et al. 2001; Derouazi et al. 2004; Baldi et al. 2005). Non-transfected cells (most commonly HEK-293, COS and BHK cells) are first cultivated usually in batch-mode to a certain cell density and then transfected with DNA encoding for a certain protein. Usually the cells are genetically not stable and soon lose their expression ability, but can produce a certain amount of protein within one batch. Originally it was used just as a preliminary test of gene expression. New developments show that large scale production up to 100 L is possible (Blasey et al. 1996; Girard et al. 2002).

Table 4.9 Comparision of batch and continuous perfusion (adapted from Griffiths 1992)

		Batch	Perfusion
Equipment	Process stability	+	−
	Sterility	+	±
Process control	Handling	+	±
	Labour	−	+
	Duration	−	+
	Development	+	−
Economics	Productivity	−	+
	Cost/unit	−	+
Licencing	Batch-definition	+	±
	Cell stability	+	?

4.5 Monitoring and Controlling in Animal Cell Culture

The production of specific products using bioprocesses requires the monitoring and controlling of a number of parameters (Table 4.10). To maintain these parameters, either of a physical or chemical nature, is crucial for a high productivity of the bioprocess. Non-optimal conditions within the bioreactor could result in a decreased product formation and a less efficient process.

The most common parameters chosen to be monitored and controlled during a bioprocess are temperature, agitation rate, pH, pO_2 and pCO_2. If one of these values differs from the specific desired set point, corrective action has to be taken (Krahe 2003; Chmiel 2006). In addition to this, other parameters like metabolite concentrations in the culture medium or the cell shape and viability can be used to provide information about the state of the process (Riley 2006). This chapter gives an overview of the methods to monitor the parameters mentioned above and how to control them appropriately during the process.

Cell line identification will not be discussed in detail in this chapter. Protocols for this purpose can be found, e.g., in J. R. W. Masters, (Masters 2000) or A. Doyle and J.B. Griffiths, (Doyle and Griffith 1998).

4.5.1 Temperature

The parameter temperature is very critical for a bioprocess and in animal cell cultures and must be maintained at the desired set point within a tight range of about $\pm 0.5\,^{\circ}C$. A higher variation in temperature often leads to production rates far from the optimum due to slow cell metabolism (temperature below lower limit) or rapid

Table 4.10 Parameters measured or controlled in bioreactors

Physical	Chemical	Biological
Temperature	pH	Biomass concentration
Pressure	Dissolved O_2	Enzyme concentration
Reactor weight	Dissolved CO_2	Biomass composition (such as DNA, RNA, protein, ATP/ADP/ AMP, NAD/NADH levels)
Liquid level	Redox potential	Viability
Foam level	Exit gas composition	Morphology
Agitator speed	Conductivity	
Power consumption	Broth composition (substrate product, ion concentrations, etc.)	
Gas flow rate		
Medium flow rate		
Culture viscosity		
Gas hold-up		

cell death (temperature above higher limit) (Hartnett 1994). To measure the temperature within the system, resistance temperature devices (RTD) are preferred (PT-100 or PT-1000) due their high accuracy and reproducibility. The device contains a metal conductor which changes resistance with changing temperature. This principle is used to determine the actual temperature within the system. Another method is the application of thermocouples, which are cheaper but also less accurate (Riley 2006; Doyle and Griffith 1998).

However, independent of the device used, the measured temperature and resulting electrical signal must be compared to the chosen set point and eventually corrective action must be taken. For this, the electrical signal obtained by the RTD or thermocouple is amplified and transmitted to a controller, where it is compared to the set point value. Depending on the actual state of the system and the desired temperature value, heating or cooling of the system is induced. For small scale and large scale bioreactors, a heating and cooling action using PID (proportional integral derivative) control provides an adequate control (Storhas 1994).

4.5.2 pH

Cells and their metabolism are very sensitive to changes in H^+-ion concentration and so pH. Therefore a control of pH in a very tight range of ± 0.1 around the set point is necessary to ensure high productivity of a bioprocess (Krahe 2003; Chmiel 2006).

Measurement of pH requires a combination of two electrodes, one measuring electrode and one reference electrode. These two electrodes are usually combined in one device, a pH probe. The electrode for measuring has a glass membrane, where a potential develops, which varies with hydrogen ion concentration within the medium to be measured. This potential is then compared to a constant potential built up at the reference electrode to determine the actual pH in the medium to be measured.

Depending on the process, there are two ways to control the pH in the culture medium (Krahe 2003; Riley 2006; Doyle and Griffith 1998). If a medium containing sodium-bicarbonate is used, then the introduction of CO_2 into the medium can be used to control the pH.

$$CO_{2\,(aq)} + H_2O <-> H^+ + HCO_3^-$$

Phosphates in the medium also aid buffering and, for some cells lines, HEPES ([N-2-hydroxyethylpiperazine-N]ethansulphonic acid) can be added to provide further buffering. If a medium containing no sodium-bicarbonate is used, a liquid base (e.g. KOH or NaOH) or acid (e.g. HCl) is used to control the culture pH. In this case, the medium should have at least some buffer capacity to avoid oscillation around the set point. The problem with this method is that the addition could result in a dilution of the medium and so nutrient levels, which then could result in lower productivity. Another problem, which could occur if base and acid are simply added by dripping on the culture's liquid surface, is cell death due to bad mixing (Langheinrich and Nienow 1999).

4.5.3 Oxygen Partial Pressure

Oxygen is usually used by animal cell cultures for the production of energy from organic carbon sources. Therefore it is a very important substrate but unfortunately, oxygen solubility in aqueous medium is relatively low and can result in oxygen limitations of the culture and thus affect cell growth rate and metabolism. The solubility of oxygen at 1 bar pressure and air saturation is only about 6–8 mg L^{-1}, depending on the liquid temperature and composition (Sect. 4.2). The oxygen consumption rate depends on the specific cell line used, growth rates, carbon sources, etc. (Doyle and Griffith 1998).

It is difficult to measure the dissolved oxygen in the cell culture medium directly; one can use the partial pressure to determine the actual dissolved oxygen concentration. At equilibrium, the partial pressure p_{O2} of oxygen in the medium is proportional to the concentration of oxygen in the vapor phase above the medium c_{O2} [Sect. 4.2 (4.15)].

The most commonly applied device to quantify the dissolved oxygen concentration in a cell culture medium is a so called polarographic Clark-electrode (Clark 1955; Cammann et al. 2002; Chmiel 2006). Here, the oxygen diffuses from the culture medium across an oxygen-selective membrane and is then reduced at a negatively polarized platinum electrode. This cathode is connected to a reference silver anode by an electrolyte solution (KCl).

$$\text{Cathode: } O_2 + 2\,H_2O + 4\,e^- + 4\,OH^-$$

$$\text{Anode: } Ag + KCl + AgCl + K^+ + e^-$$

The electrons produced by the reduction of oxygen produce current, which is directly proportional to the oxygen concentration in the medium.

Relatively new devices to determine the amount of oxygen in cell culture medium are optodes (Klimant et al. 1995; Wolfbeis 1991; Wolfbeis (2000); Wittmann et al. 2003). An optode uses dynamic quenching effects. Therefore pulsed monochromatic light is carried in an optic fiber to the oxygen sensor and excites the immobilized fluorophore (e.g., ruthenium complexes). The excited complex fluoresces and emits energy at a higher wavelength. The collision of an oxygen molecule with a fluorophore in its excited state leads to a non-radiative transfer of energy. This internal conversation decreases the fluorescence signal. The degree of quenching correlates with the oxygen concentration (Figs. 4.45–4.47). It is possible to measure either the fluorescence lifetime or the luminescence intensity. The detection of the fluorescence lifetime as oxygen dependent is not affected by external light sources, changes in the pH, ions or salinity. Changes of the optical properties of the sample like opacity, coloration or the refractive index falsify the fluorescence intensity, but not the lifetime.

The kinetics of this process follows the Stern–Volmer relationship:

$$(\tau_{10}/\tau_{1}) - 1 = K_{SV}c_{O2}$$

where τ_{10} and τ_{1} are the fluorescent lifetimes in the absence and presence, respectively, of oxygen, K_{SV} is the Stern–Volmer quenching constant and c_{O2} is the oxygen concentration.

Optodes do work in the whole region 0–100% oxygen saturation in water, but are most sensitive at low oxygen concentrations. During the measurement no oxygen is consumed, as, for e.g., by the use of a Clark-electrode (Wittmann et al. 2003).

As mentioned before, the actual value is compared to a desired set point and, if needed, corrective action is taken. To maintain the desired set point, air or pure oxygen is usually sparged directly into the culture, depending on the required oxygen transfer rates and the achievable flow rates. In the laboratory also, oxygenation by membranes is a widely used technique and ensures bubble-free aeration (Sect. 4.2).

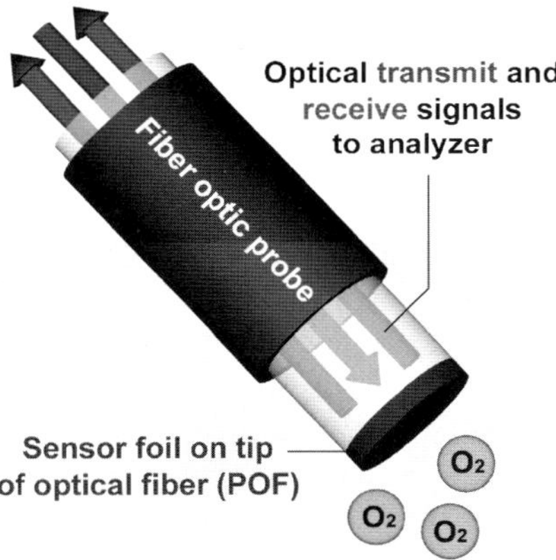

Fig. 4.45 Principle of an optode (with permission and by courtesy of Presens Sensing GmbH, Germany)

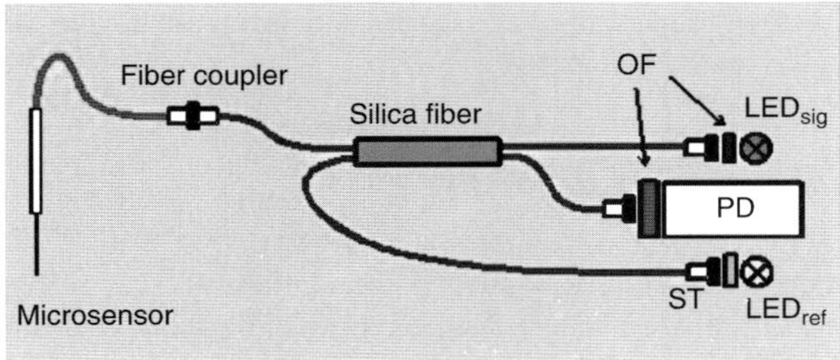

Fig. 4.46 Schematic of the assembly of measurement based on an optode (with permission and by courtesy of Presens Sensing GmbH, Germany)

Fig. 4.47 Single channel instrument for the measurement of oxygen using an optode (with permission and by courtesy of Presens Sensing GmbH, Germany)

4.5.4 Carbon Dioxide Partial Pressure

Not only the concentration of dissolved oxygen but also the concentration of dissolved carbon dioxide within the cell culture medium is very important. The concentration of carbon dioxide in the vapor leaving a bioreactor can also be used to calculate specific respiration rates and cell activity. The latter can easily be measured utilizing infrared (IR) analyzers (Riley 2006; Chmiel 2006). Dissolved carbon dioxide can now be measured in situ utilizing a fiber optic chemical sensor (Riley 2006; Wolfbeis 1991; Chu and Lo 2008; Pattison et al. 2000; Ge et al. 2003). For example, this kind of sensor uses hydroxypyrenetrisulfonate (HPTS) to quantify the dissolved carbon dioxide in the medium. The protonated and un-protonated forms have distinct excitation maxima at 396 and 460 nm, respectively, which allow a ratiometric measurement of pH and thus carbon dioxide concentration (Fig. 4.48). The company YSI Incorporated has developed an in situ device model, 8,500, utilizing this method to quantify dCO_2 (Fig. 4.49), but these devices have not been established widely in the industry yet.

4.5.5 Metabolites and Products

One goal in animal cell cultures is always to maintain suitable culture conditions with regard to nutrients to achieve fast cell growth, high cell viabilities and high product yields. With a good understanding of the cell metabolism and it needs, selective feeding strategies can be utilized to improve productivity and lower culture costs. The carbohydrate glucose and the amino acid glutamine are the two key

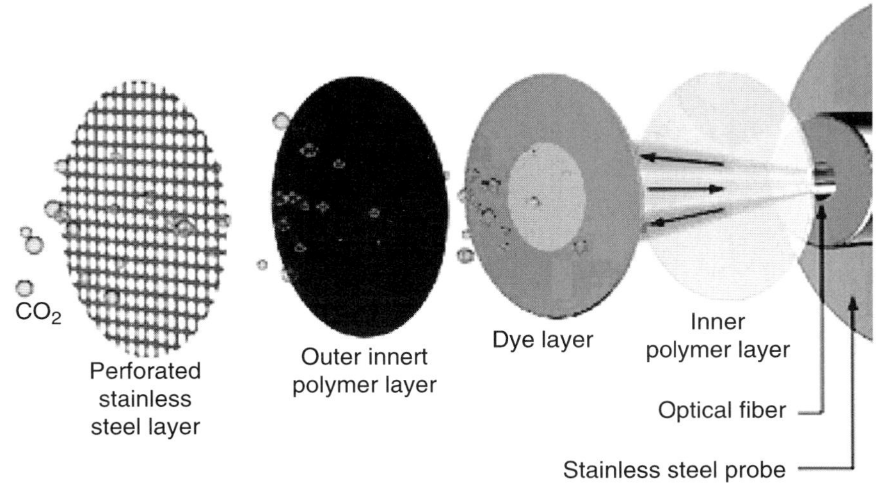

Fig. 4.48 Principle of a fiber optic chemical sensor for the dissolved carbon dioxide measurement (with permission and courtesy of YSI Incorporated, Yellow Springs, USA)

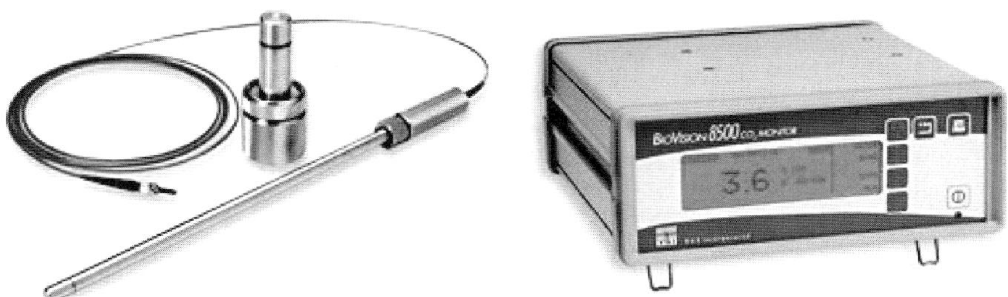

Fig. 4.49 Fiber optic sensor (**a**) and device (**b**) for the dissolved carbon dioxide measurement (with permission and courtesy of YSI Incorporated, Yellow Springs, USA)

nutrients in animal cell cultures, while ammonia and lactate are the two main metabolic byproducts. In addition to insufficient supply of nutrients, byproduct formation can have a negative effect and inhibit cell growth and product formation. Therefore it is very important to monitor and control the concentrations of both nutrients and byproducts in the medium during the process (Riley 2006; Chmiel 2006; Doyle and Griffith 1998).

One approach to quantify nutrient and byproduct concentrations is the use of biosensors. Biosensors are devices which utilize a specific enzymatic reaction to obtain an electrical signal which is proportional to the concentration of the analyte to be quantified. Due to the enzymatic nature of the sensor, a separate sensor has to be used for each analyte. These sensors are most commonly used offline, meaning a sample is taken aseptically from the culture and analyzed outside the system, but automatic on-line sampling devices are also available if preferred. In all cases the enzymatic reaction, e.g., an oxidase reaction, results in a product that can be easily detected, e.g., hydrogen peroxide.

The following example is based on the YSI Model 2700 Select Biochemistry Analyzer (YSI Incorporated; Yellow Springs, OH). Here a three-layer membrane placed on the face of the probe contains an immobilized enzyme in the middle layer. Once the sample is injected into the buffer-filled sample-chamber, the analyte diffuses through the first membrane and is oxidized after contact with the enzyme to H_2O_2 (Reaction 1) (Fig. 4.50).

$$Glucose + O_2 \xrightarrow{\ Glucose\ oxidase\ } Gluconic\ acid + H_2O_2$$
$$Lactate + O_2 \xrightarrow{\ Lactate\ oxidase\ } Pyruvate + H_2O_2$$
$$Glutamine + O_2 \xrightarrow{\ Glutamate\ oxidase\ } \alpha - ketoglutarate + NH_3 + H_2O_2$$

Hydrogen peroxide is capable of diffusing through the second membrane and is oxidized at the platinum electrode (Reaction 2).

$$H_2O_2 \xrightarrow{\ Platinum\ anode\ } 2H^+ + O_2 + 2e^-$$

Due to the free electrons produced, a current occurs, which is linearly proportional to the concentration of the analyte in the sample once a dynamic equilibrium between H_2O_2 production in and diffusion out of the enzyme layer is established. This current is then used to calculate the analyte concentration in the sample.

Ozturk used this kind of analyzer to monitor glucose and lactate during batch and perfusion culture of hybridoma cells by using cell free samples from a circulation

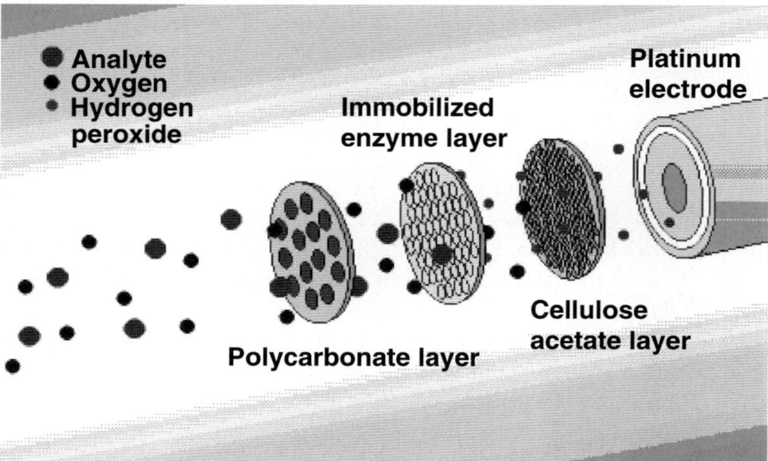

Fig. 4.50 Schematic of the sensing principle of the biosensor (with permission and courtesy of YSI Incorporated, Yellow Springs, USA)

loop of the bioreactor including a 0.45 μm hollow fiber membrane filter (Ozturk et al. 1997).

Another approach to quantify nutrient and byproduct concentrations is the flow injection analysis (FIA). A liquid sample is injected into a stream of aqueous carrier stream and forms a pulse which can then be detected, based on changes in different physical parameters (e.g., absorbance, fluorescence, chemi-luminescence, etc.). Because each sensor can only detect one analyte, the sample flow is often split into different channels for a broader range of analytes to be monitored (Riley 2006; Chmiel 2006).

4.5.6 Cell Density and Viability

Cell density can be measured utilizing different methods, either offline or online. Offline methods require aseptic sampling prior to the cell density determination. The sample is diluted appropriately and often counted with a hemocytometer (Doyle and Griffith 1998; Butler 2004).

Additionally, the dye trypan blue is generally used to determine cell viability. The dye penetrates the broken membranes of dead cells and dyes them blue, binding to the proteins present in the cytosol (Butler 2004). The membranes of viable cells inhibit the dye entering the cytosol and so these cells do not get dyed blue. A recent method to determine cell density and viability is the use of automated counters. These devices, e.g., Vi-Cell Cell Viability Analyzer (Beckman Coulter, Inc., Fullerton, CA) are usually based on the trypan blue method and can help to quantify the determined data more rapidly and are less subjective. If needed, such a device can also be connected directly to a fermenter and samples taken and analyzed automatically (autosampler). Other online methods use probes inserted into the vessel and measure the absorbance of the culture with a Vertical Cavity Surface-Emitting Laser (VSCEL) at a wavelength of 850 nm (TruCell2 probe, Finesse, Inc., Santa Clara, CA). Using this method, it is possible to determine only the total cell density within the culture, but not the percentage of dead cells.

A number of indirect methods of cell density and viability determination are also available but will not be discussed in detail in this chapter. These methods include total protein determination, e.g., the Bradford-assay, or DNA determination, e.g., the DAPI-assay. Methods to determine the percentage of viable cells within the culture include the tetrazolium assay, the lactate-dehydrogenase determination or adenylate energy charge determination. Also the rate of protein or nucleic acid synthesis can be used to determine cell viability (Doyle and Griffith 1998; Butler 2004).

In addition to this, methods for cell line identification are also available. Karyotpying, antibody labeling, isozyme analysis or DNA fingerprinting is utilized to detect cross-contaminations and mutations within a culture. Protocols for these techniques can be found in, e.g., Doyle and Griffiths, 1998.

4.5.7 Agitation

It is very important to maintain a specific rate of agitation within the system. Agitation ensures homogeneity of nutrients and oxygen in the medium and also helps to distribute additions like base or acid evenly in the culture medium (Krahe 2003; Chmiel 2006). If the agitation rate is too high, cell death could occur due to shear stress. Therefore it is very important to know what shear stress the cultivated cells can resist without a decrease in protein production or cell viability. Efficiency and shear stress can be significantly different for different types of impellers. The impeller speed can be measured by all standard speed controllers, e.g. electronic tachometer.

4.6 Questions and Problems

- Summarize briefly the main fluid-mechanical and biological aspects of cell damage.
- What kind of reaction system can be used to investigate shear effects on anchorage dependent cells? Explain the basic characteristics of this system.
- What is an acceptable wall shear stress for adherent growing cells for an exposure time of 24 h?
- Describe fluid dynamic effects on anchorage dependent cells grown on microcarriers in suspension culture.
- What is described by the Kolmogorov-eddy-length? Is this parameter suitable of estimation of shear effects on anchorage dependent cells grown on microcarriers in suspension culture?
- Which effects are responsible for cell damage in bubble columns? Which is the most important one?
- Discuss the consequences of shear effects on reactor design.

4.6.1 Problem

During a cell culture fermentation a system answer for the oxygen concentration in the culture medium was plotted by the "dynamic method" (see figure: x-axis: time in seconds, y-axis: oxygen concentration in the liquid in mg L^{-1}).

Define the dynamic method and give the mass balances.

Calculate the oxygen uptake rate (OUR) of the culture, the $k_L a$ and the oxygen saturation concentration c^* by approximation using the experimental data (Fig. 4.51).

4.6.2 Problem

The surface area of a culture grown in a spinner culture vessel is 100 cm^2 and the culture medium volume is 100 ml. The oxygen consumption by the cells is 2×10^{-8} mg cell^{-1} h^{-1}. At the end of the log phase of growth the cell concentration reached

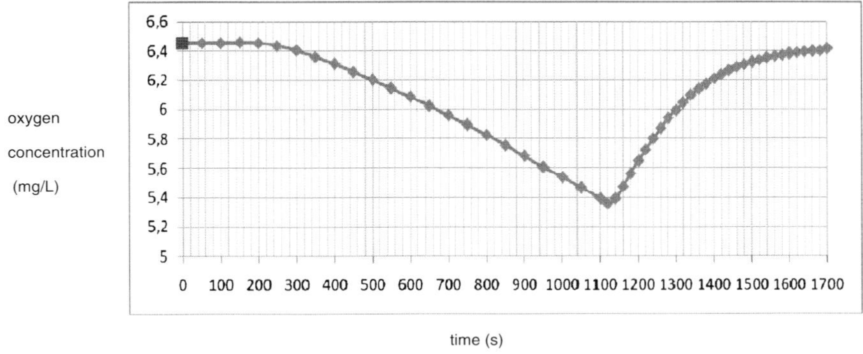

Fig. 4.51

2.5×10^6 cells mL^{-1}. Oxygen is supplied in the form of air pumped into the head space of the vessel and its rate of transfer across air/medium interface has been determined to be 0.17 mg per cm^2 and hour.

It is proposed to use a scaled up version of this spinner culture. The culture volume is to be increased tenfold. The vessel that is to be used has a cross sectional area of 200 cm^2.

What problems do you foresee in supplying oxygen to this culture (calculate the oxygen demand of the up-scaled culture) and how would you solve this problem?

- Which types of membranes can be used for bubble-free aeration of suspension reactors? Enumerate the advantages and disadvantages of bubble-free aeration systems.
- Explain techniques for immobilization and encapsulation of mammalian cells.
- Define the terms "microcarrier" and "macrocarrier" with respect to preferred reactor system.
- Name properties of an "ideal" microcarrier.
- Explain mass-transfer effects in macroporous carriers. Give the fundamental equations for a heterogeneous reaction within a spherical particle.
- What is described by the Thiele-modulus?
- List the advantages and disadvantages of batch and perfusion systems.
- Explain briefly the principles of culture modes.
- Sketch for a chemostat the relationship between cell density, substrate concentration and product concentration as a function of the dilution rate for "steady state". Explain briefly the term "wash-out point".
- Explain briefly the strategies used to determine feeding rates in fed-batch cultures
- Name important aspects, when perfusion culture has an advantage compared to batch/fed-batch culture.

- Sketch the scheme for the mass transfer across a dialysis membrane. In which respect does the permeability coefficient of a dialysis membrane depend on the molecular weight of a certain substance?

List of Symbols

A	membrane area
b	width
c	concentration
C	constant
c^*	equilibrium solubility of oxygen – oxygen saturation
c_0^*	oxygen solubility at zero solute concentration
c_1, c_2	concentrations of both sides of a dialysis membrane
c_{AL}	concentration of oxygen at steady state
c_{crit}	critical oxygen concentration
c_{iL}	concentration of ionic component i in the liquid
c_{jL}	concentration of non-ionic component j in the liquid
c_L	concentration of oxygen in solution
c_{O2}	concentration of oxygen
c_P	product concentration
c_S	substrate concentration
c_{S0}	substrate concentration in feed
D	dilution rate in chemostat and perfusion mode
D_a	dilution rate in outer chamber of membrane dialysis reactor
d_B	bubble diameter
$D_{crit.}$	critical (maximal) dilution rate
D_e	diffusive coefficient
D_i	dilution rate in inner chamber of membrane dialysis reactor
d_1	inner diameter of the membrane tube
$d_{i,cou}$	diameter of inner cylinder in couette viscometer
d_2	orifice diameter
DO	dissolved oxygen
d_o	outer diameter of the membrane tube
D_R	vessel diameter
d_R	impeller diameter
D_s	diffusion coefficient of oxygen in the membrane
d_s	Sauter mean diameter
d_T	bubble column vessel diameter
F	liquid flow rate
g	acceleration due to gravity
h	height
He	Henry constant
He_s	Henry constant of oxgen in membrane material

He_w Henry constant of oxgen in water

H_i constant for ionic component i

h_L height of fluid in bubble column

H_R filling height of the stirred reactor – impeller located at the interface

ISF   integrated shear factor

k mass transfer coefficient at the gas liquid interface

k_d cell specific death rate

K_j constant for non-ionic component j in the liquid

k_L mass transfer coefficient at the liquid-side interface

$k_L a$ mass transfer coefficient, which is the product of k_L, the overall mass transfer coefficient from the gas to the liquid phase (two film model), and a the gas–liquid interfacial area per unit of the reactor liquid volume

k_{O2} monod-constant for oxygen

k_S Monod-constant for substrate limitation

K_{SV} Stern–Volmer quenching constant

l_{Kol} Kolmogorov length scale

M_d torque

n_R rotation speed

OTR   oxygen transfer rate

OUR   oxygen uptake rate

P power input

P/V power input per unit volume

P_{mem} permeability coefficient of a dialysis membrane

p_{O2} partial pressure of oxygen in the gas phase

Q volumetric aeration rate

$Q_{O2,max}$ maximal cell specific oxygen uptake rate

q_P cell specific production rate

q_S cell specific substrate uptake rate

r radius

Re Reynolds number

$r_{i,cou}$ inner radius of a couette viscometer

r_P particle radius

r_Z cell radius

Ta   Taylor number

S   molecular flow

t time

T temperature

U flow velocity in flow chamber

U_T peripheral speed in couette viscometer

V volume

v superficial gas velocity

v_0 gas velocity at the sparger

V_k hypothetical killing volume.

w gap width

X_t total cell concentration

X cell density

X_d concentration of dead cells

X_v concentration of viable cells

Y length scale in flow chamber

Yz valency of ionic component i

α exponent (further function of superficial velocity)

β exponent independent of scale and impeller type

α_{perf} recirculation rate in perfusion culture, ratio between feed and recirculated flow

β_{perf} concentration factor in perfusion culture, ration between biomass, concentration in the reactor and in recirculated flow

γ shear rate

η_{fl} fluid viscosity

η_i effectiveness factor for internal mass transfer resistance

Φ_0 Thiele-Modulus for zero-order kinetic

μ cell specific growth rate

μ_{max} maximal cell specific growth rate

$\Delta\rho$ density difference

ε energy dissipation rate per unit mass

ν kinematic fluid viscosity

ρ_{fl} fluid density

σ surface tension

σ_z surface tension of cell membrane at cell burst

τ shear stress

τ_{crit} critical shear stress

τ_w wall shear stress

τ_{l0} fluorescent lifetimes in the absence of oxygen

τ_l fluorescent lifetimes in the presence of oxygen

References

Abu-Reesh I, Kargi F (1989) Biological responce of hybridoma cells to defined hydrodynamic shear stress. J Biotechnol 9: 167–178

Augenstein DC, Sinskey AJ, Wang DIC (1971) Effect of shear on the death o two strains of mammalian cells. Biotechnol Bioeng 40: 1004–1010

Aunins JG, Henzler HJ (1993) Aeration in cell culture bioreactors. In: Rehm HJ, Reed G (eds) Biotechnology, Vol 3, 2nd ed, pp 219–281. VCH, Weinheim

Baldi L, Muller N, Picasso S, Jacquet R, Girard P, Thanh HP, Derow E, Wurm FM (2005) Transient gene expression in suspension HEK-293 cells: application to large-scale protein production. Biotechnol Prog 21: 148–153

Barford JP, Phillips PJ, Harbour C (1992) Enhancement of productivity by yield improvements using simulation techniques. In: Murakami H et al. (eds) Animal Cell Technology: Basic & Applied Aspects. Kluwer, Netherlands, pp 397–403

Blasey HD, Aubry JP, Mazzei GJ, Bernard AR (1996) Large scale transient expression with COS cells. Cytotechnol 18: 183–192

Bliem R, Konopitzky K, Katinger H (1991) Industrial animal cell reactor systems: aspects of selection and evaluation. Adv Biochem Eng 44: 2–26

Bödeker BGD, Newcomb R, Yuan P, Braufman A, Kelsey W (1994) Production of recombinant factor VIII from perfusion cultures: I. Large-Scale fermentation. In: Spier RE, Griffiths JB, Berthold W (eds) Animal cell technology: products of today, prospects for tomorrow, pp 580–583. Butterworth-Heinemann, London

Bohmann A, Pörtner R, Märkl H (1995) Performance of a membrane dialysis bioreactor with radial-flow fixed bed for the cultivation of a hybridoma cell line. Appl Microbiol Biotechnol 43: 772–780

Born C, Zhang Z, Al-Rubeai M, Thomas CR (1992) Estimation of disruption of animal cells by laminar shear stress. Biotechnol Bioeng 40: 1004–1010

Bräutigam HJ (1985) Untersuchungen zum Einsatz von nicht porösen Kunststoffmembranen als Sauerstoffeintragssystem. Dissertation, TU Hamburg-Harburg

Butler M (2004) Animal Cell Culture and Technology – The Basics. Oxford University Press, New York

Cammann K et al. (2002) Chemical and biochemical sensors. In: Ullmann's Encyclopedia of Industrial Chemistry, Wiley-VCH, New York

Carrel A (1923) A method for the physiological study of tissues. J Exp Med 38: 407

Chalmers JJ (1994) Cells and bubbles in sparged bioreactors. Cytotechnol 15: 311–320

Cherry RS (1993) Animal cells in turbulent fluids: Details of the physical stimulus and the biological response. Biotechnol Adv 11: 279–299

Cherry RS, Papoutsakis ET (1986) Hydrodynamic effects on cells in agitated tissue culture reactors. Bioproc Eng 1: 29–41

Cherry RS, Papoutsakis ET (1988) Physical mechanisms of cell damage in microcarrier cell culture bioreactors. Biotechnol Bioeng 32: 1001–1014

Chisti Y (2000) Animal-cell damage in sparged bioreactors. Trends Biotechnol 18: 420–432

Chisti Y (2001) Hydrodynamic damage to animal cells. Crit Rev Biotechnol 21: 67–110

Chmiel H (2006) (ed). Bioprozesstechnik, 2nd ed., Spektrum Akademischer Verlag, Elsevier, Amsterdam

Chu C-S, Lo Y-L (2008) Fiber-optic carbon dioxide sensor based on fluorinated xerogels doped with HPTS, pp 120–125. Sensors Actuators B 129

Clark LC (1955) US Patent 2913386

Croughan MS, Hamel JF, Wang DIC (1987) Hydrodynamic effects on animal cells grown in microcarrier cultures. Biotechnol Bioeng 29: 130–141

Croughan MS, Hamel JF, Wang DIC (1988) Effects of microcarrier concentration in animal cell culture. Biotechnol Bioeng 32: 975–982

Croughan MS, Wang DIC (1989) Growth and death in overagitated microcarrier cell cultures. Biotechnol Bioeng 33: 731–744

Czermak P, Nehring D (1999) DE 19953137 A1

Czermak P, Nehring D (2000) DE 20053137 A2

Davis J (2007) Hollow fibre cell culture. In: Pörtner R (ed) Animal Cell Biotechnol – Methods and Protocols. Humana Press, Clifton, UK

Davis JM, Hanak JA (1997) Hollow-fiber cell culture. Meth Mol Biol 75: 77–89

de Tremblay M, Perrier M, Chavarie C, Archambault J (1993) Fed-batch culture of hybridoma cells: comparison of optimal control approach and closed loop strategies. Bioprocess Eng 9: 13–21

Derouazi M, Girard P, Van Tilborgh F, Iglesias K, Muller N, Bertschinger M, Wurm FM (2004) Serum-free large-scale transient transfection of CHO cells. Biotechnol Bioeng 87: 537–545

Dey D, Boulton-Stone JM, Emery AN, Blake JR (1997) Experimental comparisons with a numerical model of surfactant effects on the burst of a single bubble. Chem Eng Sci 52: 2769–2783

Diamond SL, Sharefkin JB, Diefenbach C, Frasier-Scott K, McIntire LV, Eskin SG (1990) Tissue plasminogen activator messenger RNA levels increase in cultured human endothelial cells exposed to laminar shear stress. J Cell Phys 143: 364–371

Doran P (2006) Bioprocess Engineering Principles. Academic Press, London

Doyle A, Griffith JB (1998) Cell and Tissue Culture: Laboratory Procedures in Biotechnology, p 192. Wiley, Chichester

Eyer K, Oeggerli A, Heinzle E (1995) On-line gas analysis in animal cell cultivation: II. Methods for oxygen uptake rate estimation and its application to controlled feeding of glutamine. Biotechnol Bioeng 45: 54–62

Fassnacht D (2001). Fixed-Bed Reactors for the Cultivation of Animal Cells. Fortschritt-Berichte VDI. VDI Verlag GmbH, Düsseldorf, Germany

Fassnacht D, Rössing S, Singh RP, Al-Rubeai M, Pörtner R (1999) Influence of *Bcl-2* on antibody productivity in high cell density perfusion cultures of hybridoma. Cytotechnol 30: 95–105

Fenge C, Klein C, Heuer C, Siegel U, Fraune E (1993) Agitation, aeration and perfusion modules for cell culture bioreactors. Cytotechnol 11: 233–244

Fenge Ch, Lüllau E (2006) Cell culture bioreactors. In: Ozturk SS, Hu WS (eds) Cell Culture Technology for Pharmaceutical and Cell-Based Therapies. Taylor & Francis, New York

Frahm B, Kirchner S, Kauling J, Brod H, Langer U, Bödeker B (2007) Dynamische Membranbegasung im Bioreaktor zur Intensivierung der Sauerstoffversorgung empfindlicher Zellinien. Chemie Ing Tech 79: 1052–1058

Frahm B, Lane P, Atzert H, Munack A, Hoffmann M, Hass VC, Pörtner R (2002) Automated, adaptive, model-based control by the open-loop-feedback-optimal (OLFO) controller for the effective fed-batch cultivation of hybridoma cells. Biotechnol Prog 18: 1095–1103

Frahm B, Lane P, Märkl H, Pörtner R (2003) Improvement of a mammalian cell culture process by adaptive, model-based dialysis fed-batch cultivation and suppression of apoptosis. Bioproc Biosys Eng 26: 1–10

Frahm B, Lane P, Munack A, Pörtner R (2005) Optimierung und Steuerung von Zellkultur-Fed-Batch-Prozessen mittels einer Kollokationsmethode. Chem Ing Tech 77: 429–435

Frangos JA, Eskin SG, McIntire LV, Ives CL (1985) Flow effects on prostacyclin production by cultured human endothelial cells. Science 227: 1477–1479

Frangos JA, McIntire LV, Eskin SG (1988) Shear stress induced stimulation of mammalian cell metabolism. Biotechnol Bioeng 32: 1053–1060

Fussenegger M, Fassnacht D, Schwartz R, Zanghi JA, Graf M, Bailey JE, Pörtner R (2000) Regulated overexpression of the survival factor bcl-2 in CHO cells increases viable cell density in batch culture and decreases DNA release in extended fixed- bed cultivation. Cytotechnol 32: 45–61

Ge X, Kostov Y, Rao G (2003) High-stability non-invasive autoclavable naked optical CO_2 sensor. Biosensors Bioelectronics 18: 857–865

Geserick C, Bonarius HP, Kongerslev L, Hauser H, Mueller PP (2000) Enhanced productivity during controlled proliferation of BHK cells in continuously perfused bioreactors. Biotechnol Bioeng 69: 266–274

Ghebeh H, Gillis J, Butler M (2002) Measurement of hydrophobic interactions of mammalian cells grown in culture. J Biotechnol 95: 39–48

Girard P, Derouazi M, Baumgartner G, Bourgeois M, Jordan M, Jacko B, Wurm F (2002) 100-liter transient transfection. Cytotechnol 38: 15–21

Glaser V (2001) Current trends and innovations in cell culture. Gen Eng N 21: 11

Gotoh T, Mochizuki G, Kikuchi K (2001) Perfluorocarbonmediated aeration applied to recombinant protein production by virus-infected insect cells. Biochem Eng J 7: 6978

Griffith B (2000) Scaling up of animal cell cultures. In: Masters JRW (ed) Animal Cell Culture. 3rd ed, p 19–66. Oxford University Press, New York

Griffiths JB (1992) Animal cell culture proceses – batch or continuous? J Biotechnol 22: 21–30

Grima EM, Christi Y, Moo-Young M (1997) Characterization of shear rates in airlift bioreactors for animal cell culture. J Biotechnol 54: 195–210

Handa A, Emery AN, Spier RE (1987) On the evaluation of gas–liquid interfacial effects on hybridoma viability in bubble column bioreactors. Dev Biol Stand 66: 241–253

Hansen HA, Madson NM, Emborg C (1993) An evaluation of fed-batch cultivation methods for mammalian cells based on model simulations. Bioproc Eng 9: 205–213

Hartnett T (1994) Instrumentation and control of bioprocesses. In: Lydersen BK, D'Elia NA, Nelson KL (eds). Bioprocess Engineering: Systems, Equipment, and Facilities. Wiley, New York

Henzler HJ, Kauling OJ (1993) Oxygenation of cell cultures. Bioproc Eng 9: 61–75

Hiller GW, Aeschlimann AD, Clark DS, Blanch HW (1991) A kinetic analysis of hybridoma growth and metabolism in continuous suspension culture on serum-free medium. Biotechnol Bioeng 38: 733–741

Howaldt M, Walz F, Kempken R (2006) Kultur von Tierzellen. In: Chmiel H (ed) Bioprozesstechnik. Elsevier – Spektrum Akademischer Verlag

Hu WS, Meier J, Wang DIC (1985) A mechanistic analysis of the inoculum requirements for the cultivation of mammalian cells on microcarriers. Biotechnol Bioeng 27: 585–595

Hu WS, Meier J, Wang DIC (1986) Use of surface aerator to improve oxygen transfer in cell culture. Biotechnol Bioeng 28: 122–125

Hübner H (2007) Cell encapsulation. In: Pörtner R (ed) Animal Cell Biotechnol – Methods and Protocols. Humana Press, Clifton, UK

Huebner H, Buchholz R (1999) Microencapsulation. In: Flickinger MC, Drew SW (ed) Encyclopedia of Bioprocess Technology: Fermentation, Biocatalysis and Bioseparation, pp 1785–1798. Wiley, Hoboken, NJ

Hülscher M (1990) Auslegung von Airlift-Schlaufenreaktoren für die Kultivierung von Tierzellen im Hinblick auf Zellschädigung. Dissertation Universität Dortmund

Hülscher M, Pauli J, Onken U (1990) Influence of protein concentration on mechanical cell damage and fluid dynamics in airlift reactors for mammalian cells. Food Biotechnol 4: 157–166

Jordan M (1993) Die Rolle von Serum bei der hydrodynamischen Belastung von tierischen Zellen im Bioreaktor; Möglichkeiten der Serum Reduktion. Dissertation an der ETH Zürich, Schweiz

Kelly BD, Chiou TW, Rosenberg M, Wang DIC (1993) Industrial animal cell culture. In: Stephanopolous G (ed) Biotechnology, Vol 3 (Bioprocessing), VCH Verlagsgesellschaft, Weinheim, Germany

Klimant I, Meyer V, Kühl M (1995) Fiber-optic oxygen microsensors, a new tool in aquatic biology. Limnol Oceanogr 40: 1159–1165

Kompala DS, Ozturk SS (2006) Optimization of high cell density perfusion bioreactors. In: Ozturk SS, Hu WS (eds) Cell Culture Technology For Pharmaceutical and Cell-Based Therapies. Taylor & Francis, New York

Krahe M (2003) Biochemical Engineering. Reprint from Ullmann's Encyclopedia of Industrial Chemistry, Vol 6. VCH Publishers, Weinheim, Germany

Kramer HW (1988) Der Einfluß von Scherkräften in Bioreaktoren auf Proliferation und Syntheseleistung tierischer Zellen. Dissertation ETH Zürich, Nr. 8565

Kretzmer G (1994) Entwicklung optimaler Prozeßbedingungen zur Produktion von rekombinanten Proteinen mit adhärenten Säugerzellen. Habilitationsschrift Universität Hannover

Kunas KT, Papoutsakis ET (1990) Damage mechanism of suspended animal cells in agitated bioreactors with and without bubble entrainment. Biotechnol Bioeng 36: 476–483

Kurokawa H, Park YS, Iijima S, Kobayashi T (1994) Growth characteristics in fed-batch culture of hybridoma cells with control of glucose and glutamine concentrations. Biotechnol Bioeng 44: 95–103

Kurosawa H, Matsumura M, Tanaka H (1989) Oxygen diffusity in gel beads containing viable cells. Biotechnol Bioeng 34: 926–932

Kyung YS, Peshwa MV, Gryte DM, Hu WS (1994) High density culture of mammalian cells with dynamic perfusion based on on-line oxygen uptake rate measurements. Cytotechnol 14: 183–190

Langheinrich C, Nienow AW (1999) Control of pH in large scale, free suspension animal cell bioreactors: alkali addition and pH excursions. Biotechnol Bioeng 66: 171–179

Lee GM, Kaminski MS, Palsson BO (1992) Observations consistent with autocrine stimulation of hybridoma cell growth and implications for large-scale antibody production. Biotechnol Lett 14: 257–262

Lee JH, Lim HC, Yoo YJ, Park YH (1999) Optimization of feed rate profile for the monoclonal antibody production. Bioproc Eng 20: 137–146

Looby D, Racher AJ, Griffiths JB, Dowsett AB (1990) The immobilization of animal cells in fixed bed and fluidized porous glass sphere reactors. In: de Bont JAM et al. (eds) Physiology of Immobilized Cells, pp 255–264. Elsevier, Amsterdam

Lüdemann I (1997) Nutrient-Split-Fütterungsstrategie für suspendierte und immobilisierte nicht-adhärente tierische Zellen. VDI-Verlag, Düsseldorf

Lüdemann I, Pörtner R, Schaefer C, Schick K, Šrámková K, Reher K, Neumaier M, Franék F, Märkl H (1996) Improvement of the culture stability of non-anchorage-dependent animal cells grown in serum-free media through immobilization. Cytotechnol 19: 111–124

Ludwig A, Kretzmer G, Schügerl K (1992) Determination of a "critical shear stress level" applied to adherent mammalian cells. Enz Microbiol Technol 14: 209–213

Lüllau E, Biselli M, Wandrey C (1994) Growth and metabolism of CHO-cells in porous glass carriers. In: Spier RE, Griffiths JB, Berthold W (eds) Animal Cell Technology: Products of Today, Prospects of Tomorrow, pp 252–255. Butterworth-Heinemann, London

Lüllau E, Kanttinen A, Hassel J, Berg M, Haag-Alvarsson A, Cederbrant K, Greenberg B, Fenge C, Schweikart F (2003) Comparison of batch and perfusion culture in combination with pilot-scale expanded bed purification for the production of soluble recombinant beta-secretase. Biotechnol Prog 19: 37–44

Lundgren B, Blüml G (1998) Microcarriers in cell culture production. In: Subramanian G. (eds) Bioseparation and Bioprocessing – A Handbook. Wiley-VCH, New York

Ma N, Chalmers JJ, Aunins JG, Zhou W, Xie L (2004) Quantitative studies of cell-bubble interactions and cell damage at different Pluronic F-68 and cell concentrations. Biotechnol Prog 20: 1183–1191

Ma N, Mollet M, Chalmers JJ (2006) Aeration, mixing and hydrodynamics in bioreactors. In: Ozturk SS, Hu WS (eds) Cell Culture Technology for Pharmaceutical and Cell-Based Therapies. Taylor & Francis, New York

Märkl H, Bronnenmeier R, Wittek B (1987) Hydrodynamische Belastbarkeit von Mikroorganismen. Chem Ing Tech 59: 907–917

Marks DM (2003) Equipment design considerations for large scale cell culture. Cytotechnol 42: 21–33

Masters JRW (2000) (ed). Animal Cell Culture. 3rd ed., Oxford University Press, New York, 2000

McQueen A, Meilhoc E, Bailey JE (1987) Flow effects on the viability and lysis of suspended mammalian cells. Biotechnol Lett 9: 831–836

Meier SJ, Hatton TA, Wang DIC (1999) Cell death from bursting bubbles: role of cell attachment to rising bubbles in sparged reactors. Biotechnol Bioeng 62: 468–478

Meissner P, Pick H, Kulangara A, Chatellard P, Friedrich K, Wurm FM (2001) Transient gene expression: recombinant protein production with suspension-adapted HEK293-EBNA cells. Biotechnol Bioeng 75: 197–203

Michaels JD, Mallik AK, Papoutsakis ET (1996) Sparging and agitation-induced injury of cultured animal cells: do cell to-bubble interactions in the bulk liquid injure cells? Biotechnol Bioeng 51: 399–409

Miller WM, Wilke CR, Blanch HW (1987) Effects of dissolved oxygen concentration on hybridoma growth and metabolism in continuous culture. J Cell Phys 132: 524–530

Miller WM, Wilke CR, Blanch HW (1988) A kinetic analysis of hybridoma growth and metabolism in batch and continuous suspension culture. Effect of nutrient concentration, dilution rate and pH. Biotechnol Bioeng 32: 947–965

Moo-Young M, Blanch HW (1981) Design of biochemical reactors – mass transfer criteria for simple and complex systems. Adv Biochem Eng 19: 1–70

Moreira JL, Cruz PE, Santana PC, Feliciano AS (1995) Influence of power input and aeration method on mass transfer in a laboratory animal cell culture vessel. J Chem Tech Biotechnol 62: 118–131

Murtas S, Capuani G, Dentini M, Manetti C, Masci G, Massimi M, Miccheli A, Crescenzi V (2005) Alginate beads as immobilization matrix for hepatocytes perfused in a bioreactor: a physico-chemical characterization. J Biomater Sci Polym Ed 6: 829–846

Nehring D, Czermak P, Luebben H, Vorlop J (2004) Experimental study of a ceramic microsparging aeration system in a pilot scale animal cell culture. Biotechnol Prog 20: 1710–1717

Nilsson K, Buzsaky F, Mosbach K (1986) Growth of anchorage dependent cells on macroporous microcarrier. Bio/Technology 4: 989–990

Ogbonna JB, Märkl H (1993) Nutrient-split feeding strategy for dialysis cultivation of *Escherichia coli*. Biotechnol Bioeng 41: 1092–1100

Olivier LA, Truskey GA (1993) A numerical analysis of forces exerted by laminar flow on spreading cells in a paralled plate flow chamber assay. Biotechnol Bioeng 42: 963–973

Oller AR, Buser CW, Tyo MA, Thilly WG (1989) Growth of mammalian cells at high oxygen concentrations. J Cell Sci 94: 43–49

Otzturk SS (2006) Cell culture technology – an overview. In: Ozturk SS, Hu WS (eds) Cell Culture Technology For Pharmaceutical and Cell-Based Therapies. Taylor & Francis, New York

Ozturk SS, Thrift JC, Blackie JD, Naveh D (1997) Real time monitoring and control of glucose and lactate concentrations in a mammalian cell perfusion reactor. Biotechnol Bioeng 53: 372–378

Ozturk SS (1996) Engineering challenges in high density cell culture systems. Cytotechnol 22: 3–16

Ozturk SS, Hu WS (eds) (2006) Cell Culture Technology For Pharmaceutical and Cell-Based Therapies. Taylor & Francis, New York

Papoutsakis ET (1991) Fluid-mechanical damage of animal cells in bioreactors. Tibtech 9: 427–437

Pattison RN, Swamy J, Mendenhall B, Hwang C, Frohlich BT (2000) Measurement and control of dissolved carbon dioxide in mammalian cell culture processes using an in situ fiber optic chemical sensor. Biotechnol Prog 16: 769–774

Pörtner R (1998) Reaktionstechnik der Kultur tierischer Zellen. Shaker, Aachen

Pörtner R, Lüdemann I, Märkl H (1997) Dialysis cultures with immobilized hybridoma cells for effective production of monoclonal antibodies. Cytotechnol 23: 39–45

Pörtner R, Märkl H (1995) Festbettreaktoren für die Kultur tierischer Zellen. BIOforum 18: 449–452

Pörtner R, Märkl H (1998) Dialysis cultures. Appl Microbiol Biotechnol 50: 403–414

Pörtner R, Platas Barradas OBJ (2007) Cultivation of mammalian cells in fixed bed reactors. In: Pörtner R (ed) Animal Cell Biotechnology – Methods and Protocols. Humana Press, Clifton, UK

Pörtner R, Platas Barradas OBJ, Fassnacht D, Nehring D, Czermak P, Märkl H (2007) Fixed bed reactors for the cultivation of mammalian cells: design, performance and scale-up. Open Biotechnol J 1: 41–46

Pörtner R, Schäfer Th (1996) Modelling hybridoma cell growth and metabolism – A comparision of selected models and data. J Biotechnol 49: 119–135

Pörtner R, Schilling A, Lüdemann I, Märkl H (1996) High density fed-batch cultures for hybridoma cells performed with the aid of a kinetic model. Bioproc Eng 15: 117–124

Pörtner R, Schwabe JO, Frahm B (2004) Evaluation of selected control strategies for fed-batch cultures of a hybridoma cell line. Biotechnol Appl Biochem 40: 47–55

Pörtner R, Shimada K, Matsumura M, Märkl H (1995) High density culture of animal cells using macroporous cellulose carriers. In: Beuvery EC et al. (eds) Animal Cell Technology: Developments Towards the 21st Century, pp 835–839. Kluwer, The Netherlands

Qi HN, Goudar CT, Michaelis JD, Henzler HJ, Jovanovic GN, Konstantinov KB (2003) Experimental and theoretical analysis of tubular membrane aeration for mammalian cell bioreactors. Biotechnol Prog 19: 1183–1189

Quicker G, Schumpe A, König B, Deckwer WD (1981) Comparison of measured and calculated oxygen solubilities in fermentation media. Biotechnol Bioeng 23: 635–650

Rampe M (2006) Scale-up und Weiterentwicklung eines effizienten Membranbegasungssystems für Bioreaktoren. Thesis, University of Applied Sciences Giessen

Riley M (2006) Instrumentation and process control. In: Ozturk SS, Hu WS (eds) Cell Culture Technology for Pharmaceutical and Cell-Based Therapies, p 276. CRC Press, Boca Raton

Ruffieux PA, von Stockar U, Marison IW (1998) Measurement of volumetric (OUR) and determination of specific (qO_2) oxygen uptake rates in animal cell cultures. J Biotechnol 63: 85–95

Ryszcuk A, Emborg C (1997) Evaluation of mammalian fed-batch cultivations by two different models. Bioproc Eng 16: 185–191

Schneider M, Reymond F, Marison IW, von Stockar U (1995) Bubble-free oxygenation by means of hydrophobic porous membranes. Enzyme Microb Technol 17: 839–847

Schumpe A, Adler I, Deckwer WD (1978) Solubility of oxygen in electrolyte solutions. Biotechnol Bioeng 20: 145–150

Schwabe JO, Pörtner R, Märkl H (1999) Improving an on-line feeding strategy for fed-batch cultures of hybridoma cells by dialysis and "nutrient-split' -feeding. Bioproc Eng 20: 475–484

Shiragami N (1997) Effect of shear rate on hybridoma cell metabolism. Enz Microb Technol 13: 913–919

Shiragami N, Unno H (1994) Effect of shear stress on activity of cellular enzyme in animal cell. Bioproc Eng 10: 43–45

Shuler ML, Kargi F (2002) Bioprocess Engineering – Basic Concepts. Prentice-Hall PTR, Englewood Cliffs, NJ

Siano SA, Muthrasan R (1991) NADH Fluorescence and oxygen uptake responses of hybridoma cultures to substrate pulse and step changes. Biotechnol Bioeng 37: 141–159

Sielaff TD, Hu MY, Amiot B, Rollins MD, Rao S, McGuire B, Bloomer JR, Hu WS, Cerra FB (1995) Gel-entrapment bioartificial liver therapy in galactosamine hepatitis. J Surg Res 59: 179–184

Singh V (1999) Disposable bioreactor for cell culture using wave-induced agitation. Cytotechnol 30: 149–158

Sinskey AJ, Fleishaker RJ, Tyo MA, Giard DJ, Wang DIC (1981) Production of cell-derived products: virus and interferon. Ann NY Acad Sci 369: 47–59

Smith CG, Greenfield PF, Randerson DH (1987) A technique for determining the shear sensitivity of mammalian cells in suspension culture. Biotechnol Lett 1: 39–44

Spraque EA, Steinbach BL, Nerem RM, Schwartz CJ (1987) Influence of a laminar steady-state fluid-imposed wall shear stress on binding, internalization, and degradation of low-density lipoproteins by cultured arterial endothelium. Circulation 76: 648–656

Stathopoulos NA, Hellums JD (1985) Shear stress effects on human embyonic kidney cells in vitro. Biotechnol Bioeng 27: 1021–1026

Stevens J (1994) Entwicklung eines neuen Verfahrens zur kontinuierlichen Fermentation von Tierzellen. Dissertation Universität Dortmund

Storhas W (1994) Bioreaktoren und periphere Einrichtungen. Vieweg

Strathmann H (1979) Trennung von molekularen Mischungen mit Hilfe synthetischer Membranen. Dr. Dietrich Steinkopf Verlag, Darmstadt

Tan WS, Dai GC, Chen YL (1994) Quantitative investigation of cell-bubble interactions using foam fractionation technique. Cytotechnol 15: 321–328

Taylor GI (1934) The formation of emulsions in definable fields of flow. Proc Roy Soc 146:501–525

Thiele EW (1939) Relation between catalytic activity and size of particle. Ind Eng Chem 31: 916–920

Tramper J, de Gooijer CD, Vlak JM (1993) Scale-up considerations and bioreactor development for animal cell cultivation. In: Goosen MFA, Daugulis AJ, Falkner P (eds): Insect Cell Culture Engineering, pp 139–177. Marcel Dekker Inc., New York

Tramper J, Smit D, Straatman J, Vlak JM (1987) Bubble column design for growth of fragile insect cell. Bioprocess Eng 2: 37–41

Tramper J, Vlak JM (1988) Bioreactor design for growth of shear-sensitive mammalian and insect cells. Adv Biotechnol Processes 7: 199–228.

Tramper J, Williams JB, Joustra D, Vlak JM (1986) Shear sensitivity of insect cells in suspension. Enz Microb Technol 8: 33–36

Truesdale GA, Downing AL, Lowden GF (1955) The solubility of oxygen in pure water and sea water. J Appl Chem 5: 53–62

van der Velden-de Groot CAM (1995) Microcarrier technology, present status and perspective. Cytotechnol 18: 51–56

van Lier FL, van Duijnhoven GC, de Vaan MM, Vlak JM, Tramper J (1994) Continuous beta-galactosidase production in insect cells with a p10 gene based baculovirus vector in a two-stage bioreactor system. Biotechnol Prog 10: 60–64

van Wezel AL (1967) Growth of cell strains and primary cells on microcarriers in homogenous culture. Nature 216: 64–65

Varley J, Birch J (1999) Reactor design for large scale suspension animal cell culture. Cytotechnol 29: 177–205

Voisard D, Meuwly F, Ruffieux PA, Baer G, Kadouri A (2003) Potential of cell retention techniques for large-scale high-density perfusion culture of suspended mammalian cells. Biotechnol Bioeng 82: 751–765

Werner RG, Walz F, Noé W, Konrad A (1992) Safety and economic aspects of continuous mammalian cell culture. J Biotechnol 22: 51–68

Wlashin KF, Hu WS (2006) Fedbatch Culture and Dynamic Nutrient Feeding. Adv. Biochem. Engin/Biotechnol 101:43–74

Wittmann C, Kim HM, John G, Heinzle E (2003) Characterization and application of an optical sensor for quantification of dissolved O2 in shake flasks. Biotechnol Lett 25: 377–380

Wolfbeis OS (2000) Fiber optic chemical sensors and biosensors. Anal Chem 72: 81R–89R

Wolfbeis OS (ed) (1991) Fiber Optic Chemical Sensors and Biosensors, Vol 1&2. CRC, Boca Raton

Wrotnowski C (2000) Cell culture now a drug discovery bottleneck. Gen Eng N 20: 15

Wu J (1995) Mechanisms of animal cell damage associated with gas bubbles and cell protection by medium additives. J Biotechnol 43: 81–94

Wu J, Goosen, Mattheus FA (1995) Evaluation of the killing volume of gas bubbles in sparged animal cell culture bioreactors. Enzyme Microb Technol 17: 1036–1042

Yoshida H, Mizutani S, Ikenaga H (1993) Production of monoclonal antibodies with a radial-flow bioreactor. In: Kaminogawa S et al. (eds) Animal Cell Technology: Basic and Applied Aspects, pp 347–353. Kluwer, The Netherlands

Zhang S, Handa-Corrigan A, Spier RE (1992a) Foaming and media surfactant effects on the cultivation of animal cells in stirred and sparged bioreators. Biotechnol 25: 289–306

Zhang Z, Al-Rubeai M, Thomas CR (1992b) Mechanical properties of hybridoma cells in batch culture. Biotechnol Lett 14, 11–16

Zhang S, Handa-Corrigan A, Spier RE (1993) A comparison of oxygenation methods for high density perfusion cultures of animal cells. Biotechnol Bioeng 41: 685–692

Zhou W, Hu WS (1994) On-line characterization of a hybridoma cell culture process. Biotechnol Bioeng 44: 170–177

Zhou W, Rehm J, Hu WS (1995) High viable cell concentration fed-batch cultures of hybridoma cells through on-line nutrient feeding. Biotechnol Bioeng 46: 579–587

Complementary Reading

Butler M (2004) Animal Cell Culture and Technology – The Basics. Oxford University Press, New York

Chisti Y (2000) Animal-cell damage in sparged bioreactors. Trends Biotechnol 18: 420–432

Chisti Y (2001) Hydrodynamic damage to animal cells. Crit Rev Biotechnol 21: 67–110

Chmiel H (ed) (2006) Bioprozesstechnik, Elsevier – Spektrum Akademischer Verlag

Doran P (2006) Bioprocess Engineering Principles. Academic Press, London

Doyle A, Griffith JB (1998) Cell and Tissue Culture: Laboratory Procedures in Biotechnology, p 192. Wiley, Chichester

Griffith B (2000) Scaling up of animal cell cultures. In: Masters JRW (ed) Animal Cell Culture. 3rd ed, p 19–66. Oxford University Press, New York

Henzler HJ, Kauling OJ (1993) Oxygenation of cell cultures. Bioproc Eng 9: 61–75

Krahe M (2003) Biochemical Engineering. Reprint from Ullmann's Encyclopedia of Industrial Chemistry; Vol 6. VCH Publishers, Weinheim, Germany

Lundgren B, Blüml G (1998) Microcarriers in cell culture production. In: G. Subramanian: Bioseparation and Bioprocessing – A Handbook. Wiley-VCH, New York

Marks DM (2003) Equipment design considerations for large scale cell culture. Cytotechnol 42: 21–33

Ozturk SS, Hu WS (eds) (2006) Cell Culture Technology For Pharmaceutical and Cell-Based Therapies. Taylor & Francis Group, New York

Pörtner R (1998) Reaktionstechnik der Kultur tierischer Zellen, Shaker, Aachen
Pörtner R (ed) (2007) Animal Cell Biotechnol – Methods and Protocols. Humana Press, Clifton, UK
Pörtner R, Schäfer Th (1996) Modelling hybridoma cell growth and metabolism – A comparision of selected models and data. J Biotechnol 49: 119–135
Shuler ML, Kargi F (2002) Bioprocess Engineering – Basic Concepts. Prentice Hall PTR, Englewood Cliffs, NJ
van der Velden-de Groot CAM (1995) Microcarrier technology, present status and perspective. Cytotechnol 18: 51–56
Varley J, Birch J (1999) Reactor design for large scale suspension animal cell culture. Cytotechnol 29: 177–205

Chapter 5
Bioreactor Design and Scale-Up

G. Catapano, P. Czermak, R. Eibl, D. Eibl, and R. Pörtner

Abstract Design and selection of cell culture bioreactors are affected by cell-specific demands, engineering aspects, as well as economic and regulatory considerations. Mainly, special demands such as gentle agitation and aeration without cell damage, a well controlled environment, low levels of toxic metabolites, high cell and product concentrations, optimized medium utilization, surface for adherent cells, and scalability have to be considered. This chapter comprises engineering aspects of bioreactor systems (design, operation, scale-up) developed or adapted for cultivation of mammalian cells, such as bioreactors for suspension culture (stirred-tank reactors, bubble columns, and air-lift reactors), fixed bed and fluidized bed reactors, hollow fiber and membrane reactors, and, finally, disposable bioreactors. Aspects relevant for selection of bioreactors are discussed. Finally, an example is given of how to grow mammalian suspension cells from cryopreserved vials to laboratory and pilot scale.

5.1 Bioreactor Design

Design and selection of cell culture bioreactors are affected by cell-specific demands, engineering aspects, as well as economic and regulatory considerations. Mainly special demands, such as gentle agitation and aeration without cell damage,

G. Catapano
Department of Chemical Engineering and Materials, University of Calabria, I-87030 Rende (CS), Italy

P. Czermak
Institute of Biopharmaceutical Technology, University of Applied Sciences Giessen-Friedberg, Giessen, Germany and Department of Chemical Engineering, Kansas State University, Durland Hall 105, KS 66506-5102, Manhattan, USA

R. Eibl, D. Eibl
Institute of Biotechnology, Zurich University of Applied Sciences, Department of Life Sciences and Facility Management, Wädenswil, Switzerland

R. Pörtner
Hamburg University of Technology (TUHH), Institute of Bioprocess and Biosystems Engineering, Denickestr. 15, D-21073 Hamburg, Germany
poertner@tuhh.de

R. Eibl et al., *Cell and Tissue Reaction Engineering: Principles and Practice,*
© Springer-Verlag Berlin Heidelberg 2009

a well-controlled environment with respect to pH, temperature, DO, dissolved-CO_2 concentration, etc., low levels of toxic metabolites (ammonia, lactate), high cell and product concentrations, optimized medium utilization, surface for adherent cells, and scalability, have to be considered (Schlegel et al. 2007). During the last few years a huge number of different bioreactor concepts have been developed. As the main problem for designing a scalable bioreactor concept, the high shear sensitivity of the cells have to be considered (compare Sect. 4.1). For example, stirred-tank bioreactors can be operated at low impeller speed and low aeration rate only, resulting in low mass transfer coefficients insufficient for optimal supply of the cells at high cell densities (compare Sect. 4.2).

The ability of cell growth at very high tissue-like cell densities ($>10^8$ cells mL^{-1}) was taken advantage of during development of cell culture bioreactors. Therefore, two strategies have been followed for increase of process intensity: (1) Volumetric scale-up of lab-scale and pilot-scale bioreactors at similar process intensity (cells/volume). Here low-density or homogeneous systems (stirred-tank, bubble column, air-lift reactor) are applied, where cells are cultivated either in suspension or, if required, on microcarriers. (2) Increase of process intensity (cells/volume).

To this category belong high-density or heterogeneous systems (fixed bed, fluidized bed, hollow fibre reactor), where cells are immobilized either in macroporous support materials or within a compartment created by membranes. Considering a maximal theoretical cell density of 5×10^8 cells mL^{-1}, a 1-L high-density culture would be equivalent to a 50 L suspension culture at 1×10^7 cells mL^{-1}. Selection of a reactor system depends to a large extent on the specific purpose such as production of a certain amount of protein (small amounts for basic scientific studies up to kilogram quantities for medical application), 3D tissue cultivation, or single- or multipurpose facility. Therefore, to date no "universal" bioreactor system has been found.

Suspension reactors (compare Sect. 5.1.1) are characterized by advantages such as the use of conventional reactor systems, availability of know-how on design and sterile operation, good mass transfer, homogeneous mixing, the possibility of sampling and determination of cell concentration, and high scale-up potential. The disadvantages are difficult oxygen supply at high cell densities, cell damage by shear and/or foam (bubble aeration), relatively high demand for control (temperature, oxygen, pH, flow rates), and the requirment of cell retention for high cell densities.

Among high-density systems, we find fixed bed and fluidized bed reactors as well as hollow fiber reactors (compare Sects. 5.1.2 and 5.1.3). Fixed bed and fluidized bed reactors have the following advantages compared to suspension reactors: high volume-specific cell density (approx. 5×10^7 cells mL^{-1} reactor) and productivity, low shear rates, simple medium exchange and cell/product separation, productivity on a high level during long-term cultures. Their disadvantages are non-homogeneous cell distribution and difficulty in the determination of cells and cell harvest.

Within hollow fiber reactors, cells are immobilized in the extracapillary space of a hollow fiber bundle. Their advantages are very high cell densities, concentrated product, lower serum demands (if required), long-term stability, easy handling, and low cost. The disadvantages are limited scale-up, mass transfer problems/concentration gradients, difficulty in the determination of cell number, proteolytic activity on the product.

During the past years a growing number of disposable or single-use bioreactor systems have appeared on the market besides hollow fiber reactors, addressing especially problems related to early process development, such as flexibility, cost effectiveness, time to market, as well as quality and regulatory issues. These systems are mostly based on bag technology. The bags with volumes up to several hundred liters are placed on tilting or rocking devices for mixing and supplying oxygen. Advantages of single-use bioreactors are reduced cleaning procedure, lack of validation issues (cleaning, sterilization, etc.), lower investment costs, easier adaptation to changing process demands, and less contamination risks.

The main characteristics of the culture systems discussed above are summarized in Table 5.1. All these systems support growth of mammalian cells in one or the other. On a lab scale, mostly disposable flask, membrane or bag systems are the methods of choice. Small suspension reactors, as well as fixed bed and fluidized bed reactors, are mostly used for research or process development. For larger amounts of products, only suspension reactors or, up to a certain scale fixed bed or fluidized bed reactors, have the required scalability.

A large number of bioreactor concepts developed during the past years have been reviewed by Fenge and Lüllau (2006), among others. Bioreactor design was addressed by Krahe (2003) with special focus on maintaining sterility within the

Table 5.1 Main characteristics of cell culture systems and bioreactors (Pörtner [1998], modified)

	T-flask/roller bottles/disposable bags	Membrane reactors (hollow fibre, miniPerm etc.)	Suspension (stirred tank, air-lift)	Fixed bed/fluidized bed
Cell density	1	3	2	2–3
Homogeneity	1	1	3	2
Shear stress	N	N	Y	N
Product concentration	1	3	2	2
Productivity	1	3	2	2
Medium efficiency	3	1	3	2
Continuous process (perfusion)	N	J	J[a]	J
Control	0	1	2	3
Downstream process	1	3	2	2
Steam sterilisable	N	N	J	J
Reusable	N	N	J	J

[a] Additional equipment for cell retention required
0: not possible, 1–3: increasing efficiency, Y: yes, N: no, (): partly

reactor. The following will focus on well-established bioreactors, but not so much on developments for special applications, which in many cases are not commercially available. Special emphasis will be given to dynamic system following the category introduced in Chapter 3. Small-scale, static culture systems are not within the focus of this chapter. The description will consider basic functional principles, questions related to design engineering, operation and scale-up advantages, and limitations. Furthermore, selection criteria based on technical, economic, and regulatory conditions are addressed.

5.1.1　Bioreactors for Suspended Cells

5.1.1.1　Stirred Tank Reactors

Design of stirred-tank bioreactors has to deal mainly with cell damage due to shear stress caused by agitation and/or aeration. Stirred-tank reactors, which are the most common type of reactor and are nowadays build up to 20,000 L scale, are especially suited for suspendable cells. Adherent cells can be grown on microcarriers and, by this, handled similar to a suspension culture (Sect. 4.3.1.2). The principal configuration of a stirred, aerated bioreactor is shown in Fig. 5.1. A cylindrical vessel is equipped with one or more impellers mounted on a rotating shaft, which is activated by an external motor. Close to the bottom, a sparger (if applied, compare Sect. 4.2) is located for supply of gas bubbles. Furthermore, up to four buffles can be installed to prevent vortexing and to improve mixing. Buffles are common for microbial

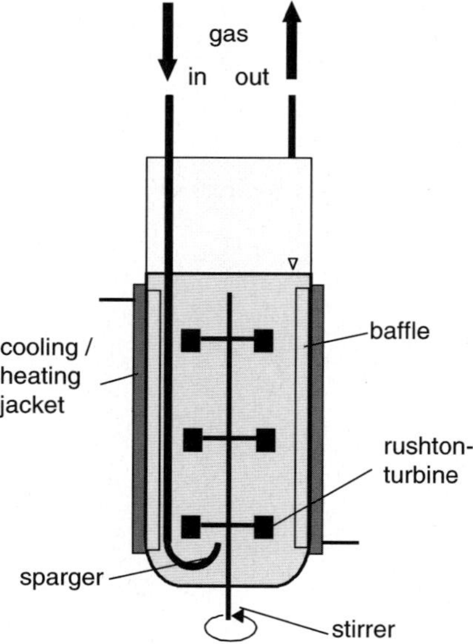

Fig. 5.1　Schematic setup of a bioreactor

fermentation operated at high impeller speed. In cell-culture reactors buffles are of minor importance, as these reactors are operated at a lower impeller speed, and are partly omitted as they may cause additional cell damage. In cell-culture reactors often a curved bottom is used instead of a flat one as for microbial fermentation to improve homogeneity in mixing.

Heating and cooling of the bioreactor can be accomplished by different means depending on the size of the reactor. Very small devices (e.g., spinner flasks) are placed in an incubator at 37 °C (5% CO_2-containing air). Small-scale reactors used outside an incubator can be equipped with an electrical heater immersed in a brooth, a heating jacket fixed to the outer vessel wall (here mostly made of glass), or a system for pumping a preheated liquid (usually tap water circulating in a loop) through a double jacket as shown in Fig. 5.1. The latter is the most common version for larger reactors (10–3500 L) as well (Krahe 2003). Very large reactors can be equipped with internal heating or cooling coils to inprove heat transfer.

Bioreactors are typically filled up to 70–80% only to provide sufficient head space for sedimentation of air drops or for removal of foam. Foam formation can be reduced by applying an antifoam agent or by means of a mechanical foam breaker (not common for cell-culture reactors).

In the laboratory scale, up to volumes of approximately 15 L, the vessel is mostly made of glass, as glass is transparent, easy to clean, inert, and steam-sterilizable. Examples for bioreactors made of glass are shown in Figs. 5.2–5.5. Glass is especially

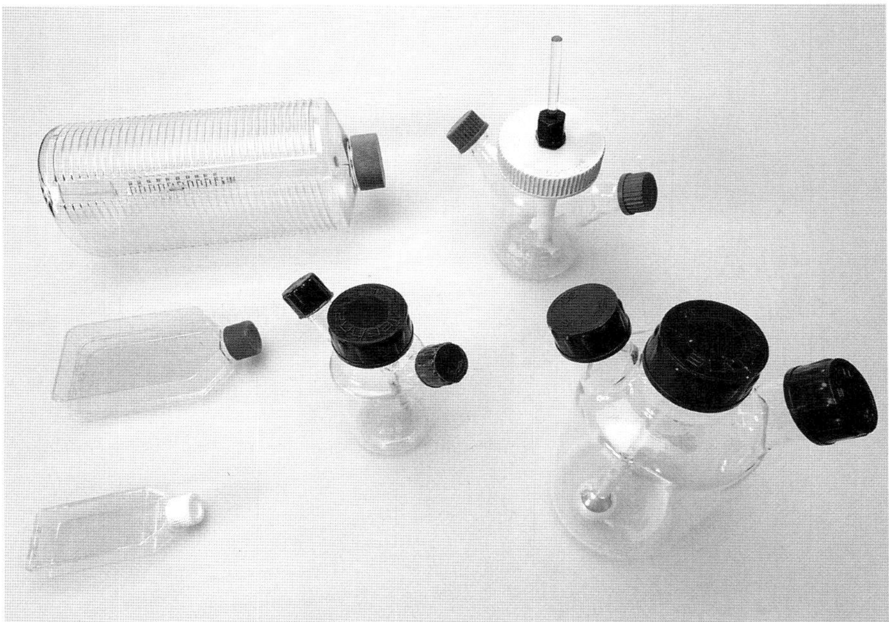

Fig. 5.2 Cultivation systems (flasks, roller bottles and spinner flasks) for use in incubator or warm room

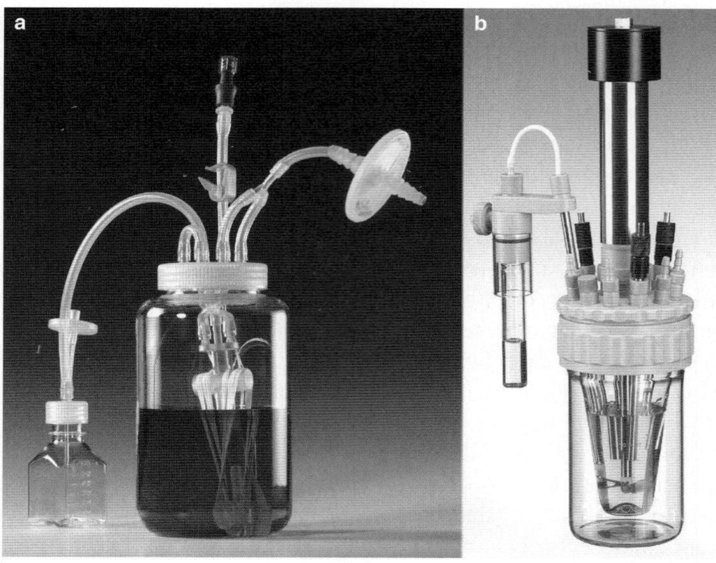

Fig. 5.3 Special bioreactors for small-scale and process development. (**a**) Superspinner, Membranrector with tumbling membrane basket for bubble free aeration, Sartorius AG, Prof. Lehmann, University of Bielefeld (courtesy of Sartorius AG, Germany); (**b**) conical reactor (courtesy of medorex, Germany)

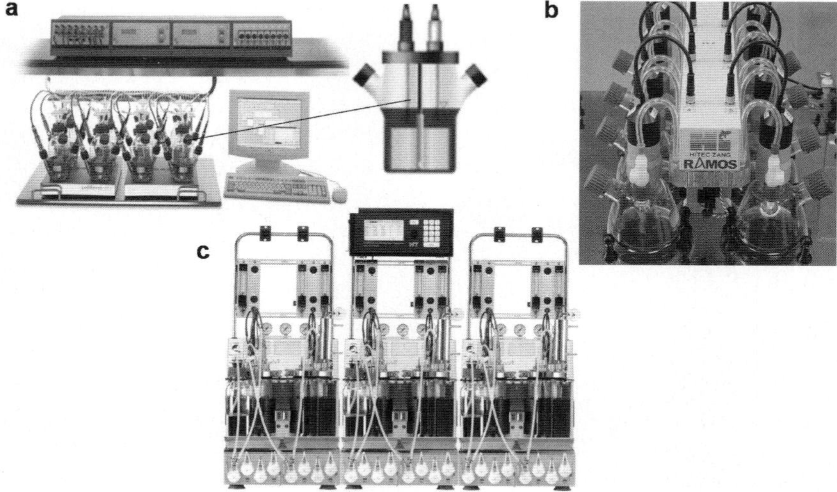

Fig. 5.4 Parallel bioreactors for process development. (**a**) cellferm-pro (courtesy of DASGIP AG, Germany), (**b**) RAMOS (courtesy of Hitec Zang GmbH, Germany), (**c**) Multifors (courtesy of Infors AG, Switzerland)

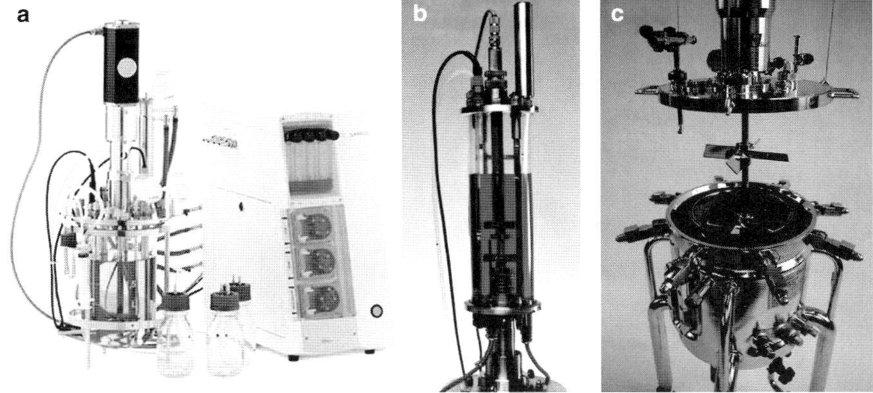

Fig. 5.5 Typical bioreactors for lab-scale use. (**a**) Cell culture reactor for autoclaving, top driven (Biostat Aplus, courtesy of Sartorius AG, Germany). (**b**) Visual safety reactor for in situ sterilisation, bottom driven (Bioengineering AG). (**c**) Stainless steal reactor for in situ sterilization, top driven (courtesy of Zeta AG, Switzerland)

preferred for bioreactors sterilized by autoclaving. Problems can arise if glass vessels are sterilized *in situ* at 121 °C and 2 bar absolute pressure, as the glass vessel might burst. Risk of injury for laboratory personnel can be partly reduced by providing a metallic safety jacket around the vessel. Alternatively, a visual safety bioreactor equipped with a special polyamide foil supported by a stainless steel base and lid was developed by Märkl (1989). Metallic shells are used to provide support to the foil during sterilisation, which are removed afterwards (compare Fig. 5.5b).

At the lab-scale reactor top, connections and impellers are usually made of austenitic steel or in some cases out of polyetheretherketone (PEEK). For reactors in pilot- or production scale the vessel is also made of austenitic steel, commonly 316 or 316 L steel (Fig. 5.5c). The expensive 316 L is preferred for good manufacturing practice (GMP) production owing to the excellent electropolishing and welding characteristics. For production processes under GMP, an electropolished surface with arithmetic mean roughness $R_a < 0.4\,\mu m$ is recommended.

The ratio of vessel height H_R to diameter D_R can vary between 1 and 3 (Doran 2006; Krahe 2003). For small-scale and lab-scale reactors, often $H_R/D_R = 1$ is preferred. At larger-scale, H_R/D_R up to 3 is common to improve retention time of gas bubbles in the reactor and to enlarge the heat exchange capacity of the vessel wall.

Appropriate mixing of the culture broth is essential to prevent inhomogeneous distribution of cells, nutrients, temperature, pH, etc. For cell culture a number of impellers have been suggested. Some common types are shown in Fig. 5.6. The rushton turbine, often used in microbial fermentation because of its excellent mixing characteristics, is regarded as less suitable for cell culture, as it requires rather high impeller speed for sufficient mixing, thereby damaging shear-sensitive cells. Alternatively, larger impellers (diameter ≥ 0.5 of vessel diameter) with axial-flow characteristics are applied (marine, paddle, or segmented impeller). As larger impellers can be operated at lower impeller speeds to provide appropriate mixing, cell

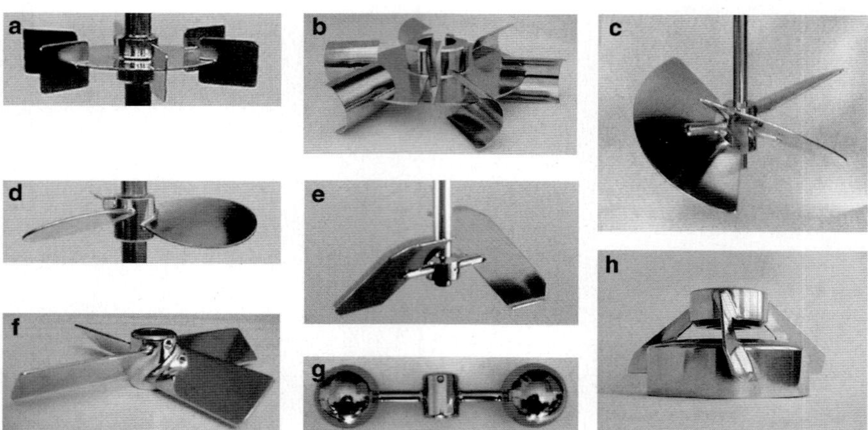

Fig. 5.6 Impeller types applied in stirred bioreactors for cell culture (courtesy of Zeta AG, Switzerland). (**a**) Rushton turbine, (**b**) hollow blade, (**c**) three blade pitched, (**d**) marine impeller, (**e**) two blade pitched, (**f**) skew blade, (**g**) sphere, (**h**) bio-m impeller

damage is significantly reduced. Data from Nagata (1975) indicated that impellers with retreated blades, large width, pitched blades, and small diameter compared to the vessel diameter are superior in producing better bulk mixing and less shear forces. Nienow (1990) draw similar conclusions on the basis of a correlation between mixing time and average energy dissipation. For bioreactors with $H_R/D_R = 1$ and equipped with a single impeller, the mixing time does not depend on the type of impeller, but on the mean energy dissipation and the impeller and vessel diameter. Therefore, impellers having a larger diameter can be operated at lower impeller speed, resulting in lower mean energy dissipation and reduced cell damage, without any impact on mixing efficiency. For vessels with high H_R/D_R ratio, multiple impellers are often combined to achieve a homogeneous mixing throughout the bioreactor at the lowest possible impeller speed. By this means, lower energy dissipation is required to achieve appropriate mixing. But around the individual impellers several compartments of flow can be formed, resulting in reduced interstage mass transfer rate. This can be overcome by increasing the stirrer speed, decreasing the distance between impellers, and applying a rushton turbine as the bottom impeller for bubble dispersion and an axial flow impeller (marine, pitched blade) as the upper impeller (Breman et al. 1996; Mishra and Joshi 1994). Details regarding choice of impeller, scale-up, and further engineering aspects are given by Ma et al. (2006) and Nienow (1990).

The rotating shaft can be driven from the bottom or from the top. The bottom drive shaft is characterized by a short length, increased space on the top head of the reactor for transfer connections and sensors, and improved lubrication of the mechanical seals by the medium/broth. By applying a top drive, abrasive media (not relevant for cell culture) can be used without damaging the bearing seal, and media do not leak in case of failure of the mechanical seal (compare Krahe 2003).

The sealing of the rotating shaft within the closed, steril vessel against the environment can be seen as the most critical part of the drive system (Krahe 2003;

Chmiel 2006). Mechanical seals, mostly double mechanical seals, are usually located outside the reactor on the shaft, and seal the rotating shaft against the static vessel. They are easy to clean but bear the risk of medium leakage, and therefore require intensive maintenance. These drawbacks can be overcome by magnetic coupling, were the torque is transmitted through the reactor wall with magnets. Here the disadvantage is the complex construction of the parts within the reactor (e.g., bushing-type bearings), which are difficult to clean. Magnetic coupling is quite common for autoclavable bioreactors. Meanwhile, even larger reactors can be equipped with this technique.

For continuous operation of stirred-tank reactors in the perfusion mode, several techniques for cell and/or product retention are in use (e.g., microfiltration, ultra-filtration, sedimentation, centrifugation, spin filters, acoustic filter, and dialysis for cell and product enrichment). These are discussed in detail in Sect. 5.1.1.3.

Examples for stirred bioreactors on different scales are introduced in the following. Small-scale spinner flasks (approximately 100 mL up to 5 L, Fig. 5.2) equipped with conical pendulum or paddle-type magnetic impellers are intended for use in a humidified CO_2 incubator or, for larger volumes, in warm rooms. They are available in glass or in plastics for single use. The flask is equipped also with ports for the inoculum, medium exchange, or sampling. The impeller speed is kept low (approximately 30–100 rpm). Oxygen is supplied via the head space through slightly opened caps. This simple concept allows for easy handling, but only low oxygen transfer rates ($0.1–4 h^{-1}$) (Spier and Griffiths 1984). As an alternative concept with improved oxygen transfer, Heidemann et al. (1994) developed the "Superspinner" (Fig. 5.3a), a standard laboratory flask equipped with a microporous, hydrophobic polypropylene membrane tube. The coiled membrane tube is mounted on a pendulum impeller, driven by a magnetic impeller, and perfused with an air/CO_2 mixture from within the incubator. A further concept with improved oxygen transfer, a conical flask equipped with a marine-type impeller, is shown in Fig. 5.3b. For screening purposes and for process development several configurations consisting of multiple bioreactors equipped further with monitoring and control units are in use (Fig. 5.4).

Examples for lab-scale bioreactors up to approximately 10 L, pilot scale up to approximately 500 L and production scale up to $20 m^3$ are shown in Figs. 5.5, 5.7, and 5.8,. The bioreactor itself has to be equipped with further devices for medium supply and harvest; buffer for pH-control; gas mixing station for supply of air, oxygen, and CO_2; sterile filter within the tube for aeration; waste air cooler, etc. An example of the setup of a lab-scale bioreactor system configured for fed-batch culture and connected to a process control system is shown in Fig. 5.9,. A practical approach for setup and instrumentation of a bioreactor system can be found in Pörtner (2007).

Appropriate design and layout of a bioreactor system for production scale is a key issue for successful production of biopharmaceuticals. Usually, a new cultivation process is investigated and optimized on a laboratory scale. These data are used to scale-up the process to the final scale. Factors that need to be considered include the following:

- How big does the vessel need to be to produce sufficient product;
- How are the materials in the vessel to be mixed efficiently but without damage to the cells in the vessel;

Fig. 5.7 Pilot-scale bioreactors for cell culture (courtesy of Bioengineering AG, Switzerland)

Fig. 5.8 Production-scale bioreactors for cell culture (courtesy of Zeta AG, Switzerland)

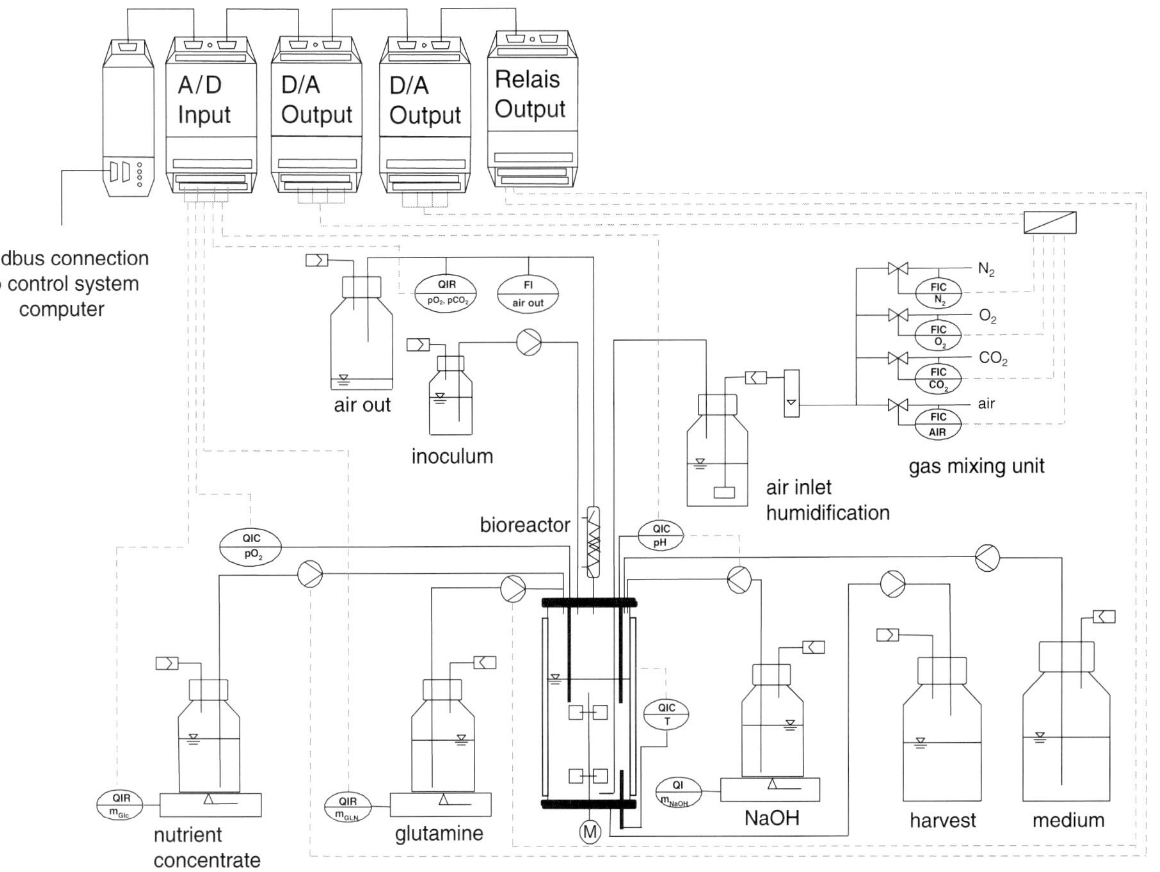

Fig. 5.9 Configuration of a lab-scale bioreactor system connected to a fieldbus control system

- What is the required aeration rate and the power input;
- What is the process time;
- How many scale-up steps are required;
- What are the costs (capital, running, depreciation) of various types of equipment, etc. (Biotechnology by open learning, 1992).

For scale-up, a wide variety of methods are in use including fundamental methods, which imply solving all the microbalances for momentum, mass, and heat transfer, semi-fundamental methods based on simplified models (e.g., plug flow, plug flow with dispersion, well mixed), rules of thumb, which are widely applied but may lead to wrong conclusions, dimensionless analysis based on dimensionless groups for parameters and the principle of similarity, or regime analysis (scale-down, use of "characteristic times"). In biochemical engineering often rules of thumb combined with extrapolation to larger scale are applied owing to the complexity of the problems involved in solving all the microbalances for momentum, mass, and heat transfer in the system.

A basic prerequisite for scaling up a bioreactor is geometric similarity. In general, two objects are said to be geometrically similar if they have the same shape. By definition, the ratio of any two linear dimensions of one object will be same for any geometrically similar object. Therefore, two bioreactors of different sizes are geometrically similar if geometrical ratios such as "vessel height to vessel diameter", "stirrer diameter to vessel diameter", "baffle width to vessel diameter", etc. are the same.

Applying rules of thumb for scale-up implies that certain criteria found to be optimal on the small scale is used for the large scale as well. Criteria applied for cell-culture reactors can be divided into two groups, those focusing primarily on mass transfer and mixing, and those considering mechanical cell damage. Consequences of the effects of cell damage on design and scale-up of cell culture bioreactors have been discussed extensively in Sect. 4.1.3. Criteria considering cell damage include impeller tip speed, mean power input per volume, and impeller Reynolds number. Tolerable tip speeds vary between 1 and $2\,\mathrm{m\ s^{-1}}$ (Fenge and Lüllau 2006). On the other hand, a tip speed of about $2\,\mathrm{m\ s^{-1}}$ is needed to ensure sufficient homogeneity (Ma et al. 2006). With respect to the mean power input per volume (mean energy dissipation rate), Chisti (2000) suggested to keep this value below $\sim 1000\,\mathrm{W\ m^{-3}}$. Further concepts such as the "integrated shear factor" (ISF) or the Kolmogorov eddy-length model have been discussed extensively in Sect. 4.1.3. Their use in practical design of cell-culture bioreactors is limited.

Scale-up criteria focusing primarily on maintaining proper rates of mass transfer take into account that scale-up is often restricted to oxygen supply and carbon dioxide removal, especially at very high cell densities. On the other hand, it became apparent that many cell lines used in industrial processes are more tolerant to mechanical agitation and that shear effects are less significant on a larger scale (compare Sect. 4.1.3). With respect to sparging, the aeration rate should be below 0.1 vvm to prevent cell damage caused by bubbles and foaming problems, even though higher aeration rates would improve mass transfer by sparging. Because of this, stirred-tank reactors are typically operated in the "flooding regime" (compare Fig. 5.10a), where gas dispersion is poor. Data on mass transfer coefficients are discussed in Sect. 4.2.

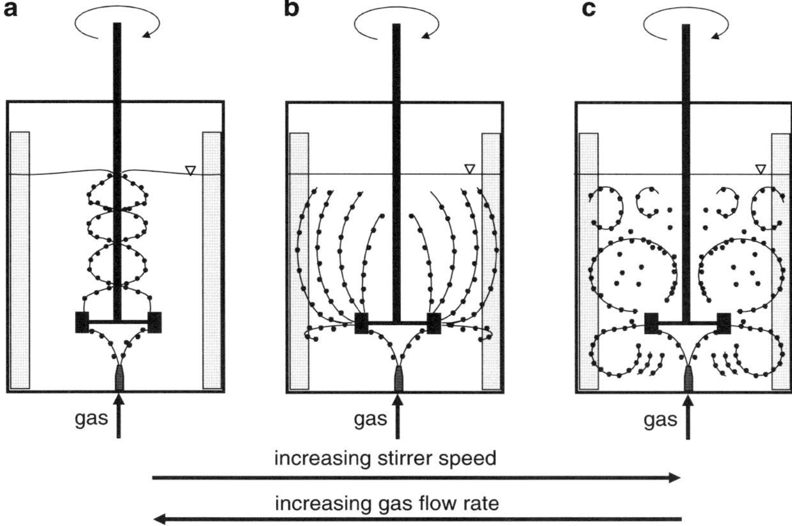

Fig. 5.10 Flow pattern in stirred, aerated bioreactors. Impact of stirrer speed and gas flow rate (from Duran 2003, modified)

Sparging and mechanical agitation determine the mixing characteristics within the bioreactor, described by the mixing time θ_{m}. Longer mixing times bear the risk of local oxygen and carbon dioxide gradients, which can be deleterious to the cells. An equation for estimating the mixing time was suggested by Nienow et al. (1996):

$$\Theta_{m} = 5.9 \left(\frac{P}{V} \right)^{-\frac{1}{3}} \left(\frac{d_{R}}{D_{R}} \right)^{-\frac{1}{3}} D_{R}^{\frac{2}{3}} \tag{5.1}$$

with the mean power input per volume P/V, stirrer diameter d_{R}, and vessel diameter D_{R}. This equation can be applied for both axial and radial flow impellers. It implies an increase in the mean power input per volume, if the mixing time is kept constant during scale-up. On the other hand, the tolerable power input per volume can be restricted owing to shear effects.

The above discussion shows that it is not possible to hold more than one of the discussed parameters (tip speed, power input, mixing time, Reynolds-number) constant at the same time during scale-up (Kossen 1994). In Table 5.2, values in every row give, for a particular scale-up criterion, the ratio of the value at $10\,m^{3}$ relative to the value at $10\,L$ (from Biotechnology by open learning – Bioreactor Design and Product Yield, 1992, modified). It becomes obvious that scale-up of bioreactors in general is not an easy task, and is "as much an art as a science" (Ma et al. 2006) and dependents a lot on experience and know-how.

Table 5.2 Ratio of the value at $10\,m^3$ relative to the value at $10\,L$; P: power, V: volume, n_R: stirrer speed, d_R: stirrer diameter, Re: stirrer Reynolds number (from: Biotechnology by open learning – Bioreactor Design and Product Yield, 1992)

Scale-up criterion	P	P/V	n_R	$n_R{}^*D$	Re
Equal P/V	10^3	1	0.22	2.15	21.5
Equal N	10^5	10^2	1	10^2	10^2
Equal tip speed	10^2	0.1	0.1	1	10
Equal Re number	0.1	10^{-4}	0.01	0.1	1

5.1.1.2 Bubble Columns and Air-Lift Reactors

Bubble columns and air-lift reactors present an alternative to stirred reactors, as aeration and mixing is achieved by gas sparging without mechanical agitation (Chmiel 2006; Duran 2006). These reactors are structurally very simple; advantages compared to stirred reactors include low capital and energy cost, lack of moving parts, and satisfactory heat and mass transfer performance, ease of scale-up, low shear characteristics, etc.

Bubble columns (Fig. 5.11a) consist of a cylindrical vessel with height much larger than the diameter, were gas is injected by a sparger located close to the bottom of the column. Bubble columns typically have no further internal structures. Hydrodynamics and mass-transfer characteristics depend on gas flow rate, sparger design, column diameter, and medium properties (e.g., viscosity). At low gas flow rates, bubbles rise with a similar upward flow velocity without backmixing in the gas phase (homogeneous flow). Liquid mixing occurs mainly in the wake of bubbling. At higher gas flow rates, the flow is regarded as heterogeneous. Bubbles and liquid tend to rise mainly in the center of the column, creating a downflow of liquid near the column wall. In this regime, flow is compartmentalized. Mixing in bubbles columns can be regarded as poor compared to stirred-vessel and air-lift reactors. Industrial applications for biotechnical processes comprise production of baker's yeast, beer, vinegar, or wastewater treatment. Examples for large-scale cultivation of mammalian cells have not been reported, mainly because of the poor mixing in bubble columns (Doran 2006; Varley and Birch 1999).

To improve mixing in aerated columns without mechanical agitation, in air-lift reactors the reactor is subdivided in two connected compartments (Fig. 5.11b–d). Gas is sparged in one part only, the "riser", creating an upflow of gas/liquid. In the second part, the "downcomer", liquid flows downward owing to the density gradient between the riser and downcomer. With respect to geometrical configuration, internal or external loop design can be distinguished. In vessels with an internal loop, riser and downcomer sections are split either by a cylindrical draft tube (Figs. 5.11b and 5.12) or by a vertical baffle (Fig. 5.11c). In external loop reactors, the riser and downcomer are two separate cylindrical columns (Fig. 5.11d). Air-lift reactors have been scaled up to $1500\,m^3$ for microbial fermentation or several tens of thousand cubic meters for wastewater treatment, demonstrating the scale-up potential of these reactors. For cultivation of mammalian cells, reactor scales up to $2000\,L$ have been reported (Varley and Birch 1999), almost solely as internal loop reactors.

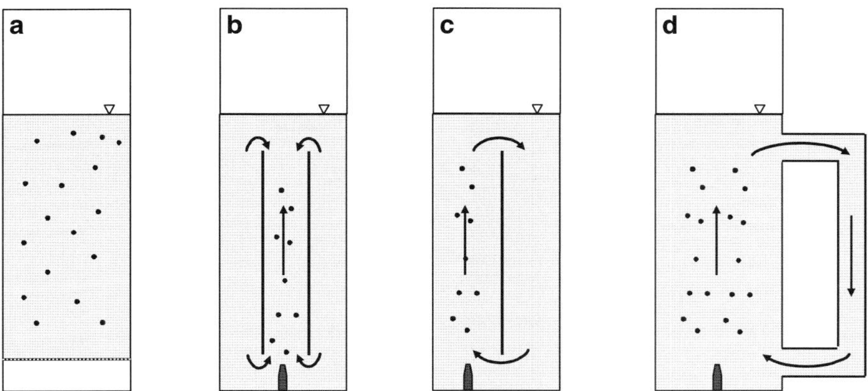

Fig. 5.11 Scheme of (**a**) bubble column, (**b**) air-lift reactor with draft tube, (**c**) air-lift fermenter with buffle for internal loop and (**d**) air-lift fermenter with external loop

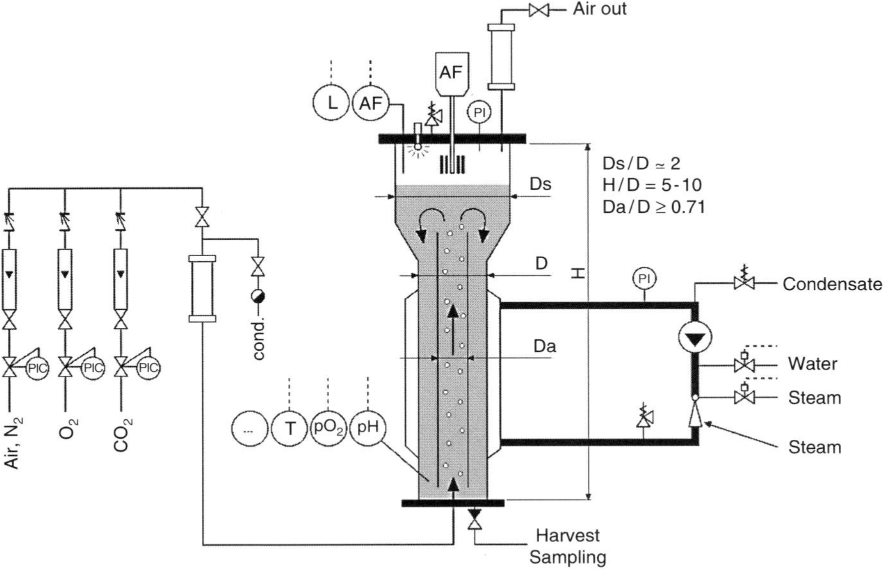

Fig. 5.12 Setup of a lab-scale air-lift reactor (courtesy of Bioengineering AG, Switzerland)

Air-lift reactors usually provide better mixing than bubble columns, and the operation range is much broader (Fig. 5.13). Hydrodynamic behavior of these reactors depend to some extent on the geometry as well as the gas flow rate, gas hold up, and circulation velocity, with the gas flow rate as the only independently controllable variable. For a review of the state-of-the-art, compare Varley and Birch (1999). Most data reported in the literature have been derived from studies using nonfermentation solutions. Reliable design criteria for microbial or even cell culture

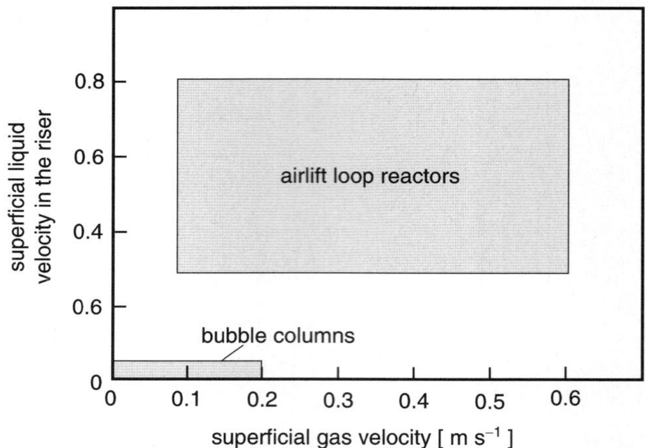

Fig. 5.13 Operating ranges of gas and liquid velocities in bubble column and air-lift reactors (from Chisti and Moo-Young 1987, modified)

in air-lift reactors are very rare. Therefore, in the following mainly qualitative design considerations are presented.

In air-lift reactors, mixing and mass transfer are generally coupled. Often, the gas (air) flow rate is set to maintain a certain dissolved oxygen concentration; this in turn determines the hydrodynamic behavior within the reactor. Furthermore, stripping of CO_2 at very high cell densities is essential. Therefore, design of air-lift reactors for cultivation of mammalian cells has to consider geometrical aspects (e.g., riser, downcomer, aspect ratio, height of liquid above and below the baffle or draft tube), sparger position and design, gas flow rate, position of electrodes (e.g., temperature, pH, DO probes), sample ports, feed and base addition, etc. None of these factors is independent, explaining to some extend the huge number of correlations for hydrodynamics in air-lift reactors. A comprehensive review of the engineering principles was provided by Varley and Birch (1999). In the following the main factors are summarized:

- Typical aspect ratios (height to diameter) are between 6:1 and 12:1;
- With respect to optimal mixing, the cross-sectional area of riser and downcomer should be similar;
- As the height of liquid above the baffle or draught tube increases, the mixing time goes through a minimum before beginning to increase. The optimum has to be determined experimentally for a certain scale;
- Recommended sparger position is just above the base of the baffle;
- The cross-sectional area just below a buffle should not exceed 1.65 times the downcomer cross-section;
- Typical gas flow rates, expressed as superficial gas velocities, are in the range of 0.001–0.01 m s^{-1} (Griffiths 1988);
- Relationships between $k_L a$ and superficial gas velocity U_g are generally of the form

$$k_L a \sim a U_g^b \tag{5.2}$$

$k_L a$ values varying between 0.7 and 20 h^{-1};

- Recommended injection point for feed or base addition is just above the gas sparger within the riser.

Scale-up of air-lift reactors for cultivation of mammalian cells is still a difficult task, despite the extensive studies on hydrodynamics and scale-up in general. This is mainly due to the fact that superficial flow velocities are in most cases well above those tolerable in air-lift reactors for mammalian cell culture owing to the high shear sensitivity of the cells. Strategies to scale-up air-lift reactors are generally based on geometric similarity, constant $k_L a$, and either constant superficial gas velocity, constant gas flow rate per unit volume of liquid, or constant mixing time (Varley and Birch 1999). Similar to stirred-tank reactors, it is not possible to keep all these parameters constant at the same time. As for large-scale mammalian cell culture, it is very important to maintain mixing times at appropriate levels. Varley and Birch (1999) recommend keeping the mixing times as constant as possible during scale-up. To decouple mixing and mass transfer, an additional gas (e.g., nitrogen) can be used. In this case, the total gas flow rate can be adjusted to maintain adequate mixing and to remove CO_2 at very high cell densities, while controlling the oxygen concentration by varying the ratio between nitrogen and oxygen in the gas. If, alternatively, the choice is between superficial gas velocity and volumetric flow rate, it is obvious that these two criteria cannot be met simultaneously.

Even though air-lift reactors are not as widespread as stirred-tank reactors, there is an increasing number of successful use of air-lift reactors for mammalian and insect cells (see reviews by Varley and Birch 1999; Fenge and Lüllau 2006). This may be due, to some extent, to historical reasons, as at times when most cell culture media contained serum, the intensive aeration required to operate air-lift reactors resulted in extensive foaming and cell death. This drawback is less profound when serum- or protein-free media are used, which is now state-of-the art. Furthermore, air-lift reactors are regarded as less flexible in terms of working volume (e.g., for fed-batch culture) or when anchorage-dependent cells have to be cultivated on microcarriers. On the other hand, they are supposed to mix the culture broth more gentle and therefore are more suited for shear-sensitive cells. With respect to high cell density perfusion culture, air-lift reactors can be equipped with external retention devices similar to stirred-tank reactors as has been shown by several authors (see review by Fenge and Lüllau 2006). Therefore, it can be concluded that air-lift reactors are a powerful alternative to stirred-tank reactors on a large scale, provided that sufficient scale-up strategies are available.

5.1.1.3 Techniques for Cell and Product Retention in Suspension Culture

Even though for suspension reactors batch, repeated batch and fed-batch are still the preferred modes of operation on an industrial scale, a number of continuous perfusion techniques have been successfully applied to increase cell and product concentration as well as time–space yield of the process. An overview on industrial application of perfusion cultures for production of monoclonal antibodies and recombinant proteins is given by Kompala and Ozturk (2006). The main goal of perfusion culture is the long-term continuous cultivation up to several months at very high cell density and high viability, preferably under physiological steady-state conditions of the cells. Owing to high volumetric perfusion rates (1–10 volumes per reactor volume per day), the residence time of the product is low, an advantages for fragile, glycosylated proteins.

Basically, perfusion refers to the continuous exchange of spent medium from the culture broth while retaining the cells in the bioreactor. The setup of a bioreactor systems for perfusion culture is shown in Fig. 5.14 and consists of a feed tank, a bioreactor (usually a stirred tank), an internal or external device for cell retention, a harvest tank for the spent medium, and several pumps for the medium supply and removal as well as for medium circulation in case of an external cell retention device. External devices allow an easy exchange of the device upon failure and are advantageous with respect to scale-up, as in some cases this can be done by applying parallel units. The main drawback of these techniques is the need to use pumps for recycling the cells, which might cause mechanical damage to the cells (especially for fragile cells, cell aggregates, or microcarrier cultures) and might hold the risk for contamination. Furthermore, control of important process parameters, such as DO, pH, and temperature, is more difficult.

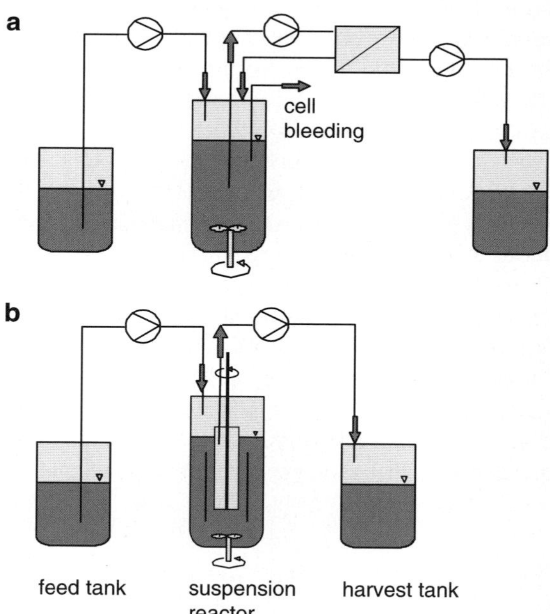

Fig. 5.14 Scheme of perfusion culture for in suspension reactors. (**a**) External cell retention and recirculation of cells back to the bioreactor, (**b**) internal cell retention (e.g., spin filter)

The numerous cell retention strategies suggested so far have been reviewed extensively (see Voisard et al. 2003; Fenge and Lüllau 2006). Here, only those having a technological or commercial importance will be discussed. The main problems in the design of cell retention devices are the size of suspension cells ($10–30\,\mu m$), the low density difference between the cells and the culture medium (approximately 5%), and the high shear sensitivity of the cells, which might lead to cell death or even cell disruption while circulation through the retention loop (Lüllau and Fenge 2006). Basically, almost all devices applied for cell retention rely on the particle size and/or the density as the essential separation characteristic of the particles (cells). Other physical or chemical characteristics of particles (electrical charge, dielectric constant, or surface properties) are of minor importance (Voisard et al. 2003).

Retention by Size

Retention of cells depending on their size can be performed by means of filtration or dialysis. Filtration refers to mechanical separation of solid particles from a fluid by a physical barrier; dialysis is based on diffusion of solutes across a semipermeable membrane. Owing to their greater importance in cell culture technology, filtration techniques will be discussed first. In this respect, basically separation by various types of membrane filtration principles and spin filters has to be addressed. The major drawback of all filtration devices is their tendency to foul and clog. All suggested design concepts attempt to avoid or delay these phenomena in one way or the other. Owing to the intended very long run times of cell culture processes, solving this problem is essential and decides its acceptance for industrial purposes.

For membrane filtration (Fig. 5.15a) of cells, usually microfiltration membranes having a pore size range of $0.1–10\,\mu m$ made of cellulose ester, nylon, poly(ethylensulfone), polypropylene, or poly(vinyl difluoride) (PVDF) are used. The degree of cell retention is high, but fouling and clogging of the membrane due to the content of cell debris and macromolecules in the culture supernatant, i.e. proteins contained in the medium (serum) or produced by the cells and released DNA, may cause problems in lengthy operation (Belfort 1989). As shown in Fig. 5.16, the flux through membrane is significantly reduced. A technique widely applied in particle separation is *cross-flow filtration*, where the particle suspension is forced to flow tangentially to the membrane (mostly flat-sheet membranes) to reduce deposition of particles or macromolecules in pores or on the membrane surface. Despite many efforts to adapt this technology to perfusion culture of mammalian cells (including modul design, e.g., spiral filters or controlled-shear filters, or application of backflushing or electrofiltration, reviewed by Voisard et al. 2003), mostly small-scale applications have been reported. Alternatively *hollow fiber modules* (diameter of hollow membrane fiber approximately $0.5–1\,mm$) have been used, both with internal and external devices, where fouling is minimized by applying high liquid flow rates at the membrane surface. This technology has gained some commercial importance and modules in the range of $3\,m^2$ of membrane area are available intended for bioreactors ranging from 90 to $250\,L$ (cited by Voisard et al. 2003).

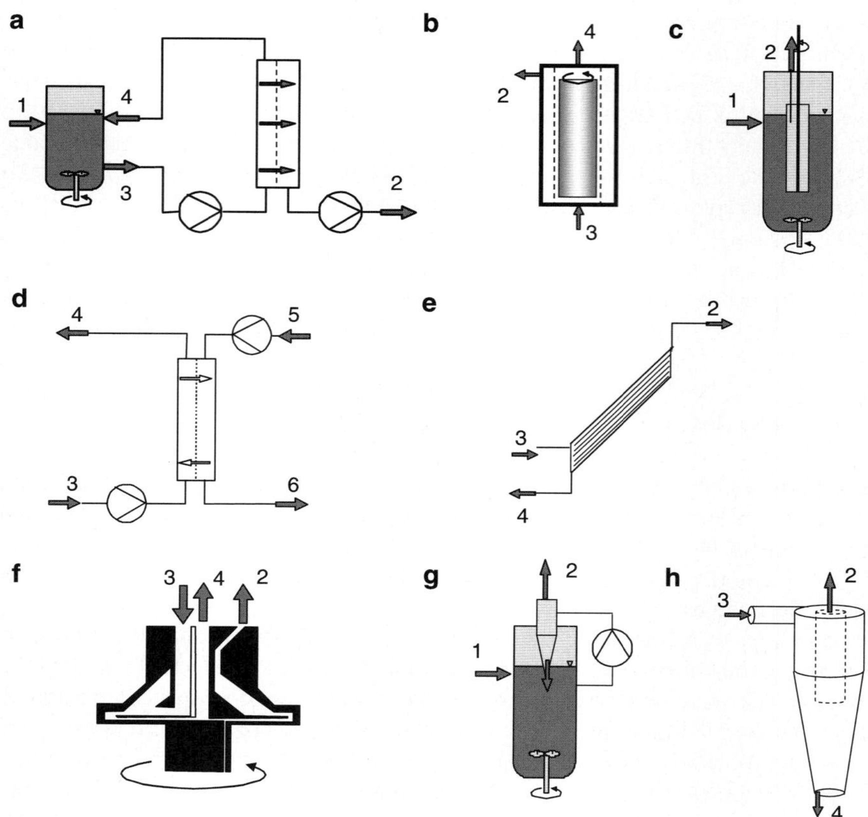

Fig. 5.15 Cell and product retention in suspension reactors. (**a**) Membrane filtration, internal or external, (**b**) vortex-flow filter (controlled shear), external, (**c**) spin or rotor filter, internal or external, (**d**) dialysis, internal or external, (**e**) settler, external or internal, (**f**) centrifuge, external, (**g**) accustic filter, internal, (**h**) hydrocyclone. (1) feed, (2) cell-free harvest, (3) medium from bioreactor, (4) concentrated cells back to bioreactor, (5) dialyzing fluid in, (6) dialyzing fluid out

The *Alternating Tangential Flow* (ATF) technology comprises a hollow fiber module connected to a bioreactor by a single port. The cell suspension is alternately pumped into the filtration module and back to the bioreactor by a fast diaphragm pump (period of 10 s). This provides an efficient tangential flow under extremely low shear operation as compared to the shear going through a pump head. The system can be adapted to suspension cells as well as to anchorage-dependent cells grown on microcarriers. In conclusion, for microfiltration the following critical parameters can be named: membrane material, flow rate (especially the high flow rates in the external loop may be detrimental to cells), module geometry, membrane fouling, and cell bleeding. Reactor scales between 1 L to several hundred liters as well as perfusion rates up to 1200 L d^{-1} (theoretically for the available ATF technology) have been reported (reviewed by Lüllau and Fenge 2006; Voisard et al. 2003).

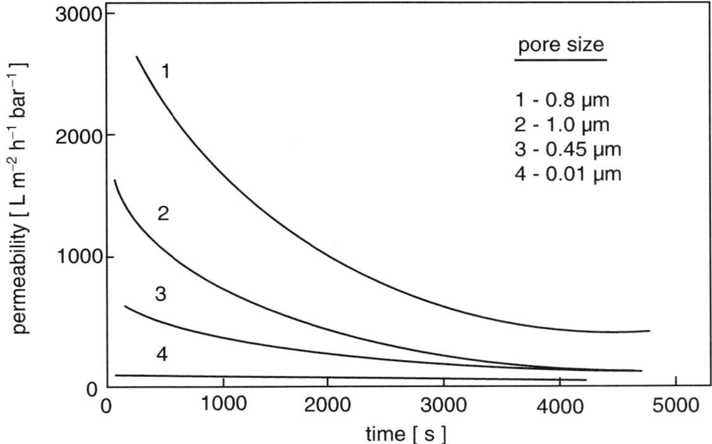

Fig. 5.16 Permeability of different ultrafiltration membranes. Filtration of bovine serum albumine (1 g L⁻¹ in pH 6.7 phosphate buffered water solution, Re = 100, Δp = 1 bar) (data from McDonogh et al. 1992)

The *vortex-flow filter* (Fig. 5.15b) applies a membrane filtration technique to increase the shear forces on the membrane without inducing high pressure drop in the flow channels, as is inevitable in cross-flow and hollow-fiber modules. It consists of three concentric cylinders, an outer cylindrical housing, an intermediate cylinder supporting the membrane (e.g., 200–400 cm²), and an inner cylindrical rotor. The gap within the annular chamber between the inner rotor and the cylindrical membrane has a width of 2 mm to induce Taylor vortex formation to prevent fouling. The cell suspension flows through this annular gap from the bottom to the top (retentate). The permeate leaves the chamber at the top of the outer cylindrical housing. Here, the critical parameters are the pore size of the membrane, spinning velocity, and scale-up. Perfusion rates up to 100 L d⁻¹ have been reported (cited by Voisard et al. 2003).

Spin-filters (also called rotor filters or rotating sieves, Fig. 5.15c) for cell retention were introduced very early by Himmelfarb et al. (1969). Basically, particles are retained sterically by a rotating mesh (metal wire or polymer) having a pore size larger than the average cell size. Mesh pore size is usually 5–20 µm for suspendable cells, and larger (50 µm, 120 µm) for microcarrier culture or cell aggregates. Spin filters can be operated internally on the stirrer axis or on a separate axis, mostly surrounded by a draft tube to induce a vertical flow and to homogenize the hydrodynamic conditions at the screen surface. Alternatively, they can be external, operated on a central axis. In certain setups, Taylor vortex formation is induced to prevent fouling. Critical parameters are the pore size, spinning velocity/screen tangential speed (approximately 1700–2200 cm min⁻¹) as well as clogging when (1) the required filtration flux exceeds the retention capacity of the filter screen and cells accumulate in the pores and (2) after long-term operation the pores are filled by cells that have grown and attached to the mesh. Reported reactor scales are in the range of 1–500 L for internal, and up to 4 L for external arrangement (data from Lüllau and Fenge 2006).

All filtration techniques mentioned so far are mainly intended to retain the cells. Products, even high-molecular-weight proteins, are usually not enriched. *Dialysis* (Fig. 5.15d), also a membrane technique, allows for simultaneous enrichment of cells and high molecular products. Dialysis is defined as a process in which solutes diffuse from a high concentration solution to a low concentration solution across a semipermeable membrane until equilibrium is reached (Strathmann 1979; Moser 1985). Contrary to micro- or ultrafiltration the nonporous dialysis membranes are not perfused. This technique is very reliable, as dialysis membranes do not change their permeability properties even after long process times (Bohmann et al. 1995). Since the nonporous membrane selectively allows low-molecular-weight molecules to pass while retaining those with a higher molecular weight and cells, dialysis can effectively be used as a separation process based on size exclusion. For molecules having a very high molecular weight, the dialysis membrane can be regarded as impermeable. Usually this is described by the "cut-off" value of the dialysis membrane. Typical values are in the range of 10–100 kDa. In a dialysis process, the transport S_i of a certain component i through the membrane is controlled by the transmembrane concentration difference Δc_i, the active membrane area A, and the permeability coefficient P_i of a single component i:

$$S_i = P_{mem,i} A \Delta c_i \tag{5.3}$$

The permeability coefficient $P_{mem,i}$ covers the transport resistance through the membrane. It is membrane- and substrate specific, and depends on the resistance within the boundary layers on both sides of the membrane. With increasing molecular weight, the permeability coefficient decreases similar to the diffusion coefficient (Fig. 5.17). Dialysis membranes can be applied internally or externally (for a review, see Pörtner and Märkl 1998) with different process strategies (dialysis batch, dialysis fed-batch, "nutrient-split" fed batch, continuous, etc.) that allow high cell densities. Examples are discussed by Pörtner and Märkl (1998), Bohmann et al. (1995), and Frahm et al. (2003) (compare Sect. 4.4.2.4). Critical parameters of dialysis reactors are the required membrane area per volume, sophisticated process strategies for optimal time–space yield, and technical problems on large scale (pumps for medium circulation, etc.). Size of reactor systems applying dialysis technique varies from several milliliters up to approximately $1 \, m^3$. A dialysis reactor for lab-scale use is shown in Fig. 5.18.

Retention by Density

As mentioned earlier, cell retention by density is hampered by the very small density difference between cells and culture medium. Inevitably, gravitational settling velocities are low (approximately 1–15 cm h^{-1}). Strategies for improving the separation efficiency can be derived from (5.4):

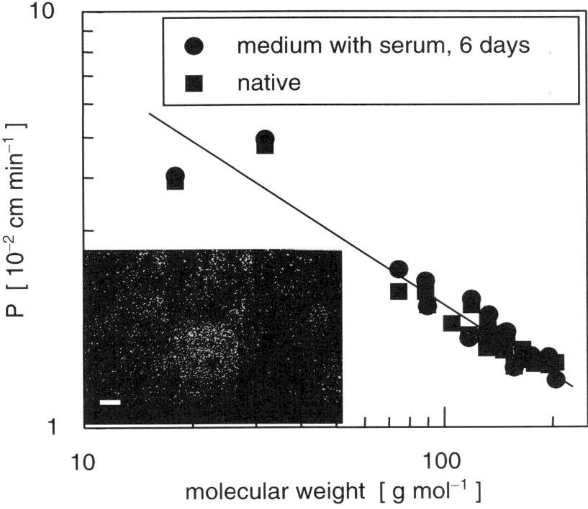

Fig. 5.17 Permeability coefficient P of a dialysis membrane, membrane: Cuprophan (Membrana, Wuppertal), cut-off: 10,000 Da, determined in membrane-dialysis reactor: cell culture medium, inner stirrer: 70 min^{-1}, outer stirrer: 700 min^{-1}

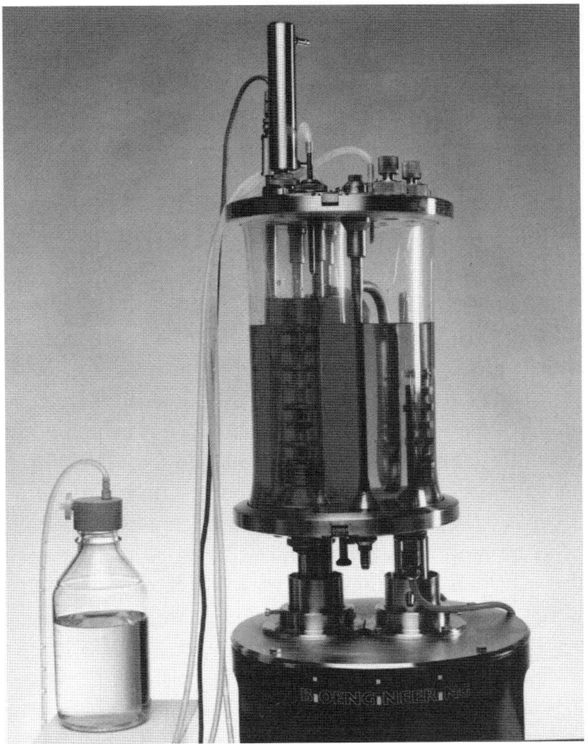

Fig. 5.18 Membrane-dialysis reactor (Prof. Märkl, TU Hamburg-Harburg & Bioengineering AG)

$$U_p = \frac{2(\rho_p - \rho_{fl})g r_p^{2}}{9\eta_{fl}} \tag{5.4}$$

where U_p is the particle velocity in the gravitational field, ρ_{fl} is the fluid density, g is the gravitational force, r_p is the particle radius, and η_{fl} is the fluid viscosity.

Retention techniques have better potential for long-term operation compared to filtration techniques. The most relevant techniques including gravity settlers, centrifuges, acoustic filters, and hydrocyclones will be briefly introduced in the following.

Gravity settlers are technically very simple and robust devices for cell retention by gravitational forces without moving parts that damage the cells. Basically, the low gravitational retention efficiency is increased by reducing the distance for cell sedimentation by the following two mechanism (Voisard et al. 2003): (1) Cells are allowed to settle in a vertical countercurrent flow at a liquid velocity lower than the average settling velocity of the viable cells. This technique requires large areas. (2) The cell suspension flows between inclined parallel plates in a laminar regime (Fig. 5.15e). Cells sediment, settle on the lower plate, and slide down. By this, they generate a contercurrent flow, thereby enhancing the settling efficiency (Boycott effect). The latter technique is mostly applied. Typical parameters of inclined settlers are a plate gap of 0.5–0.7 cm, plate width of 5–15 cm, plate length of 15–40 cm, and plate angle of 20–30°. Gravity settlers are mostly applied in external retention loops, even though special bioreactor designs including internal settling zones have been suggested (reviewed by Voisard et al. 2003). They have been shown to retain viable cells more efficiently than dead cells with a cell retention efficiency of typically 95%. The main drawback, oxygen limitation within the nonaerated settler due to long residence times (approximately 1–2 h), can be overcome by decreasing the temperature within an external retention loop to room temperature or even lower to reduce the metabolic activity of the cells. The device itself is easily scalable. But as the range of operation parameters for optimal cell retention is quite narrow, the performance characteristics are regarded as inflexible and the scale-up of operation parameters is difficult. Critical parameters are the settling surface area, harvest flow rate, and temperature gradient between the reactor and the external settler. Reported reactor scales for settlers are up to several hundred liter perfusion bioreactors (Fenge and Lüllau 2006; Kompala and Ozturk 2006). Perfusion rates in the range of 2000 L d^{-1} and more seem possible (Kompala and Ozturk 2006; Kaiser 2006).

In *centrifuges* (Fig. 5.15f) a centripetal acceleration is used to separate substances of higher and lower densities. By this, the separation efficiency is significantly improved compared to gravitational settling. A number of centrifuges for perfusion cultures have been suggested (reviewed by Fenge and Lüllau 2006; Voisard et al. 2003). Here, two concepts of commercial relevance for centrifuges used in an external loop will be mentioned briefly. A device offered under the trade name Centritech (pneumatic scale) comprises a disposable sterile insert designed as an inverted question mark, which permits one end of the tube to be rotated while the other end is fixed. This anti-twist mechanism eliminates the need for rotating seals. Devices able to perfuse up to 2800 L d^{-1} applying maximal centrifugal forces up to 320 g are available. Another centrifuge for large scale is offered by Westfalia

Separator AG. This disc-stack centrifuge is able to operate with perfusion rates up to 3000 L d^{-1} applying maximal centrifugal forces up to 560 g at 200 rpm (Voisard et al. 2003). The main advantages of centrifuges are (1) no clogging or fouling, (2) a separation rate controlled by the g-force and feed flow rate, and (3) separation of viable and dead cells by adapting the g-force. Critical for industrial application are insufficient robustness and reliability. Alternatively, the so-called "centrifugal bioreactors" have been developed, in which the cells are retained within the bioreactor by internal centrifugation (Kompala and Ozturk 2006). Despite the promising results for lab-scale applications, the system needs further improvement.

In *acoustic filters* (Fig. 5.15g), an ultrasonic separation of cells and the medium is achieved. Basically, the particle size is increased by exposing the cell suspension to an acoustic resonance field, where three forces act on the cells: (1) the primary radiation force due to the compressibility and density difference between cells and the medium, driving the cells toward the antinodes of the resonance field; (2) the secondary radiation force due to interaction between the cells inducing cell aggregation; and (3) the Bernoulli force driving cell aggregates to the local maxima of the acoustic velocity amplitude within the velocity antinodes planes. Critical aspects of this technique are the narrow range of operation parameters for optimal cell retention, harvest limited by cell concentration and input power, and difficult scale-up. Acoustic filters also allow selective retention of viable and dead cells. Devices for perfusion rates up to 250 L d^{-1} (Applicon) are offered.

Hydrocyclones (Fig. 5.15h) apply centrifugal forces by injecting a cell suspension tangential to the wall with typical pressure drops of 0.5–4 bar. Concentrated cell suspension exits in the underflow, and the clarified medium exits in the overflow of the device. For perfusion culture of mammalian cells, hydrocyclones are especially attractive because of the small volume, high separation efficiency, and low shear forces. Successful perfusion cultures operated at approximately 500 L d^{-1} have been reported (Jockwer et al. 2001). Critical parameters are scale-up, operation parameters, and lack of available data for design.

Potential for Large-Scale Perfusion

For evaluation of cell retention techniques, first of all aspects such as the retention capacity and length of operation time in perfusion mode have to be considered. Furthermore, simplicity of operation, robustness, scalability, and costs have to be considered (Voisard et al. 2003; Kompala and Ozturk 2006). For both cell retention by size as well as by density, feasibility has been demonstrated for a large number of different mammalian and insect cells (reviewed by Voisard et al. 2003; Fenge and Lüllau 2006) at different scales. Filtration-based systems allow almost complete cell retention (excluding spin filters), but are limited in long-term operation because of membrane clogging. The time to clog can vary between 10 and 100 days. Therefore, only vortex-flow filters and spin filters are applicable for larger scales. Devices using cell retention by density can be referred to as *open perfusion systems*, which by definition do not clog. Therefore, culture duration is not an issue. Here, the degree of cell retention has to be considered, which decreases with increasing perfusion rate. According to Kompala and Ozturk (2006), perfusion rate should be below $2\,d^{-1}$ for cell retention rates >90%.

The potential of available cell retention techniques for use in large-scale manufacturing processes has been discussed at length by Voisard et al. (2003). On the basis of an extensive literature survey, four ranges of perfusion capacities were defined (compare Table 5.3). The data provided by Voisard et al. were complemented by newer references (reviewed by Kompala and Ozturk 2006). This has given rise to slightly different conclusions, especially with respect to inclined settlers. Whereas Voisard et al. limits inclined settlers to range up to $100\,L\,d^{-1}$ because of a limited scale-up potential, more recent publications demonstrate successful operation up to $2000\,L\,d^{-1}$ (Kompala and Ozturk 2006) or even more. As discussed by Voisard et al. (2003), in the perfusion range of $0\text{–}5\,L\,d^{-1}$ typically cell retention techniques based on filtration are used, mostly spin filters, vortex-flow filters and hollow fibers. In the range $50\text{–}250\,L\,d^{-1}$ hollow fiber techniques are preferred, with alternating tangential filtration, vortex-flow filters inclined settlers, and centrifuges as potential alternatives. Between 250 and $1000\,L\,d^{-1}$ the following techniques can be applied: centrifugation, settling, spin filters, acoustic filters, and hydrocyclones. In the largest range, $1000\text{–}5000\,L\,d^{-1}$, only centrifuges and inclined settlers have the potential for scale-up.

The retention techniques available to day are well established, and the number of successful applications on industrial scale is increasing. Nevertheless, it seems that the industry still hesitates to apply continuous perfusion techniques instead of simpler batch- and fed-batch strategies. This is partly due to the fact that an ideal cell retention device does not exist and therefore prediction of performance of a respective retention device for a specific cell line is difficult. As a consequence, perfusion culture is still regarded as a technology associated with uncertainty and risks for manufacturing operation. Perfusion cultures are especially suited for unstable, non-growth-associated recombinant products (e.g., Factor VIII, retrovirus particles for gene therapy). To date, this group of products represents a minority

Table 5.3 Typical range of operation for cell retention techniques (data from Voisard et al. 2003, modified)

Perfusion capacity [L d^{-1}]	0–50	50–250	250–1000	1000–5000
Membrane filtration (hollow-fibre)				
Alternating tangential filtration				
Vortex-flow filtration				
Spin filter				
Dialysis[a]				
Settling				
Centrifugation				
Acoustic				
Hydrocyclone				

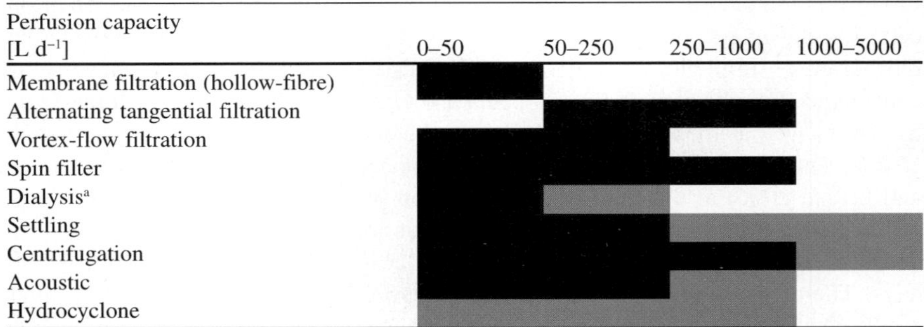

[a]Dialysis allows cell and product retention and is mainly used as dialysis batch or dialysis fed batch
Black: data available, grey: no data available, potential

within the therapeutic proteins. It is expected that this will change in the near future and industrial demand for perfusion cultures will increase.

5.1.2 Fixed Bed and Fluidized Bed Bioreactors: Design, Performance and Scale-Up

5.1.2.1 Engineering Principles

Fixed bed and fluidized bed bioreactors have gained growing attention in the cultivation of mammalian cells since they allow cultivation of immobilized cells within macroporous carriers at very high cell densities. A fixed bed bioreactor (also referred to as "packed bed bioreactor") consists of a vessel filled or packed with carrier material used as support for immobilization of cells (Fig. 5.19). Whereas in fixed bed bioreactors the stationary bed of carriers is not agitated, in fluidized bed bioreactors the carriers are kept floating ("fluidization"). Both types of bioreactors have shown potential in the establishment of optimum culture conditions with a wide versatility for many cell culture purposes (reviewed by Fenge and Lüllau (2006), Meuwly et al. (2007), Pörtner et al. (2007)). In the following, at first some general engineering principles of fixed bed and fluidized bed bioreactors will be introduced, followed by requirements and developments specific for mammalian cell culture (Sects. 5.1.2.2 and 5.1.2.3). Finally, some examples for cultivation of different types of mammalian cells in fixed bed bioreactors will be presented (Sect. 5.1.2.4).

In fixed bed bioreactors, a column filled with solid particles is perfused by a liquid flow. The packing materials create a pressure drop, which is often described by the Carmen–Kozeny equation, which relates pressure drop Δp in a fixed bed of solid particles to the liquid flow rate in the laminar flow range (Re <10):

$$\Delta p = -\frac{H_{\mathrm{FB}} K (1-\varepsilon)^2 (\rho_s - \rho_{\mathrm{fl}})}{\varepsilon^3} \eta_{\mathrm{fl}} U_{fl} S_v^2, \tag{5.5}$$

where H_{FB} is the height of the bed, K is a constant depending on particle shape and surface properties, ε is the void fraction, η_{fl} is the viscosity of the fluid, U_{fl} is the fluid velocity, and S_v is the surface area per unit volume of particles. The Reynolds number Re is defined as

$$\mathrm{Re} = \frac{U_{fl} D_{\mathrm{FB}} \rho_{\mathrm{fl}}}{\eta_{\mathrm{fl}}} \tag{5.6}$$

where D_{FB} is the diameter of the bed and ρ_{fl} is the fluid density. According to this model, a linear increase of the pressure drop with increasing flow velocity or, for a constant cross-sectional area of the column, increasing flow rate, has to be expected. This model was initially designed for solid particles. However, as the carriers used in mammalian cell culture are usually macroporous, intraparticle convection is expected to influence the hydraulic behavior of the system. In this respect two cases

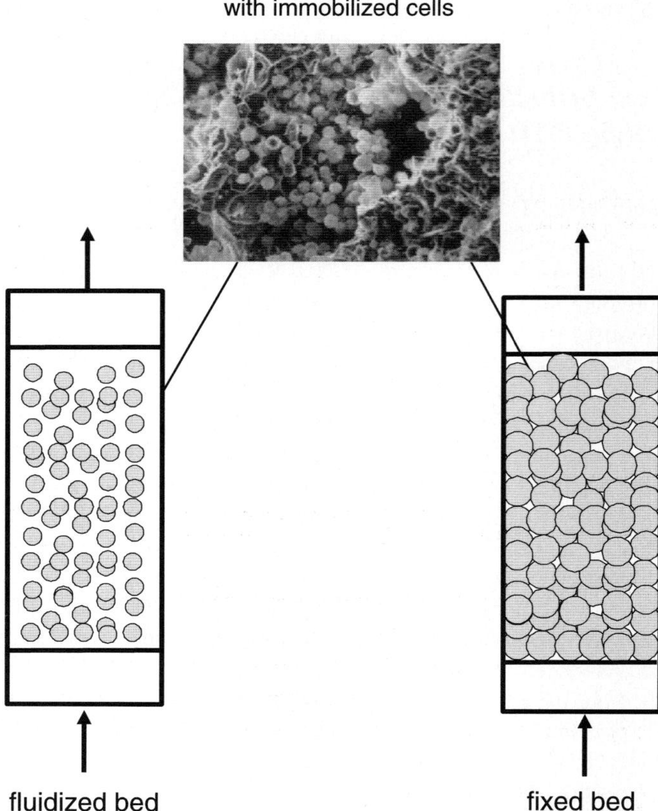

Fig. 5.19 Fixed bed and fluidized bed reactors

have to be considered: (1) carriers with very high porosity and open pores, which will not be closed by cell growth, and (2) carriers with a lower porosity and/or non-open pores, or carriers totally covered with cells in a tissue-like cell density. Examples for cell growth on different carriers were discussed earlier (Sect. 4.3.1.3). For case (1) a simple model was developed by Park and Stephanopoulos (1993) to describe fluid flow through a fixed bed packed with porous particles, based on the Blake–Kozeny equation. Here, it can be assumed that most of the fluid will flow through the interstitial spaces of the bed, but in response to the pressure drop some medium will also flow within the pores of the carriers. The consequences on fixed bed performance have been discussed by Warnock et al. (2005). For case (2) this approach cannot be applied, as the macroporous carriers are not actively perfused. But because of the pressure gradient between the interstitial area and the inner parts of the carrier caused by the interstitial flow, convection within the particles has to be considered. A practical approach to describe an increase of mass transfer due to this type of intraparticle convection effects has been applied by Fassnacht and Pörtner (1999) by using a modified diffusion coefficient. This model was used successfully to simulate oxygen limitation within the bed (see below).

As mentioned earlier, in fluidized bed bioreactors the carriers are kept in a "fluidized" stage. If an initially stationary bed of solid particles is operated in upflow mode with particles of appropriate size and density, at first the bed remains as it is and the pressure drop of the bed increases as discussed earlier (Fig. 5.20). At a sufficiently high liquid flow rate, the bed expands to the upward motion of the particles. It is brought to a "fluidized" state as soon as the volumetric flow rate of the fluid exceeds a certain limiting value, the minimum fluidization flow rate (Werther 2007). In the fluidized bed, the particles are held suspended by the fluid stream. During the fluidization, the pressure drop Δp of the fluid on passing through the fluidized bed is constant (Fig. 5.20) and equal to the weight of the solids minus the buoyancy, divided by the cross-sectional area A of the fluidized bed vessel following

$$\Delta p = \frac{H_{FB}(1-\varepsilon)(\rho_s - \rho_{fl})g}{A} \tag{5.7}$$

with the porosity ε of the fluidized bed as the void volume of the fluidized bed (volume in interstices between particles, not including any pore volume in the interior of the particles) divided by the total bed volume; ρ_s the solids apparent density; ρ_{fl} the fluid density; and H_{FB} the height of the fluidized bed. Because particles in fluidized beds are floating, channeling and clogging of the bed are avoided. At a certain flow rate, the particles cannot be kept in the fluidized state and are washed out.

Macroporous carriers for application in fixed bed and fluidized bed bioreactors have been discussed extensively in Sect. 4.3.1.3. The properties of the cells and the mode of cultivation carried out define the type of carrier to be chosen. Thus, carriers of diameters below 0.3 mm (microcarriers) are used mainly for the cultivation of adherent cells in suspension, carriers between 0.3 and 1 mm are appropriate for fluidized beds, and carrier sizes above 1 mm are used in fixed bed systems (preferably 3–5 mm) to prevent blocking of the free channels between the carriers by cell

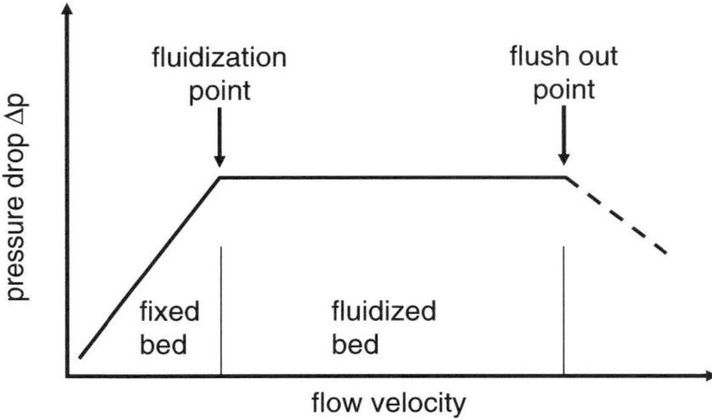

Fig. 5.20 Principles of fluidization in a fluidized bed reactor

growth. The density of the carrier has to be appropriate for the respective culture system. Whereas for suspension culture mostly carriers with a specific gravity slightly higher than the density of the medium are used, a significantly higher specific gravity of >1.5 of the carrier is typically chosen to achieve homogeneous fluidization in a fluidized bed bioreactor. In case of fixed bed bioreactors, the specific gravity is not that much of a problem, as in most cases the carriers are retained in the fixed bed column by a porous mesh or other means.

5.1.2.2   Fixed Bed Bioreactors for Cultivation of Mammalian Cells

A number of different design concepts for fixed bed bioreactor systems have been proposed in the literature (Meuwly et al. 2007; Fenge and Lüllau 2006; Pörtner and Platas Barradas 2007). Most of them are laboratory homemade systems. Only a few systems have been commercialized. The general problem for design and operation of fixed bed bioreactors is to fulfil certain, sometimes competing, process requirements such as meeting the high demand of oxygen required by the cells growing in a very high cell density, appropriate supply of nutrients, and preventing cell damage due to too high shear rates. Almost all fixed bed bioreactor system consist basically of a column filled or packed with carrier material used as support, where the cells grow immobilized, and a medium reservoir for conditioning of the culture medium (Pörtner 1998; Pörtner et al. 1999; Meuwly et al. 2007). The medium is recirculated in a loop through the fixed bed and back to the reservoir for conditioning, especially for oxygen enrichment, while the cells (adherent or non-adherent) are retained in the fixed bed. Oxygen supply is mostly done by simple bubble aeration, as the cells do not come in to contact with the bubbles and therefore cell damage due to bubbles is not an issue. In the conditioning reservoir, the exhausted, product-containing medium is exchanged batchwise or continuously (Pörtner 1998). Usually, further control elements such as sensors for pO_2, pH, and temperature are installed in the reservoir as well for process control.

Most common are external or internal loops for medium recirculation (Figs. 5.21–5.23). In case of an external loop (compare concepts "A" in Fig. 5.21), the enriched medium is pumped from the reservoir, often called "conditioning vessel", with a certain flow rate or flow velocity, which can exert a shear stress on the circulating cells, through the fixed bed and back. This two-vessel arrangement is quite popular on a laboratory scale, as it is easy to set up, but might lead to sterilization problems on a large scale. As an alternative, internal medium circulation was suggested (compare concepts "B" and "C" in Fig. 5.21). A further problem with respect to scale-up of fluidized bed bioreactors also is the appropriate supply of oxygen. In an axial stream of the medium through the fixed bed (also referred to as "axial-flow") without any further aeration, the concentration of oxygen in the medium decreases from the bottom of the bed to the top. The degree of oxygen depletion depends on the number of immobilized cells X, the cell specific oxygen consumption rate q_{O2}, the mean flow velocity through the bed U_{fl}, and the height of the bed H_{FB} according to

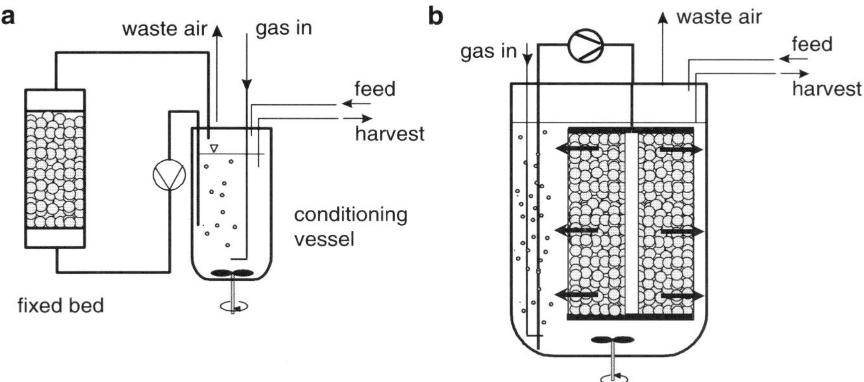

Fig. 5.21 Examples for design concepts of fixed bed reactor. (**a**) axial-flow fixed bed with external conditioning vessel (fixed bed volume VFB ~ 50–100 mL, data from Medorex (D) (Pörtner et al. 2007)), (**b**) radial-flow integrated in conditioning vessel, external circulation of medium (VFB ~ 1–25 L, data from Medorex (D) and Bioengineering (CH)), (**c**) axial-flow fixed bed integrated in conditioning vessel, internal circulation of medium VFB ~ 0.7–5 L (data from NewBrunswick (US))

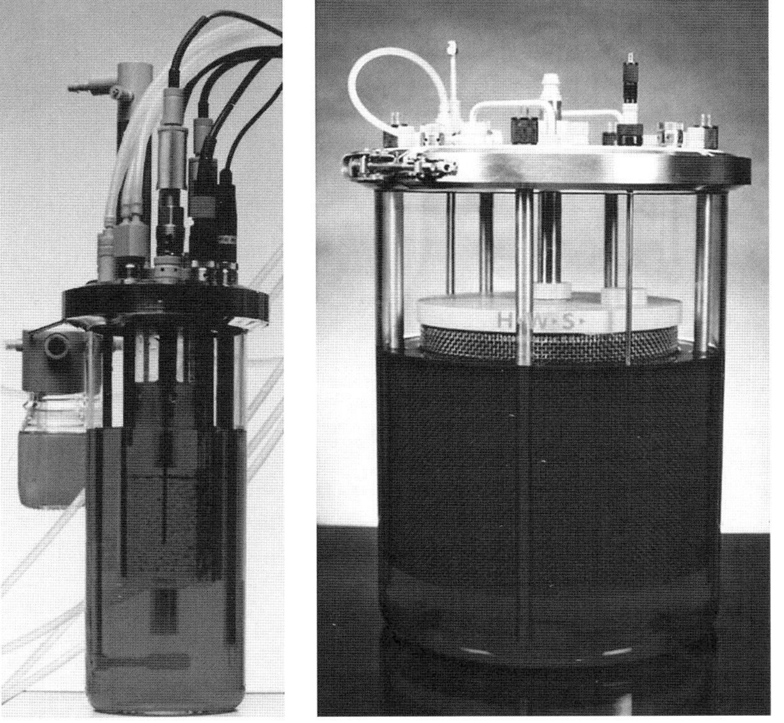

Fig. 5.22 Examples of design of fixed bed reactors for mammalian cell culture. Left: 100 mL, axial flow, right: 5 L, radial-flow (Courtesy Medorex, Germany)

Fig. 5.23 CelliGen 310 basket fixed bed reactor equipped with FibraCell macroporous carriers (courtesy of New Brunswick Scientific)

$$c_{o2}(\text{in}) - c_{o2}(\text{out}) = \frac{q_{o2}XH_{FB}}{U_{fl}} \qquad (5.8)$$

with $c_{O2}(\text{in})$ the oxygen concentration at the inlet and $c_{O2}(\text{out})$ the concentration at the outlet of the bed. The flow velocity through the bed is limited because of several reasons. Of course, damaging shear rates should be avoided. But as most of the cells are entrapped within the macroporous structures of the carriers, this criterion is not that important. More critical is a washout of cells at too high flow velocities, especially for non-adherent cells. A maximal flow velocity below $U_{fl} = 1\,\mathrm{mm\,s^{-1}}$ has been suggested by Bohmann et al. (1992) for hybridoma cells to prevent washout. As reviewed by Meuwly et al. (2007), most of the studies published during the last 10 years operate in this range. If a cell density of 5×10^7 cells $\mathrm{mL^{-1}}$ of fixed bed, a cell-specific oxygen uptake rate of 1×10^{-10} mmol $\mathrm{cell^{-1}\,h^{-1}}$, aeration with air ($c_{O2}$(in) $= 0.2\,\mathrm{mmol\,L^{-1}}$), and a minimal oxygen concentration of c_{O2}(out) $= 0.04\,\mathrm{mmol\,L^{-1}}$ are assumed, the length of an axial-flow fixed bed should be in the range of approximately 10 cm. This simple estimation underlines the fact that the length of a fixed bed is limited owing to oxygen limitation. This can be overcome to some extent by aerating the medium with pure oxygen (see below). For further scale-up, Meuwly et al. (2007) suggest increasing the diameter of the column up to 2 m, similar to chromatographic columns. This would correspond to a bed volume of up to 900 L for a height of 30 cm. This concept has not been realized to date. Alternatively, radial-flow fixed beds have been studied (compare concept "B" in Fig. 5.21) (Yoshida et al. 1993; Bohmann et al. 1995; Pörtner et al. 1997; Pörtner et al. 1998; Fassnacht 2001; Kwon et al. 2005; Hongo et al. 2005; Iwahori et al. 2005). The medium is pumped from the medium reservoir either through the center of the cylindrical bed along the radius or vice versa. As the length for oxygen depletion depends on the radius of the bed, the height of the bed can be enlarged without the risk of insufficient oxygen supply. Fassnacht et al. (2001) scaled up a radial-flow fixed bed reactor packed with SIRAN carriers to 5.6 L bed volume. Operated with perfusion rates upto 20 L per liter fixed bed and day would result in approxinately 100 L of medium supernatant per day. A 10-day operation would be equivalent to a 1000 L suspension reactor. Investigations on the flow profile in radial-flow columns up to 90 cm height indicate that at least up to this height problems due to inhomogeneous flow through the bed can be neglected (own unpublished data). Actually, this type of bioreactor has been build at least up to 25 L fixed bed volume for cultivation of mammalian cells. For an easier handling, sterilization, and scale-up, with further improved temperature control in a more compact unit, the fixed bed is integrated with the conditioning vessel. Design criteria and model predictions can be found in Fassnacht and Pörtner (1999).

Owing to the low shear stress, immobilization enables the application of serum- or protein-free medium (compare Table 5.4); it allows an easier medium exchange and reduces the steps of downstream processing. Of special importance is the improved culture stability through immobilization (Fussenegger et al. 2000; Lüdemann et al. 1996; Fassnacht et al. 1999; Moro et al. 1994). With a simple, and easy-to-handle bioreactor design, fixed bed bioreactor systems enable a high-volume specific cell density and productivity (10^8 cells $\mathrm{mL^{-1}}$) and a versatile range for production of biopharmaceuticals over long periods of time (up to several months) (Meuwly et al. 2007; Fenge and Lüllau 2006). Whereas the earlier studies focused on production of biopharmaceuticals as an alternative to suspension bioreactors or hollow fiber systems, new applications can be found in other areas including gene and cell therapy or tissue engineering. Some of the recent developments in fixed bed bioreactor technology are presented in Table 5.5.

Table 5.4 Cultivation of a mouse/mouse hybridoma cell line and a human/mouse transfectoma cell line in different media and reactor systems (adapted from Lüdemann et al. 1996)

Cell line	Hybridoma		Transfectoma	
Medium	HS	IR	HS	IR
T25-flask (static)	++	++	++	++
500-ml-spinner (stirred)	++	++	–	–
2-L-stirred reactor	++	(+)–	n.d.	n.d.
fixed bed reactor (SIRAN®)	++	++	n.d.	++

Medium: IMDM/Ham's F12 plus additives, supplemented with 3% (v/w) horse serum (*HS-medium*), or an iron-rich supplement (*IR-Medium*, protein free)
++: cultivation successful; (+)–: cultivation possible, but not reproducible; –: cultivation not possible; n.d.: not done

Table 5.5 Successful applications of fixed bed bioreactors for cultivation of mammalian cells

Type of cell	Product/purpose	References
Hybridoma	Monoclonal antibodies	Ong et al. (1994), Looby et al. (1990), Bohmann et al. (1995), Golmakany et al. (2005), Yang et al. (2004), Wang et al. (1992), Fassnacht et al. (1999)
Transfectoma	Chimeric fragmented antibodies	Lüdemann et al. (1996), Pörtner et al. (1998)
rBHK	Glycosylated antibody-cytokine	Burger et al. (1999)
CHO	NMR-studies	Ducommun et al. (2002), Matsushita et al. (1990), Thelwall et al. (1998)
rCHO	rec. Erythropoietin, rec. human placental alkaline phosphatase (SEAP)	Deng et al. (1997), Fussenegger et al. (2000)
VERO		Matsushita et al. (1990)
L293	Rec. protein	Own data
HeLa	Vaccine	Hu et al. (2000)
NIH3T3		own data
Insect cells	Baculovirus recombinant protein	Lu et al. (2005), Kwon et al. (2005)
Packaging cell lines	Retrovirus production	Nehring et al. (2006), Merten et al. (2001), Sendresen et al. (2001), McTaggart and Al-Rubeai (2000), Forestell et al. (1997)
Immortalized hepathocytes	Bioartificial liver	Fassnacht et al. (1998), Kataoka et al. (2005), Hongo et al. (2005), Iwahori et al. (2005), Akiyama et al. (2004), Kawada et al. (1998), Ijima et al. (1998)
Primary hepathocytes	Bioartificial liver	Kurosawa et al. (2000), Yamashita et al. (2001), Gion et al. (1999)
Human kidney cells	Membrane vesicles	Own data
Human melanoma cells	Growth factors	Own data

5.1.2.3 Fluidized Bed Bioreactors for Cultivation of Mammalian Cells

Fluidized bed bioreactors have been described quite extensively in the literature for a variety of adherent and non-adherent cell lines (reviewed by Fenge and Lüllau 2006). Similar to fixed bed bioreactors, most studies were performed in laboratory homemade systems. Only a few systems have been commercialized. The basic design concepts of fluidized bed bioreactor systems are summarized in Fig. 5.24. The general problem for design and operation of a fluidized bed, similar to fixed bed bioreactors, is to fulfil certain, sometimes competing process requirements such as meeting the high demand for oxygen required by the cells growing in a very high cell density, appropriate supply of nutrients, and maintaining a fluidized stage of the carriers (gravity of the carrier, fluidization velocity). Again, almost all design concepts are based on a loop for medium recirculation, either external or internal. Within this loop, oxygen is supplied to achieve a certain oxygen

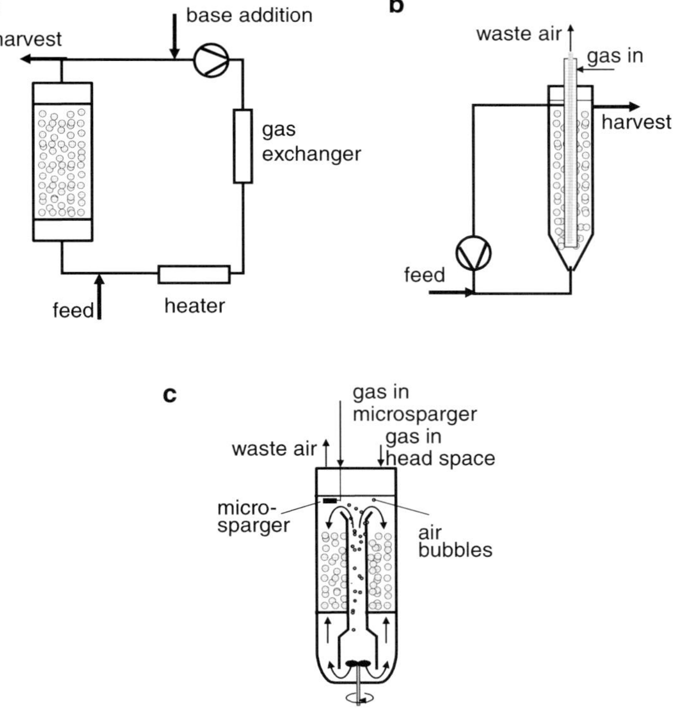

Fig. 5.24 Examples for design concepts of fluidized bed reactor. (**a**) fluidized bed with external loop, gas exchanger, and heater (Verax system, fluidized bed volume V_{FB} ~ 0.4–24 L [Fenge and Lüllau 2006]), (**b**) fluidized bed with inline aeration module for bubble and gradient free oxygen supply and external loop for medium circulation (V_{FB} ~ 0.05–0.5 L, data from Celonic (Germany)), (**c**) fluidized bed with internal medium circulation and microsparger aeration (V_{FB} ~ 0.1–40 L, data from GE Healthcare (Austria) [Lundgren and Blüml 1998])

concentration at the entrance of the bed, fresh medium is fed, and spent medium is harvested. Usually, further control elements such as sensors for pO_2, pH, and temperature are installed in the loop for process control. In case of an external loop (compare concepts "A" and "B" in Fig. 5.24), the medium has to be pumped with a high flow rate, which can exert a shear stress on circulating cells. Furthermore, sterilization problems might occur in large scales. As an alternative, internal medium circulation was suggested (compare concept "C" in Fig. 5.24). A further problem with respect to scale-up of fluidized bed bioreactors is also the appropriate supply of oxygen. In an upflow stream of the medium without any further aeration, the concentration of oxygen in the medium decreases from the bottom of the bed to the top. Even though the flow velocity in fluidized bed bioreactors is usually high compared to fixed bed bioreactors, the maximal flow velocity is limited by the gravity of the carrier, and therefore the bioreactor height, and the scalability is limited. This problem of an oxygen gradient along the height of the bioreactor can be overcome to some extent by integrating a membrane module directly into the fluidized bed (compare concept "B" in Fig. 5.24). The most important design concepts are introduced in the following.

Concept "A" refers to a technology introduced by the company Verax Corporation in late 1980s. The fluidized bed is integrated in an external loop for medium circulation. For cell immobilization a special macroporous carrier was designed, which consisted of a sponge-like collagen structure (diameter 500 µm) of interconnected pores (20–40 µm). To increase the specific gravity (>1.6) the carrier was weighted with noncytotoxic steel. By this, flow velocities up to 70 cm min^{-1} could be applied to improve mass transfer, especially oxygen supply, in the fluidized bed. It was one of the first macroporous carriers supporting cell growth >10^8 cells per milliliter of carrier. The design principles of this bioreactor, which was available from 0.4 to 24 L and which is no longer on the market, and examples for cultivation of hybridoma and rCHO cells are comprehensively described by Dean et al. (1987) and Ray et al. (1989, 1990).

Based on this concept a number of fluidized bed bioreactor systems have been suggested during the 1990s (reviewed by Fenge und Lüllau 2006). One focus of research on improving and optimizing this bioreactor concept was to find a technically feasible solution to overcome one of the major disadvantages of fluidized beds: the progressive depletion of oxygen along the height of the expanded bed. The group of Wandrey and Biselli (Hambach et al. 1992; Born et al. 1995) integrated a membrane oxygenation module and an in-line gasification tube module, respectively, directly into the fluidized bed for bubble-free aeration (Concept "B" in Fig. 5.24). By this, a uniform distribution of oxygen along the height of the fluidized bed column could be achieved. But the oxygen transfer rate of membranes is quite low (compare Sect. 4.2.1). For the silicone membranes (thickness 0.8 mm) used here, a specific membrane area of approximately 100 m^2 m^{-3} is required to support 10^8 cells per milliliter carrier or 5×10^7 cells per milliliter fluidized bed. For small-scale bioreactors this is feasible, and a commercial bioreactor with an in-line gasification tube module is offered by Celonic (Fig. 5.25) with working volume up to 1150 mL and microcarrier bed volume up to 500 mL using a SIRAN carriers for cell immobilization (Celonic 2007).

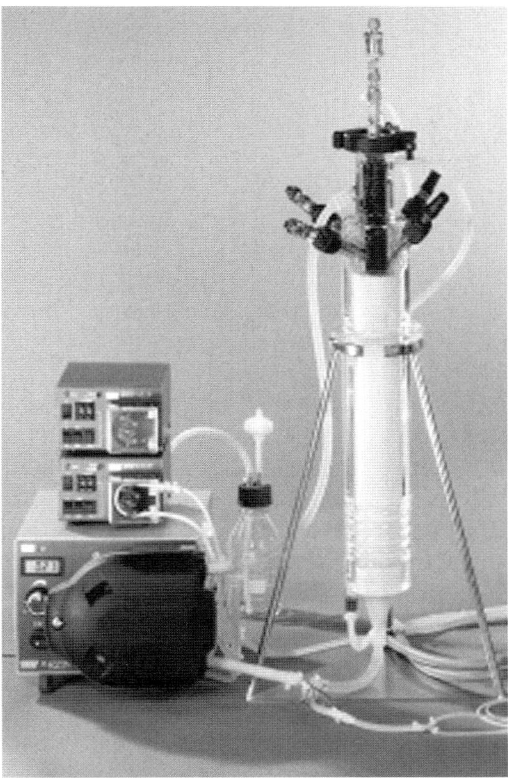

Fig. 5.25 Fluidized bed reactor E500 (courtesy of Cellonic GmbH, Germany). Culture volume: 300–1150 ml; Microcarrier bed volume: 50 mL to 500 mL

An alternative fluidized bed concept (Concept "C" in Fig. 5.24) with an internal medium recirculation was developed by the group of Katinger and Blüml (Reiter et al. 1991; Lundgren and Blüml 1998). The Cytopilot (now offered by GE Healthcare, Fig. 5.26) comprises two chambers divided by a distributor plate. An internal draft tube for medium recirculation runs through both chambers. Liquid agitated by an impeller in the lower chamber is conveyed via the distributor plate into the upper chamber. Settled microcarriers are fluidized by the hydrodynamic pressure. The bed expands depending on the impeller speed. The culture medium circulates back down through the draft tube into the lower chamber. Oxygen bubbles are sparged into the culture medium in the carrier-free zone in the upper bioreactor chamber. Coalescent bubbles immediately rise to the surface and only the suspended liquid saturated with microbubbles flows down to the impeller. Fluidized bed volumes between 0.1 and 40 L have been reported (Lundgren and Blüml 1998). For the Cytopilot mini (bed volume 500 mL), a circulation flow rate of about 0.5–2.0 L m^{-1} is recommended (Manual "Cytopilot"). The Cytopilot is often used with a certain carrier developed by the group of Katinger and Blüml (Lundgren and Blüml 1998).

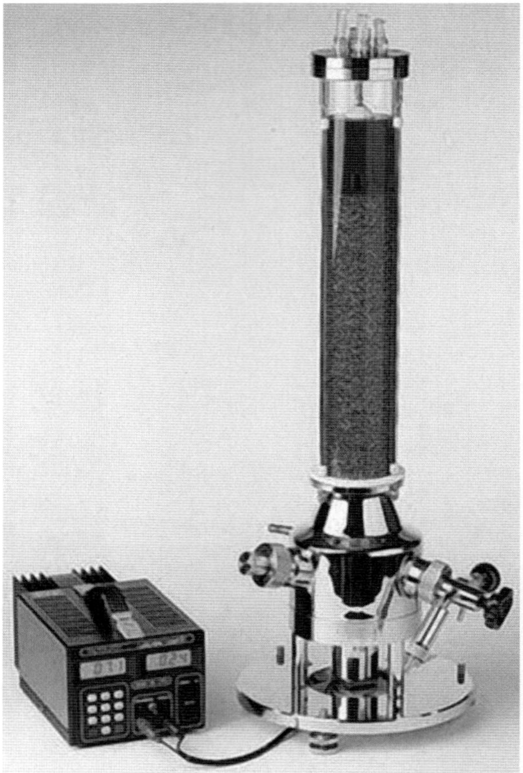

Fig. 5.26 Cytopilot mini fluidized bed reactor (courtesy of GE Healthcare)

This carrier, commercially available as Cytoline (GE Healthcare, compare Sect. 4.3.1.2), consists of polyethylene and is weighted with chalk and silicates to adjust the buoyant density of the carrier to the intended application (stirred tank, fixed bed, fluidized bed).

Fluidized bed bioreactors have been scaled up to pilot scale (e.g., 40 L bed volume for the Cytopilot) (Lundgren and Blüml 1998). Nevertheless, most applications are in the laboratory scale (<1.5 L bed volume).

5.1.2.4 Examples for Cultivation of Mammalian Cells in Fixed Bed Bioreactors

Fixed bed bioreactors have demonstrated their potential in the establishment of optimum culture conditions with wide versatility for many cell culture purposes. Some of the recent developments were mentioned earlier. The examples presented in the following include the increase of antibody productivity by hybridoma cells, high-performance cultivation of a hepatoblastoma cell line, and the cultivation of cells for production of retroviral vectors. Furthermore, criteria for selection of process strategies and scale-up concepts are addressed.

The heterogeneity inside the carriers, i.e., concentration and cell gradients over the radius of the carrier (compare Sect. 4.3.1.4), can limit productivity. These gradients are due to the substrate or oxygen consumption of the immobilized cells. If the diffusion rate of substrate into the carrier is slow compared to the reaction rate, then limitation occurs. The opposite can happen with products that accumulate to inhibiting concentrations inside the carrier (e.g., carbon dioxide). Fassnacht and Pörtner (1999) used an oxygen reaction–diffusion model to describe and optimize the performance of axial-flow as well as radial-flow fixed beds. Furthermore, they could show that an increased oxygen level could significantly improve the performance of a fixed bed bioreactor. A 100 mL fixed bed bioreactor was operated with a hybridom cell line at high dilution rates between 10 and 20 L medium per liter fixed bed per day (Fig. 5.27). At these extremely high perfusion rates, the volume-specific glucose uptake rate as well as the production rate for monoclonal antibodies were up to 2 times higher at oxygen levels of 300% air saturation compared to air saturation. Although dissolved oxygen concentrations higher than air saturation are usually considered to be toxic for mammalian cells in suspension cultures (Simpson et al. 1997), it was found that this is not the case for fixed bed systems.

In the field of tissue engineering as well as in cell culture technology, high cell densities during the culture are usually required. In order to develop a bioartificial liver (BAL), for example, many authors have attempted to optimize a culture of hepatocytes at high cell densities in different cultivation systems trying to mimic

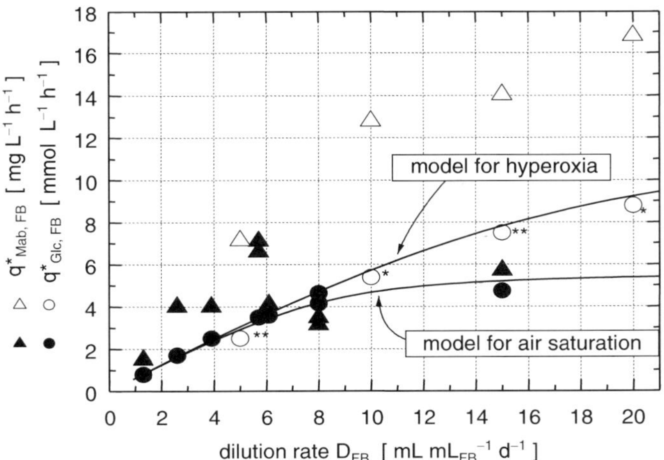

Fig. 5.27 Steady-state data of several fixed bed experiments (fixed bed volume 100 mL) of the hybridoma line IV F19.23 operated at various dilution rates D_{FB}. The diagram shows the glucose consumption $q^*_{Glc,FB}$ and antibody production rates $q^*_{MAb,FB}$ for the case of dissolved oxygen concentrations of air saturation (solid symbols), 300% air saturation (open symbols, one asterisk), and oxygen saturation (open symbols, two asterisks). The lines indicate the modeled glucose uptake rates (from Fassnacht and Pörtner 1999)

tissue-like growth (Matsushita et al. 1990; Kurosawa et al. 2000; Hongo et al. 2005; Fassnacht et al. 1999; Miyoshi et al. 1994; Kataoka et al. 2005; Iwahori et al. 2005; Akiyama et al. 2004; Kawada et al. 1998; Ijima et al. 1998; Yamashita et al. 2001; Gion et al. 1999). However, compared to cell lines with higher adaptability to cultures, such as hybridom and other cell lines, hepatocytes are considered one of the most difficult cell types to be cultivated in an artificial environment.

The fact that the human hepatocellular carcinoma cell lines such as Hep-G2 secrete the major plasma proteins and the hepatitis B surface antigen has directed the attention towards these cell lines for research purposes. They have potential use as screening systems, where cultures of these cells are put under the effect of different substances such as new drugs or toxic metabolites. The future of these techniques is in the performance of clinical trials in culture systems of human cells, eliminating the need of research in whole animals, where the complicated process of blood screening and ethics during research are of major importance.

An example for fixed bed cultivation of the hepatoblastoma cell line Hep-G2 is shown in Fig. 5.28, were different culture conditions were applied during 24 days of cultivation. Continuous perfusion of medium in the conditioning vessel was started when the culture approached glucose-limiting concentrations. A higher glutamine concentration in culture increased slightly the substrate uptake rate during continuous culture. An oxygen inlet of 200% air saturation and a dilution rate of $D_{FB} = 8\,d^{-1}$ after 20 days of culture increased significantly the substrate consumption rates. This successful example underlines again the importance of proper selection of process parameters such as perfusion rates and oxygen level.

A number of examples for successful applications of fixed bed bioreactors for production of high titer retroviral vectors for gene therapy can be found in the literature (Nehring et al. 2006; Merten et al. 2001; Hu et al. 2000; Sendresen et al. 2001; McTaggart and Al-Rubeai 2000). Since pseudotype vectors are promising for gene transfer in many gene therapy approaches, low vector concentration in standard batch cultures and high-temperature-dependent decay are limiting factors in sufficiently large-scale production. To overcome these obstacles, the kinetic relations of the bioreaction, cell growth, and vector formation in different culture modes need to be understood. Furthermore, not only generic optimization of the packaging cell line and the vector stability but also effective optimization of process modes is needed. Nehring et al. (2006) used a mathematical model developed on the basis of experimental data measured in culture dishes. The kinetics of cell growth, nutrient consumption, and vector production and decay was applied to analyze different process modes and to optimize the cultivation and harvest strategy during fixed bed cultures. The optimization with the aid of the model led to the conclusion that optimizing the harvest cycles and feed flows resulted in a higher yield in comparison to a batch of fed-batch culture.

Retroviral pseudotype vector, derived from the murine leukaemia virus carrying the HIV-1 envelop protein MLV (HIV-1), was produced using a 200 mL fixed bed bioreactor for high cell density cultivation on macroporous carriers. Reasonable optimal parameters achieved by the model were applied to run the cultivation. After starting the cultivation in batch mode, the bioreactor was either run in perfusion,

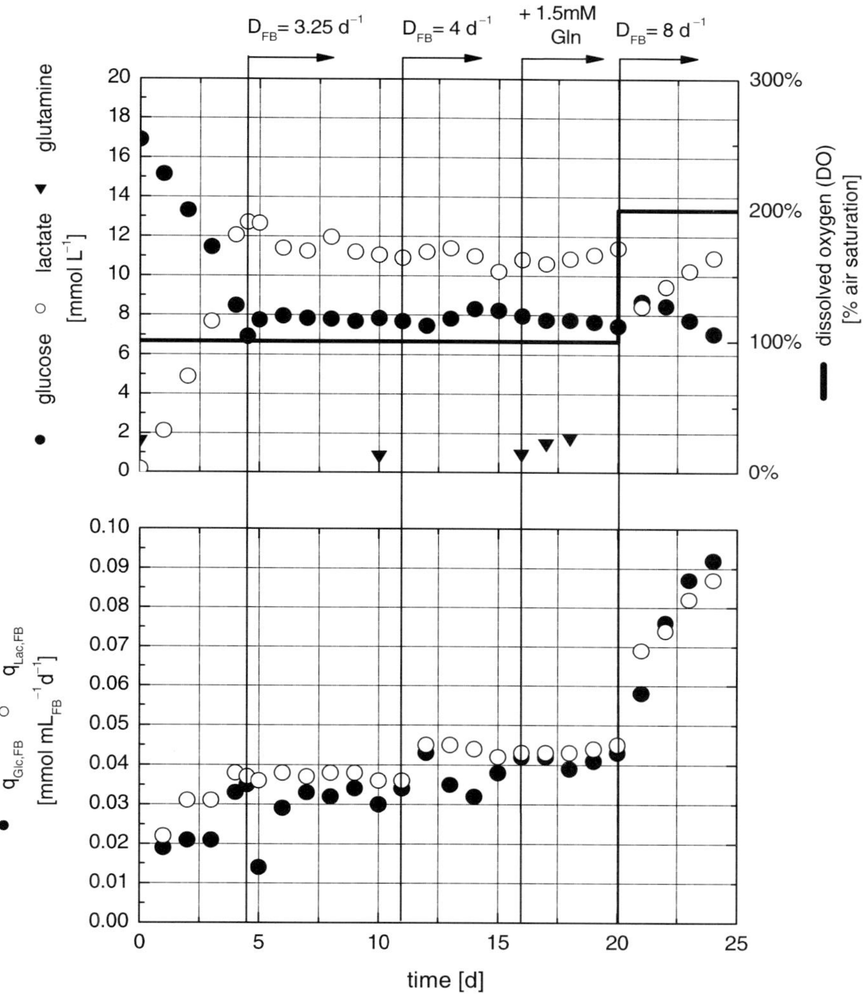

Fig. 5.28 Cultivation of the Hep-G2 in a 100 mL axial-flow fixed bed reactor. Top: Glucose (•), lactate (○) and glutamine (▲) concentrations as well as the relative oxygen percentage at the air inlet (—). Down: Glucose uptake rate $q_{Glc, FB}$ (•) and the lactate production rate $q_{Lac, FB}$ (○) during the culture. The vertical lines indicate the different dilution rates and conditions during the 24 days of culture (with kind permission of Bentham Science)

perfused fed-batch or repeated-batch (Fig. 5.29). The different culture modes are compared in Table 5.6. Whereas the highest maximal end point titer was measured in perfusion mode, the highest medium-related productivity was achieved in repeated-batch mode. Sensitivity analysis performed on the basis of the results of the optimization showed that under optimal conditions the final concentration as well as the yield of the entire process is more sensitive to the parameters of process

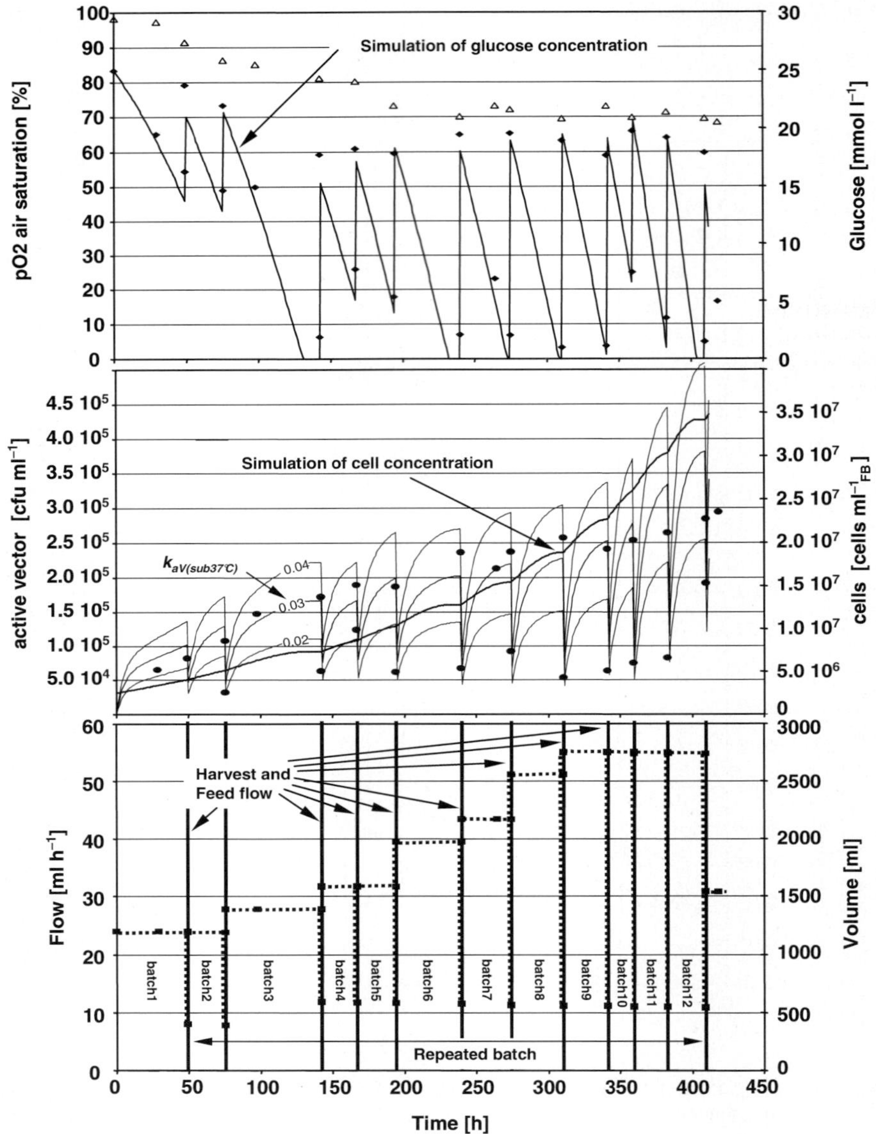

Fig. 5.29 Fixed bed cultivation of the retroviral packaging cell line TELCeB6/pTr712-K52S producing permanent MLV(HIV-1) vector particles containing the transfer vector MFGlnslacZ (data from Nehring et al. 2006). Time course of cultivation in a 200 mL fixed bed reactor showing experimental data of pO2 (Δ), glucose (♦), and active vector (•) concentrations measured in conditioning chamber. During the first 48.5 h the cultivation was kept in a batch mode. After this, the conditioning chamber was fed with a stepwise, increasing feed. The vector-containing medium was harvested three times at 192.5, 295, and 341 h. A volume of approximately 670 mL was left in the conditioning chamber after each harvest. The feed flow reached its maximum value of 50.8 mL h⁻¹ after 360 h. The entire medium volume altered between 4200 and 670 mL while the $pO2$ (Δ) was above 60% until the end. The maximum vector titer of 1.45×105 cfu mL⁻¹ was measured after 314.7 h. Simulation was performed on the basis of an unstructured kinetic model. For calculation of vector titer, different decay constants kaV were applied (with kind permission of Elsevier)

Table 5.6 Fixed bed cultivation of retroviral packaging cell line TELCeB6/pTr712-K52S (for details compare Fig. 5.29). Comparison of different process strategies (data from Nehring et al. 2006)

	Perfusion	Repeated fed batch	Repeated batch	Flask culture
Total time (h)	412	414	418	24
Total medium amount (L)	12.66	11.74	18.27	0.01
Final cell number (10^9 cells)	2.64	2.08	5.42	0.00748
Total vector production (10^8 cfu)	27.6	9.43	46.4	0.0140
Maximal vector titer (10^6 cfu mL^{-1})	6.7	2.28	11.1	0.0583

operation than to the bioreaction owing to the fast degradation of replication of competent pseudo-type vectors.

Scale-up is a major demand for high-yield, stable bioprocess systems in mammalian-cell-culture-based biopharmaceutical production (Fassnacht and Pörtner 1999; Cong et al. 2001; Ducommun et al. 2002; Golmakany et al. 2005). Using the concept of radial flow geometry, Fassnacht and Pörtner (1999) could show at least a scale-up from 100 mL to 5 L fixed bed volume. As perfusion rates of 10 L medium per liter fixed bed per day correspond to a medium flow rate of approximately 50 L per day for a 5 L bioreactor, this bioreactor size can be regarded as pilot scale. Further optimization of the fixed bed process can be achieved by reducing the inoculum volume (Fassnacht 2001). For this, it is crucial to reduce the initial cell density to a minimum without affecting exponential growth, as well as to increase the bed volume in the bioreactor during the fermentation process. The combination of both a low initial-cell-density inoculum and a cultivation strategy in which the surface for cell growth can be increased progressively might result in a decrease of the preculture steps needed before starting up a cell culture process, where inoculation means not only the cell density to be inoculated but also the working volume of the bioreactor to be inoculated. From batch cultures performed in a 10 mL fixed bed bioreactor, a cell density of 5×10^4 cells per milliliter of medium volume was found as optimal with respect to growth rate within the fixed bed (own data, not published). In suspension culture, usually $1–2 \times 10^5$ cells per milliliter are required.

To further reduce the inoculum volume in a fixed bed bioreactor, the "Inoculation during Cultivation" strategy can be applied (compare Fig. 5.30). This strategy consists in partially filling the conditioning vessel with the medium at the beginning of the cultivation; only a fraction of the fixed bed is submerged and therefore active. Only this submerged part of the fixed bed is then inoculated with a low initial cell density. After a short cultivation period, the initial cell density is increased and the medium level within the conditioning vessel is raised to another desired height. This activates a fresh part of the fixed bed, which is then inoculated by the small number of cells that are washed out of the carriers during cultivation. The maximal cell density is then finally reached during the culture.

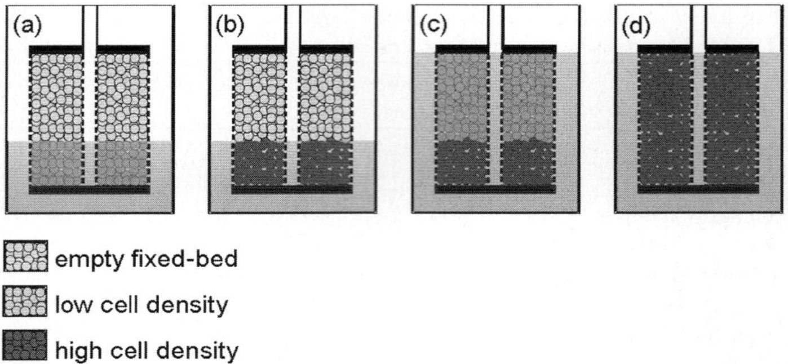

Fig. 5.30 The "inoculation during cultivation" procedure (from Fassnacht 2001). Only a small part of the fixed bed is inoculated at the beginning of an experiment to keep the preculture as small as possible (**a**). After proliferation (**b**), the medium level is raised in the conditioning vessel, and a fresh part of the fixed bed is automatically seeded by circulating cells (**c**) which then proliferate again (**d**)

This strategy was successfully applied in a 5.6 L radial flow fixed bed bioreactor. The start-up phase was divided into three steps, each step covering one-third of the fixed bed in the bioreactor. With a very low initial cell density of 5×10^4 cells per milliliter of medium for the used type of fixed bed bioreactor, harvested from two roller bottles, it was possible to start the cultivation. The bioreactor has proved already to achieve cell densities higher than of 5×10^7 cells per milliliter fixed bed, meaning an increase in cell density of more than 1300 times. This is at least one order higher compared to suspension bioreactors run in fed-batch mode or continuous perfusion. Therefore, with this pilot-scale bioreactor system a significant reduction of the preinoculation steps was achieved. The reduction of the number of cells to be inoculated into a fixed bed bioreactor by means of the optimization of the initial cell density and by means of scaling-up strategies will lead to the reduction of the inoculation steps during the cultivation cascades in industrial processes.

5.1.2.5 Summary of Fixed Bed and Fluidized Bed Bioreactors

Fixed bed and fluidized bed bioreactors have proved their potential for cultivation of mammalian cells for a broad range of applications, not only for production of biopharmaceuticals but also for tissue engineering and gene therapy. With respect to production of biopharmaceuticals, advantages can be seen in (1) high volume-specific cell density, (2) high volume-specific productivity during long-term cultures (compare Table 5.7), (3) low shear forces, and (4) simple medium exchange and cell/product separation. As disadvantages the following have to be mentioned: (1) concentration gradients, (2) inhomogeneous cell distribution, (3) impossibility to directly count cells. Reliable determination of cell density in immobilized cultures used to be a drawback of fixed bed and fluidized bed bioreactors, as most

Table 5.7 Volume specific productivity for monoclonal antibodies of hybridom cells grown continuously in fixed bed or fluidized bed bioreactors and in batch suspension culture

Reference	Culture system	Carrier	Antibody productivity $(mg\ L^{-1}\ d^{-1})$	Antibody productivity related to productivity in batch suspension culture
Fassnacht (2001)	Batch suspension		5	1
	Fixed bed	SIRAN, 3–5 mm	450[a]	90
Celonic (2007)	Batch suspension		8	1
	Fluidized bed	SIRAN, 0.7 mm	435[b]	54
Lundgren and Blüml (1998)	Batch suspension		2.5	1
	Fluidized bed	Cytoline 1	40[b]	16
	Fluidized bed	Cytoline 2	110[b]	44

[a]Productivity related to the fixed bed volume
[b]Productivity related to the carrier volume

carriers applied for these bioreactors cannot be solubilized as can be done enzymatically with carriers consisting of proteinaceous materials (e.g. collagen). This problem could be overcome to some extent by monitoring cell growth by the use of dielectric spectroscopy (Noll and Wandrey 1997; Ducommun et al. 2001). Both fixed bed and fluidized bed reactors have been scaled up at least to a pilot scale.

Compared to suspension bioreactors, still as the preferred type of bioreactor system on an industrial scale, fixed bed bioreactors can be regarded as a reliable, multipurpose production system. Whereas cultivation of production cell lines in suspension often requires extensive optimization of medium and cell line (e.g., tolerance to shear stress) or process conditions (e.g., stirrer speed, aeration rate), most cell lines can be cultivated in fixed bed bioreactors without further process development. Therefore, advantages of this technology can be seen especially in those cases in which unoptimized cell lines or media are used, e.g., for production of small quantities of a product for cell line selection or clinical trials. A further advantage can be seen in the reduced preculture steps compared to suspension bioreactors. Applications in the field of tissue engineering and gene therapy, in which usually adherent cells are applied and tissue-like structures are required, will further promote especially the use of fixed bed bioreactors.

5.1.3 Membrane Bioreactors

5.1.3.1 Introduction

The recent progress in genetic engineering and the advances in analytical methodologies have boosted the growth of the biotechnology industry in the past decade, a growth that is expected to continue steadily in the next 10–20 years. An increasing number of protein therapeutics (e.g., proteins, vaccines, monoclonal antibodies, hormones, growth factors, etc.) are entering, or will soon enter, the marketplace, which are produced in large-scale mammalian cell culture. Among those licensed lately, typical examples are provided by Herceptin, a monoclonal antibody (mAb) for treating breast cancer; Enbrel, an immunoglobulin-tumour necrosis factor (TNF) receptor fusion protein for treating rheumatoid arthritis; Vaqta, an inactivated Hepatitis A vaccine; and OKT3, an immunosuppressant (Chu and Robinson 2003). Critical issues to large-scale mammalian culture are oxygen supply, waste product accumulation, shear sensitivity of animal cells, possibility of growing adherent cells, and sophisticated control schemes (Glacken et al. 1983). The reactor technology of choice for large-scale cell culture in most cases tends to settle on the standard stirred-tank bioreactor (i.e., the chemostat) (Sect. 4.4.2). Solution to its inherent drawbacks is approached by adapting anchorage-dependent cells in preference to suspension culture (e.g., by growing cells in adhesion on microcarriers), by minimizing shear damage through the use of additives (e.g., Pluronics F68) and bubbleless gas spargers, and by maximizing oxygen supply with the use of more efficient spargers and better bioreactor fluid dynamics. Suspension cultures in stirred bioreactors or production animals are easy to operate and permit easy cell harvesting, but operate at low cell and product concentrations and require extensive downstream processing for product purification. Moreover, the risk persists of contamination by adventitious agents through the raw materials, as in the case of viruses, antigenic proteins, nucleic acids, and bioactive substances used to prime the peritoneum of mice prior to the inoculation of hybridoma cells for mAb production. The selling price of a product has been reported to be inversely proportional to its concentration in the culture medium in or leaving the bioreactor (Dwyer 1984). This is an incentive to develop bioreactor design and operation modes yielding higher product concentrations in the culture medium than those in stirred-tank bioreactors.

To this purpose, membranes can be effectively used to segregate or entrap cells in bioreactors (hence the name *membrane bioreactors*) operated either in static conditions or in perfusion mode. In membrane bioreactors (MBRs), cells are separated from nutrients and reactants by semipermeable barriers (i.e., membranes) that are totally impermeable to cells, thereby permitting their physical entrapment, yet are freely permeable to nutrients and products, and that may function as scaffolds for anchorage-dependent cells. Membrane bioreactors are advantageous because they permit cell culture at high densities (greater than 10^8 cells per milliliter) nearing those in vivo, may provide a privileged physical environment for cell culture by combining low shear forces, to avoid damage to shear-sensitive cells, with continuous nutrients supply and waste removal from the cell compartment, and permit the use of protein-free culture media. These features ultimately yield large productivity per bioreactor volume, and high product concentrations associated with lower downstream processing costs (Schonberg and Belfort 1987).

In the 1970s, Porter and Michaels first proposed the use of membranes to enhance cell reactor (or fermentor) productivity (Porter and Michaels 1970) and Knazek and Gullino showed that mammalian cells could be grown around and outside hollow fiber membranes (Knazek et al. 1972). The metabolic behavior of adherent cells, the fluid dynamics of the cell compartment, and ultimately the membrane bioreactor performance are strongly influenced by the properties of the membrane used. For this reason, in the following the properties of the membranes used in MBRs are briefly discussed. More detailed information can be found in available membrane textbooks.

5.1.3.2 Membranes for MBRs

A semipermeable membrane is generally defined as "a phase acting as a barrier that prevents the net motion of mass but permits the regulated and restricted passage of one or more species". Many different semipermeable barriers meet this definition. Therefore, commercial membranes are generally classified according to

- The chemical nature of the material they are made of. Membranes may be made of inorganic or polymeric material. The polymer may be of natural origin (e.g., cellulose) or synthetic (e.g., polypropylene);
- The morphology of the wall. The wall of the membrane may have a uniform morphology and porosity (then it is said to be symmetric or isotropic) or may have a 0.1–0.5 µm thick dense layer (e.g., on the lumen side) mechanically supported by a large pore substructure (then it is said to be asymmetric or anisotropic). A symmetric membrane may have pores not visible by SEM (then it is said to be nonporous or dense), or clearly visible pores (then it is said to be porous);
- The separation mechanism (e.g., size vs. charge);
- The force driving the transmembrane transport. When a transmembrane concentration difference promotes solute transport across the membrane, one may call it a *dialysis membrane*. When the transmembrane solute transport is promoted by a transmembrane pressure difference, one may call it a *filtration membrane*. When it rejects solutes smaller than ca. 150 kDa, one may talk of an *ultrafiltration membrane*. When it has the largest pores greater than or equal to ca. 0.1 mm diameter, one may talk of a *microfiltration membrane*;
- The material physical-chemical properties (e.g., charged vs. neutral, hydrophobic vs. hydrophilic).

Transport Across Membranes

Solute transport across the wall of semipermeable membranes is generally promoted by a transmembrane chemical potential difference. In most cases of practical interest, solute transport is driven by existing transmembrane concentration or pressure gradients according to a diffusive and/or convective mechanism. According to the irreversible thermodynamics of two-component mixtures (i.e., where the species in large excess is termed *solvent*, and solute is termed the *species* at much lower concentration), the net mass (or volume) of solvent and solute permeating the membrane per unit time and membrane surface area flux (i.e., the fluxes J_v and J_s, respectively) can be expressede as follows:

$$J_v = L_p \left(\Delta P - \sigma \Delta \pi \right) [=] \text{m s}^{-1} \tag{5.9}$$

$$J_s = P_M \Delta C + J_v \left(1 - \sigma \right) C_{\text{avg}} [=] \text{gmol s}^{-1} \text{ m}^2 \tag{5.10}$$

where L_p is the membrane hydraulic permeability towards the solvent (m³ s⁻¹ m⁻² mmHg⁻¹); P_m is the solute diffusive permeability in the membrane (m s⁻¹); C_{avg} is the solute weighted average concentration between upstream and downstream membrane sides; and σ is the dimensionless Stavermann's reflection coefficient expressing the solute capacity to permeate the membrane. $\sigma = 0$ means that the solute freely permeates the membrane, and $\sigma = 1$ means that the solute is fully rejected by the membrane. Solvent transport models and experimental evidence show that L_p is inversely proportional to the permeating fluid viscosity and proportional to the squared average pore radius. It follows that L_p of a given membrane towards protein-rich media is lower than that of pure water because of the higher viscosity of the former. P_m is the reciprocal resistance to solute transport across the membrane for pure concentration-driven diffusive transport (i.e., no net solvent flux). According to Fick's law, it is proportional to the solute effective diffusivity in polymeric membranes, D_m, which decreases with the solute molecular weight more rapidly than that predicted by the Einstein–Stokes equation

$$D_s = \frac{R_u T}{6 \pi \mu a N_A} \tag{5.11}$$

for diffusion in liquids. In fact, the solute mobility in the membrane depends also on the membrane pore structure and distribution, and the solute-to-pore radius ratio. Its value is best estimated from experimental information reported in the literature. Diffusive fluxes are often referred to concentration gradients in the liquids on the two sides of the membrane. However, solutes may dissolve differently in the membrane than in the liquids on the two sides of the membrane depending on their physico-chemical properties (e.g., charge, hydrophilicity, etc.). In fact, solutes bearing a charge may partition differently between the liquid and a charged polymeric membrane. High-molecular-weight (HMW) solutes might even not enter small membrane pores, which would preclude their transfer. Independent of the nature of the interactions between the solute and the membrane, solute partitioning between the liquid and the membrane is often accounted for by means of a solute partition coefficient, K, defined as the solute concentration ratio in the membrane and in the liquid, as follows:

$$K = \frac{C_M^w}{C_L^w} [=] - . \tag{5.12}$$

K is often assumed to be the same on both sides of the membrane. Neglecting curvature effects and accounting for the effects mentioned above, the solute diffusive permeability in the membrane may be expressed as follows:

$$P_{M} = \frac{1}{R_{M}} = \frac{D_{M}K}{\delta}[=]\ \text{m s}^{-1} \tag{5.13}$$

where δ is the membrane wall thickness. Equations (5.9) and (5.10) show that the convective contribution to solute transport across the membrane depends on the membrane hydraulic permeability, L_p, and the capacity of the solute to permeate the membrane (i.e., σ). The membrane hydraulic permeability is determined by the pore density and pore size distribution, and is better determined experimentally in pure filtration experiments. The complement to 1 of membrane sieving coefficient for a solute at very high solvent flux, $(1 - S)$, is generally used in the place of σ. It is often termed the *rejection coefficient, $R = 1 - S$*, and measures the membrane capacity to retain or reject a solute. S is defined as the solute concentration ratio at the liquid–membrane interface downstream (i.e., on the permeate side) and upstream (i.e., on the retentate side) of the membrane in pure filtration experiments, in the absence of mass transport resistance on both membrane–liquid interfaces. The value of S generally accounts also for possible solute partitioning, as follows:

$$S = \frac{C_{P}^{W}}{C_{R}^{W}} \cong (1-\sigma)[=]-. \tag{5.14}$$

In the case that the solute concentration in the retentate stream (i.e., the fluid that has to be filtered) changes along the membrane length from the module inlet to the outlet, S is often estimated as follows:

$$S \cong \frac{2C_{P}^{w}}{C_{R,i}^{w} + C_{R,o}^{w}}. \tag{5.15}$$

To optimize the bioreactor performance, it is important to know the membrane capacity to reject solutes of different molecular weights (MWs) or sizes. This information is generally provided in terms of the sieving coefficient spectra, where S is reported for solutes of increasing MWs. The slope of these curves is determined by the membrane pore size distribution, steeper slopes indicating more selective membranes. More slanted slopes indicate poorly selective membranes that are not able to separate solutes with similar MWs. Commercially, the separation properties of a membrane are often expressed in terms of its nominal molecular weight cut-off (NMWCO), which for membrane bioreactors generally indicates the MW of a solute with $S = 0.95$. It is important to mention that the actual transfer rate of partially rejected solutes may also be affected by concentration polarization (i.e., solute accumulation at the liquid–membrane interface) and fouling (i.e., irreversible or reversible solute adsorption on/in the membrane surface or pores) phenomena. A more detailed description of membrane preparation, properties and of concentration polarization, and fouling phenomena is beyond the scope of this chapter and can be found in textbooks on membrane science (Mulder 1990).

Membranes for MBRs

In MBRs, polymeric membranes are generally used that have been developed and marketed for other technical and medical applications. Symmetric microfiltration membranes with a maximal pore size of a fraction of a micron, mostly made of polyolefines (e.g., polypropylene), are used when only the cells in a suspension have to be retained in the cell compartment. Serum proteins are then freely exchanged with the medium compartment and secreted proteins may be recovered in the medium compartment. Asymmetric ultrafiltration membranes with an inner dense layer, often made of a polysulfone or cellulose derivatives, are used in flat-sheet or hollow-fiber configuration when cells and HMW proteins have to be retained in the cell compartment. The use of ultrafiltration membranes makes it possible to prime the cell compartment with medium supplemented with the serum and to feed the bioreactor with serum-free medium, since serum proteins cannot leave the cell compartment. In membrane-compartmentalized cell reactors (MCCRs) (Sect. 5.1.3.3), cells are generally cultured outside hollow fiber membranes. Endothelial cells are more often cultured under a shear-stress field in adhesion on the inner membrane surface. Cell presence and high protein concentrations on the outer membrane surface may cause severe fouling, which may cripple the membrane exchange capacity.

5.1.3.3 Membrane Bioreactor Design

A continuous stirred-tank bioreactor (Sect. 4.4.2) can be connected in closed loop with a membrane module in a configuration often referred to as *cell recycle membrane reactor* (CRMR) or *membrane fermentor*, to continuously filter the suspension containing medium, nutrients, products, and cells, and return the cells to the bioreactor tank while removing LMW products at a flow rate equaling that of the feed (Fig. 5.31). A number of configurations have been proposed differing in the membrane geometry (e.g., flat slab, tubular or hollow fiber), transport and separation properties (i.e. membranes for dialysis, ultrafiltration or microfiltration), and the operation mode. Cells may also be grown compartmentalized by means of membranes within the shell of a hollow fiber membrane module in which medium flows in the membrane lumen (i.e., under dynamic culture conditions), or between flat-sheet membranes of different properties under static conditions (Fig. 5.32). This configuration may be referred to as *membrane compartmentalized cell reactor* (MCCR). In the latter configuration, medium controlled for its temperature, pH, and dissolved oxygen concentration may either flow through the membrane module only once (i.e. in single-pass mode), it may be continuously recirculated to the stirred tank in closed loop, or only a fraction of it may be continuously removed at a flow rate equaling that in the feed, as shown in Fig. 5.33.

The design of membrane bioreactors is generally approached by treating the cell compartment as pseudo-homogeneous. The mass balance equations are written for a finite or differential control volume whose size is generally assumed to be much

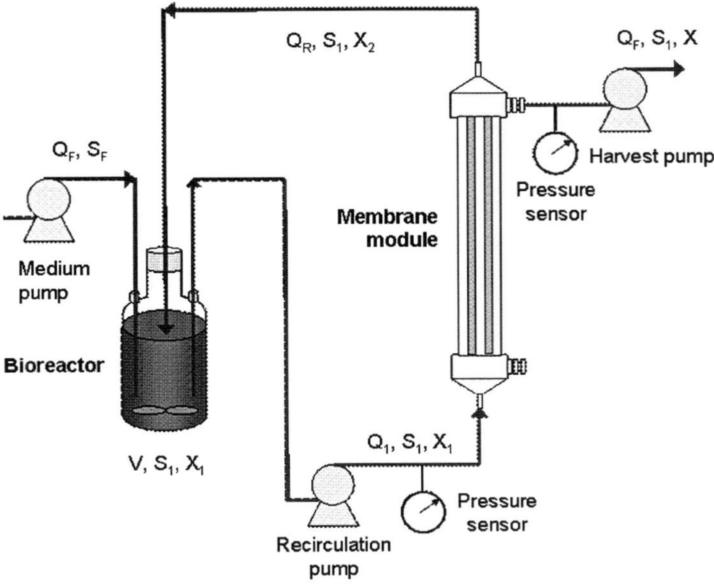

Fig. 5.31 Scheme of a cell recycle membrane reactor

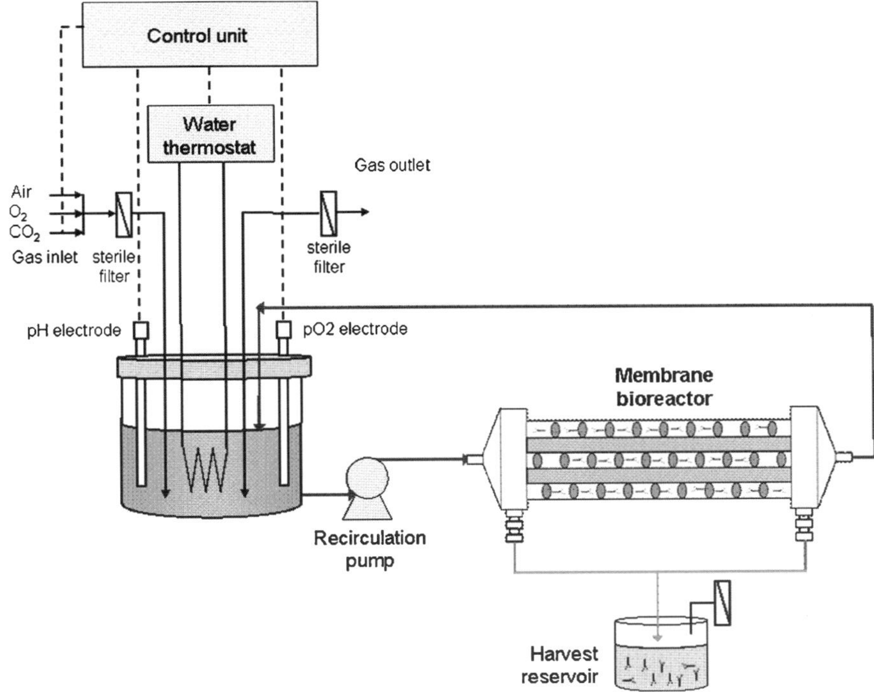

Fig. 5.32 Scheme of a membrane compartmentalized cell reactor

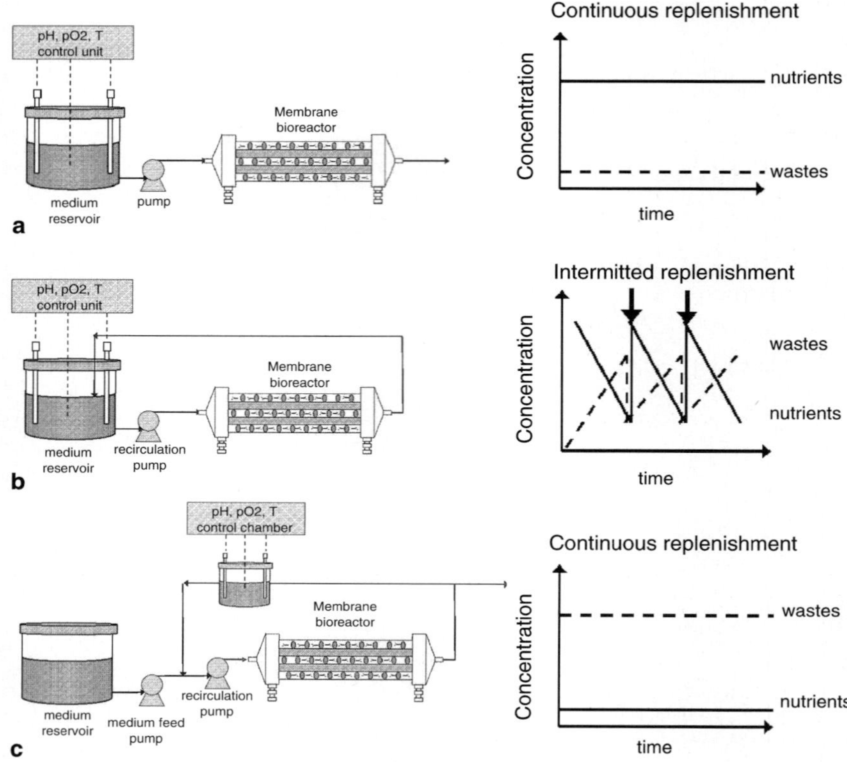

Fig. 5.33 Scheme of different modes of medium supply to a MCCR, together with nutrients and medium concentrations shown on the right side: a) single pass mode; b) closed loop; c) open loop (after Schoenherr and van Gelder, 1988)

greater than that of the cells in it, so that they are considered on par with other solutes and the suspension is approximated to a homogeneous liquid. However, cells and the suspension heterogeneity are taken into account when the transport parameters are estimated (i.e., the suspension viscosity or rheological behavior, and solute diffusivity in it), and by introducing the metabolic reaction kinetics terms in the species mass balance equations. Design of membrane bioreactors is also generally approached under the assumption that the bioreactor is isothermal and operated under steady-state, or pseudo-steady-state, conditions. The simultaneous transport of nutrients and waste products to and away from the cells, the metabolic consumption of nutrients, and the formation of new cells and metabolic products affect the actual concentration of species in the bioreactor and the bioreactor performance to an extent that depends on the bioreactor configuration and operation mode. For this reason, in the following the designs of cell recycle membrane reactors and membrane compartmentalized cell reactors are discussed separately.

Cell Recycle Membrane Reactors (CRMRs)

CRMRs are built by connecting a stirred-tank bioreactor, in which the cells are allowed to grow in a controlled environment, in closed loop with a membrane separation module, as shown in Fig. 5.31. The tanks are designed and provided with stirrers that generally ensure a uniform distribution of cells, solutes, pH, and temperature in the suspension volume. The complete mixing makes process control and maintenance of the stability of the cell slurry relatively easy. When the operating conditions are stable and cells have grown to the desired density and metabolic behavior (e.g., after inducing a metabolic shift) in batch operation, the cell suspension is continuously recirculated from the tank into the membrane module, where a fraction of the liquid medium is filtered out together with valuable metabolic products, and the concentrated cell suspension is pumped back into the tank. Ultrafiltration or microfiltration membranes are generally used to retain the cells in the vessel, and liquid permeation across the membrane is promoted by a transmembrane hydrostatic pressure difference. Nutrients are generally fed into the vessel at a volumetric flow rate equaling that of the permeate to keep the liquid level in the tank constant. Also, in the biological treatment of municipal and industrial wastewaters in large stirred-tank bioreactors, membranes are used as barriers to bacteria and suspended solids to produce a low-turbidity effluent with very low bacterial counts. In this case, the membranes are often submerged in the activated sludge to combine biotransformation with the solid–liquid separation step, as shown in Fig. 5.34 for the Puron® process by Koch. It is worth remembering that cells may be retained in the bioreactor vessel also with centrifuges and acoustic or spin filters, although with some cell loss in the effluent. In this chapter reference is made only to membrane bioreactors. Dialysis membranes are seldom used to selectively remove LMW solutes from the circulating solute-rich medium by diffusion across the membrane against a solute-poor solution. In this last case, the bioreactor is batch-operated.

CRMR Design

CRMRs may attain a steady state only when a part of the cells is able to permeate the membranes used or is continuously removed with the medium. In this case, the bioreactor design equations may be obtained by writing the mass balance equations for cells and solutes about the finite stirred-tank volume and the membrane module. Assuming that the tank is completely mixed, and that one nutrient is limiting the cell growth, the mass balance equations may be written as follows:

cells
$$Q_R X_2 + V X_1 \mu = Q_1 X_1, \tag{5.16}$$

nutrient
$$Q_F S_F + Q_R S_1 = \frac{1}{Y_{X/S}} X_1 \mu V + Q_1 S_1. \tag{5.17}$$

Assuming that suspension density remains nearly constant upon filtration, the mass balance equations about the membrane module yield:

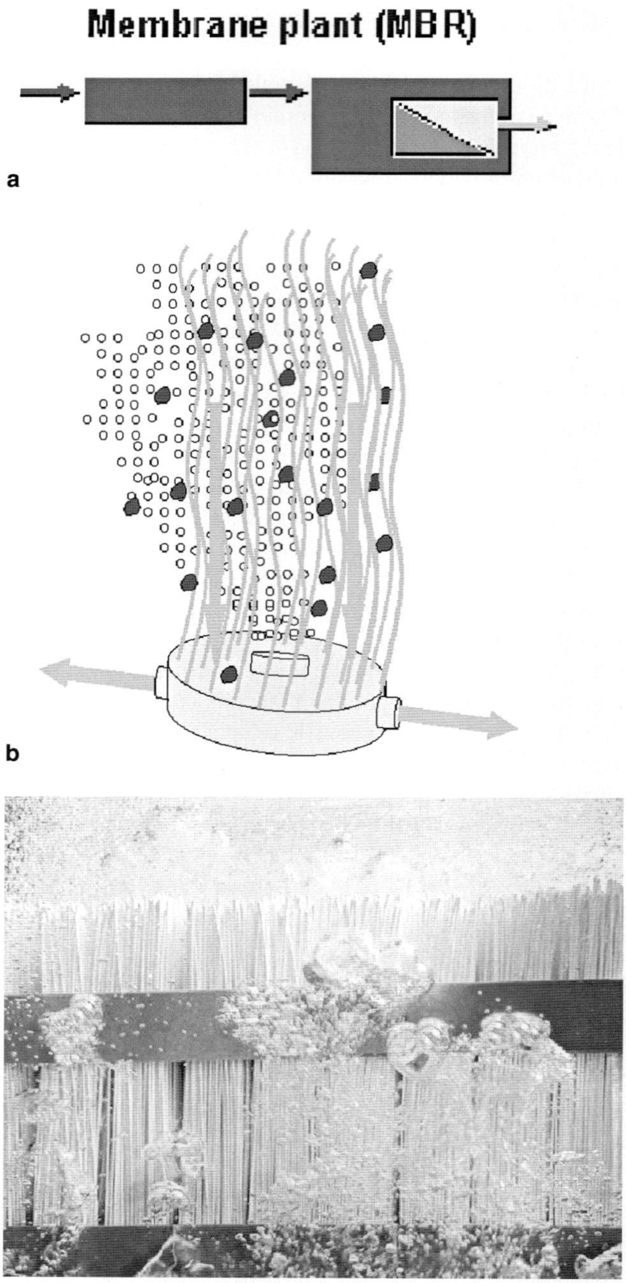

Fig. 5.34 (**a**) Scheme of submerged membrane bioreactor for wastewater treatment; (**b**) Membranes sealed at the top submerged in the activated sludge. Filtrate permeates into the membrane lumen and flows down to the bottom collector, while air bubbles rise from the center of the membrane bundle according to the Puron® scheme, Koch; (**c**) Loose fibres in the bundle of submerged membrane modules by Koch according to the Puron® scheme (Reprinted with permission)

cells
$$Q_1 X_1 = Q_F X + Q_R X_2 \tag{5.18}$$

whole liquid
$$Q_F + Q_R = Q_1. \tag{5.19}$$

Upon introducing the recycle ratio $R_R = Q_R/Q_F$, the cell concentration factor $C = X_2/X_1$ and the dilution rate $D_R = Q_F/V$, and substituting $Q_1 X_1$ from (5.18) in (5.16), and (5.19) for Q_1, equations (5.16)–(5.19) may be rearranged to give:

cells
$$D_R R_R C X_1 + X_1 \mu - D_R X_1 \left(1 + R_R\right) = 0 \text{ and} \tag{5.20}$$

nutrient
$$X_1 \mu = D_R Y_{X/S} \left(S_F - S_1\right). \tag{5.21}$$

Combination of (5.20) and (5.21) yields:

$$X_1 = Y_{X/S}\left(S_F - S_1\right)\frac{1}{\left(1 + R_R - R_R C\right)} = Y_{X/S}\left(S_F - S_1\right)\frac{1}{1 - R_R\left(C - 1\right)}. \tag{5.22}$$

Equation (5.22) shows that cell concentration in a CRMR is greater than that in a continuous stirred bioreactor by the factor $\dfrac{1}{1 - R_R(C-1)}$, both C and R_R being generally greater than 1. Substitution of (5.22) in (5.21) and further rearrangement yield:

$$D_R = \frac{X_1 \mu}{Y_{X/S}\left(S_F - S_1\right)} = \frac{\mu}{1 - R_R\left(C - 1\right)}. \tag{5.23}$$

Equation (5.23) shows that cells growing at a given rate in cell recycle reactors permit the treatment of larger volumetric feed flow rates per unit reactor volume than continuous stirred-tank bioreactors. Table 5.8 shows that continuous operation with cell recycle permits operation at higher cell concentration than batch stirred-tank bioreactors also, thereby reducing the vessel volume necessary for a given productivity and the bioreactor capital costs.

If the cell-specific growth rate can be expressed according to a Monod rate equation (Sect. 2.5) it is easy to show that the critical dilution rate at which cell washout occurs is greater than that of continuous stirred-tank bioreactors by the factor $[1 - R_R(C - 1)]$, as follows:

$$D_{R,crit} = \frac{\mu_{max} S_F}{\left(K_S + S_F\right)\left[1 - R_R\left(C - 1\right)\right]}. \tag{5.24}$$

When membranes completely reject the cells, cells cannot be washed out. However, the bioreactor does not attain a steady state and has to be designed accounting for its transient behaviour. Owing to the nonlinearity of the metabolic kinetics, the resulting unsteady-state mass balance equations are better solved numerically.

High cell concentrations increase the cell suspension viscosity and the pumping costs and limit the rate at which the medium may filter across the membrane. In fact, as rejected cells accumulate at the upstream membrane surface, concentration polarization and fouling phenomena make the permeate flux to become invariant

Table 5.8 Comparison of bioreactors in different configurations: CRMR – cell recycle membrane reactor; CSTR – continuous stirred tank reactor; MCCR – membrane compartmentalized cell reactor; P – product concentration; Pr – productivity; S_F – substrate feed concentration; X – cell concentration

Cell strain	Substrate	Reactor	Pr (g L^{-1} h^{-1})	S_F (g L^{-1})	P (g L^{-1})	X (g L^{-1})	References
Kluyveromices fragilis	Lactose	Batch	3	50	22	3.8	Cheryan and Mehaia (1983)
	Lactose	CRMR	25	50	15	3.8	Cheryan and Mehaia (1983)
	Lactose	CRMR	65	50	10	41	Cheryan and Mehaia (1983)
	Lactose	CRMR	240	150	40	90	Cheryan and Mehaia (1983)
	Lactose	MCCR	63	50	15–25	82	Mehaia and Cheryan (1984)
	Whey lactose	MCCR	35	45	5–10	117	Mehaia and Cheryan (1984)
Lactobacillus delbreuckii	Glucose	Batch	1–2		45	7–8	Vick Roy et al. (1983)
	Glucose	Batch w/ Dialysis	2–3		35	11	Vick Roy et al. (1983)
	Glucose	CSTR	7		38	7–8	Vick Roy et al. (1983)
	Glucose	CRMR	76	30	35	54	Vick Roy et al. (1983)
	Glucose	MCCR	100		2	350	Vick Roy et al. (1983)
	Glucose	Ca-Alg Gel beads	3		46	67	Vick Roy et al. (1983)
S. cerevisiae	Glucose	Dialysis	2.1–5.8				Kyung and Gerhardt (1984)
	Glucose	Batch Multiple	1.8–25				Kyung and Gerhardt (1984)
	Glucose	Vacuum Continuous	44				Kyung and Gerhardt (1984)
	Glucose	CRMR	53				Kyung and Gerhardt (1984)

with the transmembrane pressure (hence polarized regime). In the polarized regime, permeate flow rate (i.e., Q_F) may only be increased by promoting cell transport away from the membrane surface or by using larger membrane surface areas. The actual permeate flux at which a membrane module may be operated in the polarized regime can be effectively related to the operating conditions (e.g., channel geometry and recirculation flow rate) and cell concentration according to Michael's gel theory (Mulder 1990), as follows:

$$J_v = k_x \, Log(X_{gel} \, / \, X),\qquad(5.25)$$

where k_x is the overall cell mass transport coefficient, which can be obtained according to the semiempirical correlations available in the technical literature, X is the cell concentration, and X_{gel} is the cell concentration when a gel/cake forms. More sophisticated theories to account for concentration polarization and fouling effects in the filtration of suspensions can be found in textbooks on membrane science.

Figure 5.35 and Table 5.8 show the typical dependence of CRMR performance on dilution rate and cell concentration, and how it compares with other bioreactors. Table 5.8 shows also that the productivity of the CRMR is generally higher than that of other bioreactors. Figure 5.35 shows that with increasing D_R, productivity increases at first, attains a maximum value, and then decreases. Product concentration is comparable to that of other bioreactors, but decreases with increasing D_R. Substrate concentration continuously increases with increasing D_R. These features suggest that when searching for optimal production conditions, high productivity may be achieved at the cost of lower product and higher substrate concentrations in the permeate stream. This increases the product recovery and purification costs, and sets limits to productivity in those applications in which substrate presence in the product is not tolerated, as is the case of the wastewater treatment in the dairy industry. However, high dilution rates can maintain the concentration of inhibitory metabolic products

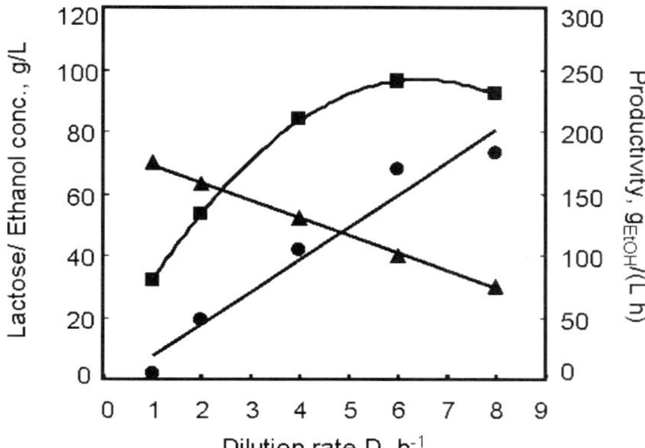

Fig. 5.35 Fermentation performance of a cell recycle membrane reactor with cells of *Kluyveromices fragilis*: cell concentration, $X_o = 90 \, gL^{-1}$; lactose concentration, $S_f = 150 \, gL^{-1}$: (•) lactose, (▲) ethanol, (■) productivity (Adapted from Cheryan and Mehaia 1983)

low (e.g., both primary and secondary metabolites) and may lead to significant enhancement of bioreactor performance. That said, the above suggests that on a case-by-case basis the economics of the whole production process generally determines the optimal design and operation choice.

Membrane Compartmentalized Cell Reactors (MCCRs)

Product purity may be increased while offering an effective protection from shear damage by culturing cells compartmentalized between flat sheet membranes (Kyung and Gerhardt 1984), or in the shell space of a hollow fiber membrane bioreactor (Mehaia and Cheryan 1984) while medium flows in the fibre bore, as originally proposed by Knazek et al. (1972). Unlike suspension cultures in which the medium bathing the cells contains uniform concentrations of soluble nutrients, products, and effectors (e.g., growth factors), semipermeable membranes separate the cells from the medium, and often from a gas stream, and divide the bioreactor into compartments featuring different environments. In fact, ultrafiltration membranes restrict the transport of HMW proteins and permit feeding the bioreactor with serum-free medium once serum proteins have been supplemented to the cell compartment. Irrespective of the membranes used, LMW nutrients and waste products are generally freely transported across the membranes to or away from the cells, respectively. They are transported across the wall of dialysis membranes at lower rates than across ultrafiltration membranes. In the rest of this chapter reference is made to cells compartmentalized outside hollow fiber membranes arranged parallel in a bundle inserted in a cylindrical housing and potted at both ends (i.e., shell-and-tube configuration), as shown in Fig. 5.36. In fact, this configuration maximizes the membrane surface area to bioreactor volume ratio, which is particularly attractive when anchorage-dependent cells are used. The compartment where cells are grown is referred to as *the extracapillary space* (ECS), whereas the compartment where culture medium is fed is referred to as the *intracapillary space* (ICS). Extension to membranes with other geometries is rather straightforward.

MCCRs have been used for bacterial and yeast cells, but their typical features have been exploited for mammalian cell culture for the production of biologicals. They can support culture at near *in vivo* density of both anchorage-dependent and independent cells. The continuous perfusion of the ICS with medium simulates the *in vivo* environment by continuously supplying nutrients to, and removing waste metabolites from, the cells. This permits maintainence of cell viability and production for long periods of time. These bioreactors allow more efficient use of expensive HMW medium components (in fact, dialysis and ultrafiltration membranes of suitable NMWCO retain HMW proteins in the ECS), the production of highly concentrated products, and a large productivity per reactor volume. However, LMW medium components (e.g., growth factors) and autocrine factors partition across the membrane on the basis of their transport and separation properties, cellular consumption (or production), and the solute mobility, and have to be continuously supplied. Moreover, when cell density is high and approaches that of organs ($>10^8$ cell mL^{-1}) transport processes

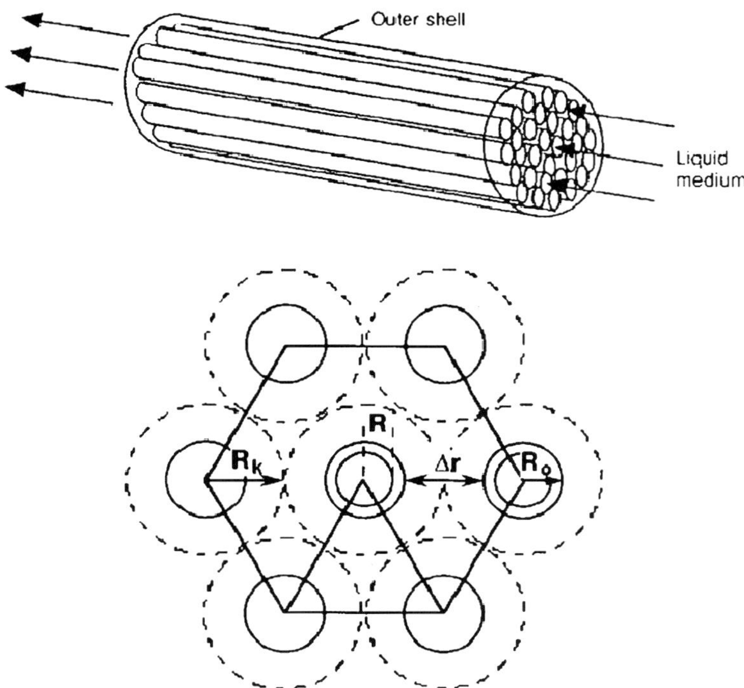

Fig. 5.36 Scheme of multiple hollow fiber membranes surrounded by cells laid in a Krogh cylinders (Reprinted with permission from Brotherton and Chau 1996)

may limit the reactor productivity and even cause loss of cell viability. For this reason, in addition to maximizing the productivity of the species of interest, the design of MCCRs has to ensure that nutrients and substrate gradients do not cause cell death.

MCCR Design

In MCCRs in which cells are cultured in the ECS of a hollow fiber membrane module, nutrients are fed to the membrane lumen and are transported by a diffusive and/ or convective mechanism across the membrane and to the cells. Metabolic products are produced in the ECS and are transported into the ICS. Convective transport is to a large extent due to the occurrence of a secondary flow in the ECS (often termed Starling flow (Starling 1896)). In fact, at the entrance of the bioreactor the pressure in the membrane lumen is higher than that in the ECS. This causes a net transport of the medium from the ICS to the ECS, which carries solutes with it. At the exit of the bioreactor, the lumen pressure is lower than that in the ECS owing to the axial pressure drop, and medium from the ECS is reabsorbed into the ICS. The effect of convective transport on bioreactor performance, as well as on nutrients and waste product concentration profiles in the ECS, depends on membrane hydraulic permeability, the inlet pressure, as well as the cell density and metabolic activity in the ECS. The opposing requirements to maximize the bioreactor productivity and to prevent nutrient and oxygen starvation suggest the need of a bioreactor optimization strategy. In fact, productivity increases with the number of cells cultured in the ECS, which suggests a design of the bioreactor with large interfiber spaces to maximize the ECS volume to host the cells. To ensure that all cells are supplied with nutrients and oxygen, the interfiber space is made as small as possible to reduce the solute pathway length in the cell mass in the ECS. Optimization strategies take into account economic considerations, which are beyond the scope of this chapter. However, they are better based on a sound description of transport and metabolic reaction phenomena in the bioreactor for the specific cell/production system considered. These bioreactors consist of at least three compartments, namely, the liquid, the membrane, and the cell compartment. For an isothermal, axisymmetric bioreactor at steady state, balance equations ought to be written about a two-dimensional differential control volume, and should state the continuity and the conservation of momentum and mass of the relevant species (i.e., the limiting substrate, nutrients, cells, and relevant metabolic products) in each compartment. The design equations proposed in the literature generally introduce some simplifying assumptions to ease the problem. Solutes are assumed to be transported across the membrane only along its radius. Transport in the cell compartment is generally treated only either immediately after cells have been seeded, or when cells have grown to tissue-like density. In the former case, solute transport in the ECS is unobstructed by the cells, and the cell compartment is treated as a continuum. In the latter case, solute transport in the ECS is obstructed by the cells. The cell compartment is generally treated as an isotropic (same properties along the coordinate axes) porous medium, and transport is described in terms of Darcy's law (Apelblat et al. 1974):

$$v_i = -\frac{K_{pm}}{\eta_{fl}} \frac{\partial P}{\partial x_i} \quad \text{with } i = 1,2 \text{ and } x1 = z; x2 = r. \tag{5.26}$$

where K_{pm} is the intrinsic permeability of the porous medium, and η_{fl} is medium viscosity in the ECS. In all cases, the cells are assumed to surround each fiber forming a uniform cylindrical annulus (the Krogh cylinder), and the Krogh cylinders of neighbouring fibers are assumed to touch each other (Fig. 5.36). The outer surface of the Krogh cylinder is assumed adiabatic to mass transport, therefore neglecting mass transport in the interstices among neighboring cylinders.

Cell presence is accounted for in the value of the transport parameters (e.g., the solute effective diffusivity, or K_{pm}) and by the metabolic reaction term in the mass balance equations. Cells are generally assumed to grow according to a Monod kinetics, although first- and zero-order kinetics are often used to ease the mathematical treatment. Even under these simplifying assumptions, many partial differential mass balance equations are required to describe the transport, which are often nonlinear and are better solved numerically.

Dimensional analysis of these equations when solute transfer among compartments occurs by a convection–diffusion mechanism shows that the solute concentration profiles in each compartment (hence the bioreactor performance) are determined by the following dimensionless groups: namely, the reduced axial Peclet number, $Pe_{ax} = \dfrac{u_o L}{D_s}\left(\dfrac{Ri}{L}\right)^2$; the inlet Reynolds number, $\mathrm{Re}_{in} = \dfrac{u_o R_i}{\eta_{fl}}$; the fibre pressure modulus, $\alpha = L\left(\dfrac{16\eta_{fl} L_p}{R_i^3}\right)^{\frac{1}{2}}$; the local wall Peclet number, $Pe_w(z) = \dfrac{v_w(z) R_i}{D_s}$; the initial or feed squared Thiele modulus, $\Phi^2 = \dfrac{\mu_{max} X_o}{Y_{X/S} S_F D_s}(R_K - R_o)^2$ and the dimensionless yield coefficient, $\Psi = \dfrac{S_F Y_{X/S}}{X_o}$ (Brotherton and Chau 1996).

The reduced axial Peclet number compares the convective to diffusive rate of solute transport along the fiber length in the lumen, $Pe_{ax} >> 1$ indicating that solute axial diffusion may be neglected relative to axial convection. The inlet Reynolds number provides information on the magnitude of convective transport in the fiber lumen at the bioreactor entrance. The fiber pressure modulus measures the magnitude of the convective, transmembrane Starling flow into the ECS. Large pressure moduli (i.e., $\alpha \geq 1$) indicate that a large fraction of the flow entering the bioreactor leaves the lumen, enters the ECS in the inlet half, and re-enters the lumen in the outlet half. Small pressure moduli (i.e., $\alpha << 1$) indicate that Starling flow in the ECS is insignificant, and that solute transfer from the ICS to the ECS, and vice versa, occurs mainly by diffusion. The local wall Peclet number compares the local ratio of radial convective to diffusive solute transport in the lumen. The Thiele modulus compares the maximal rate of substrate utilization to the maximal rate of substrate diffusion in the cell compartment. Information on how design and operating conditions affect the bioreactor performance may be effectively gathered by considering the case of a bioreactor under pseudo-steady-state conditions (i.e., constant cell concentration in the ECS) with Monod kinetics of substrate utilization at start-up (i.e., transport unobstructed by cells), and when cell density nears that in a tissue (i.e., transport obstructed by cells) (Brotherton and Chau 1996). Commercial membranes and membrane modules used for MCCRs feature geometric characteristics (i.e., length, L, membrane inner diameter, R_i) and membrane permeability (i.e., L_p) ranging from those typical for small bench-top units equipped with ultrafiltration membranes (i.e., $L = 0.1\,\mathrm{m}$, $R_i = 1.10^{-4}\,\mathrm{m}$, $L_p = 10^{-10}\,\mathrm{m^2\ s\ kg^{-1}}$) to those typical of large units for wastewater treatment equipped with microfiltration membranes (i.e. $L = 1\,\mathrm{m}$, $R_i = 2.5 \times 10^{-4}\,\mathrm{m}$, $L_p = 10^{-8}\,\mathrm{m^2\ s\ kg^{-1}}$). Correspondingly, the α values for aqueous media (i.e., $\eta_{fl} \approx 10^{-3}\,\mathrm{kg\ m^{-1}\ s^{-1}}$) may be expected to range from ca. 0.1–3. In the case of closed-shell bioreactors, Fig. 5.37 shows that the pressure profiles in the ICS drop linearly along the bioreactor length, whereas the pressure in the ECS remains about constant at half the inlet lumen pressure (Kelsey et al. 1990). The closed-shell constraint leads to equal ICS and ECS pressures halfway down the bioreactor length. At the highest pressure modulus value (i.e., $\alpha = 3$), the axial pressure profile in the ICS is only slightly nonlinear. Figure 5.38 shows the typical fluid streamlines (i.e., the trajectories of fluid elements) generated by the pressure profiles and module geometry. Similar streamlines are obtained for all α values, but the radial and axial velocity values are rather different. In fact, at $\alpha = 3$ about 60% of the flow rate

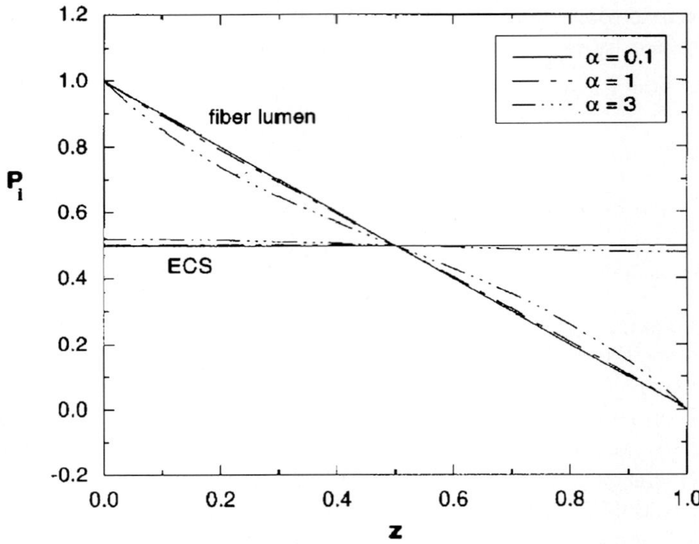

Fig. 5.37 Axial pressure profiles in the ICS and ECS of a MCCR at various a: (---) $\alpha = 0.1$; (-- - --) $\alpha = 1$; (--- - - ---) $\alpha = 3$ (Reprinted with permission from Brotherton and Chau 1996)

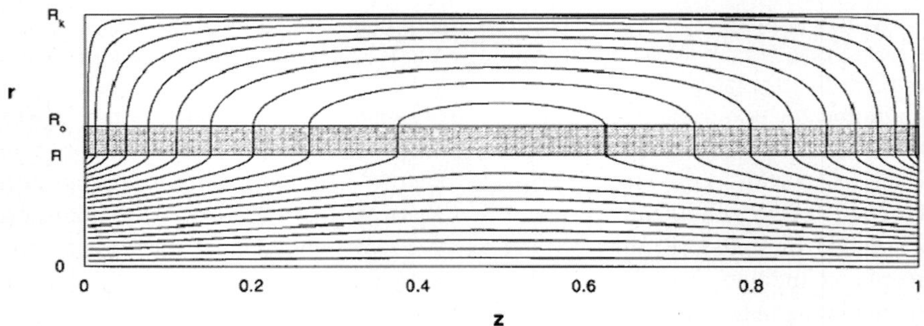

Fig. 5.38 Fluid streamlines within a MCCR for $\alpha = 3$ (Reprinted with permission from Brotherton and Chau 1996)

entering the membrane lumen flows into the ECS in the inlet bioreactor half and re-enters the ICS in the outlet bioreactor half. This amount decreases to about 40% at $\alpha = 1$ and to less than 5% at $\alpha = 0.1$. At the smallest pressure modulus, the effect of convection on solute transfer between ICS and ECS is negligible, and solutes are mainly transported by diffusion. The extent of the effect of the actual feed flow rate on radial convection at various pressure moduli is shown in Fig. 5.39 in terms of the axial profile of the local wall Peclet number normalized by the inlet Reynolds number. In bioreactors in which dialysis or low ultrafiltration membranes are used so that α is small (e.g. $\alpha = 0.1$), Pe is as low as 0.3 and it does not exceed 30 as Re increases by 2 orders of magnitude from 5 to 500 as the Pe/Re values suggest.

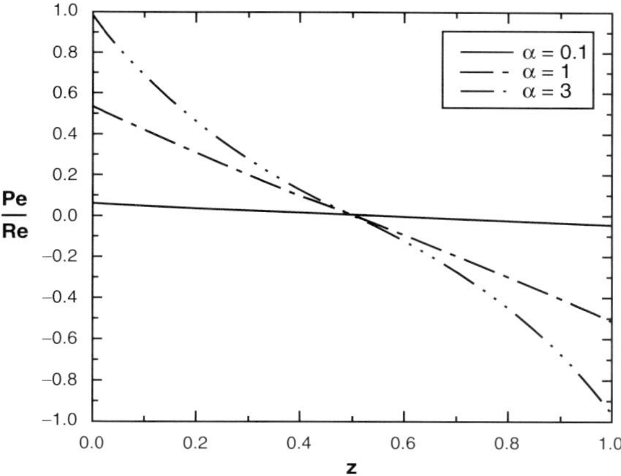

Fig. 5.39 Axial profiles of the wall Peclet number normalized by the inlet Reynolds number in a MCCR for fluid streamlines within a MCCR for at various a: (---) $\alpha = 0.1$; (-- - --) $\alpha = 1$; (--- - - ---) $\alpha = 3$ (Reprinted with permission from Brotherton and Chau 1996)

When very permeable microfiltration membranes are used, α increases to as much as 3, and the maximal Pe attains values as high as 500 as Re = 500, thereby ensuring the occurrence of large convective medium flows towards the ECS. Pseudo-steady-state spatial concentration profiles have been reported in the literature for the transport of dissolved oxygen in MCCRs, which can be effectively used to illustrate how the dimensionless groups affect the supply of limiting substrate to cells (Brotherton and Chau 1996). Cell concentrations in the ECS of 10^7 cells mL^{-1} and 10^9 cells mL^{-1} are representative of conditions at bioreactor start-up and when the cells have grown to tissue-like density, respectively. Typical values for the parameters determining the Thiele modulus for dissolved oxygen are $\mu_{max} = 6.4 \times 10^{-6}$ s^{-1}; $S_F = 0.2$ mol m^{-3}; $D_s = 2 \times 10^9$ m^2 s^{-1}; and $Y_X/S = 10^{11}$ cells mol^{-1} (Wei and Russ 1977). For thicknesses of the Krogh cylinder from 100 to 500 µm, and cell concentrations from start-up to tissue density, the order of magnitude of the squared initial Thiele modulus ranges from 1 to 10. At start-up, MCCRs equipped with permeable microfiltration membranes ensure adequate ECS perfusion and good nutrient transport to all cells, as Fig. 5.40 shows. When less permeable membranes are used (e.g., low-flux ultrafiltration or dialysis membranes), already at low squared initial Thiele moduli (i.e. Φo = 1) nutrient concentration at the exit end of the ECS is less than 20% of the feed value and the bioreactor is under diffusion control (Fig. 5.41a). In the presence of more metabolically active cells, or at higher cell concentrations (but assuming that the ECS can still be considered a continuum), cells are severely deprived of nutrients (Fig. 5.41b). Increasing the Reynolds number by an order of magnitude alleviate nutrient deprivation also with poorly permeable membranes (Fig. 5.41c). Figure 5.42 shows that, also when very permeable membranes (i.e., $\alpha = 3$) are used, operation at high cell concentrations and low Reynolds number causes severe nutrient deprivation. Figure 5.43 shows that when cells at high concentrations organize in aggregates featuring tissue-like hydraulic

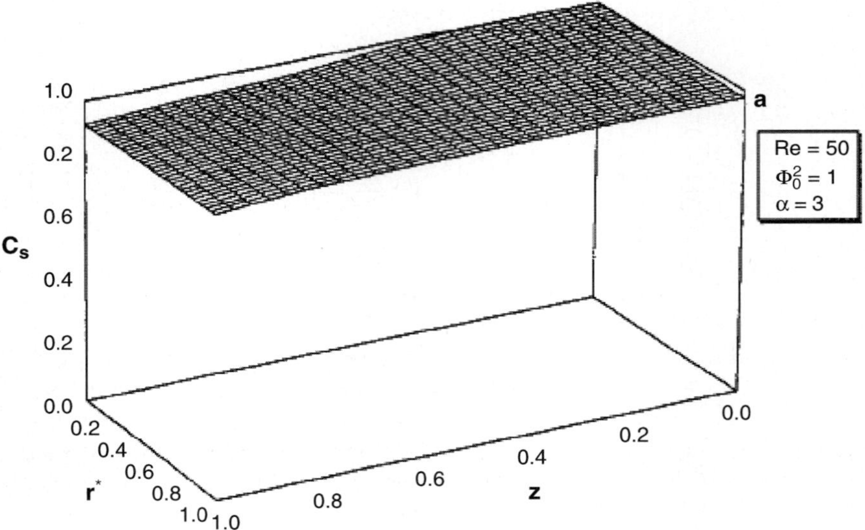

Fig. 5.40 Pseudo-steady-state oxygen profiles in the ECS of a MCCR, with $\Phi_o^{\,2} = 1$, Re = 50, for $\alpha = 3$. The coordinate r^* is defined as zero at the fiber outlet wall radius and unity at the Krogh cylinder radius (Reprinted with permission from Brotherton and Chau 1996)

resistance, practically all cells are cultured under even more severe nutrient deprivation also when the MCCR is equipped with microfiltration membranes.

The behavior briefly discussed above has been qualitatively confirmed in experiments. Table 5.8 reports some other typical features of MCCRs. In particular, it shows that higher productivity and yield may be obtained than with stirred-tank reactors. Figure 5.44 shows that for whey permeate fermentation to ethanol by *Kluyveromices fragilis*, similar to cell recycle membrane reactors, high productivities are obtained from an MCCR at the cost of low substrate conversions at least at low dilution rates. MCCRs are also very effective in the culture of cells inhibited by the metabolic products. In fact, LMW products diffuse or are transported away from the cells as they are produced and are diluted out in the much larger volume of the stirred tank, thereby minimizing their effect on cells cultured in the ECS.

5.1.3.4   Commercial Membrane Bioreactors and Applications

CRMRs are generally commercialized by firms with expertise in membrane filtration processes for cell recovery or downstream processing of biological products and in bioreactors for cell culture, or fermentors. In fact, a CRMR can be easily set up by coupling commercial stirred-tank bioreactors (e.g., a chemostat) through the bioreactor ports to a commercial ultrafiltration or microfiltration membrane module via an external circulation loop in which the cell suspension is kept flowing by means of pumps. This approach makes it easy for these firms to manufacture

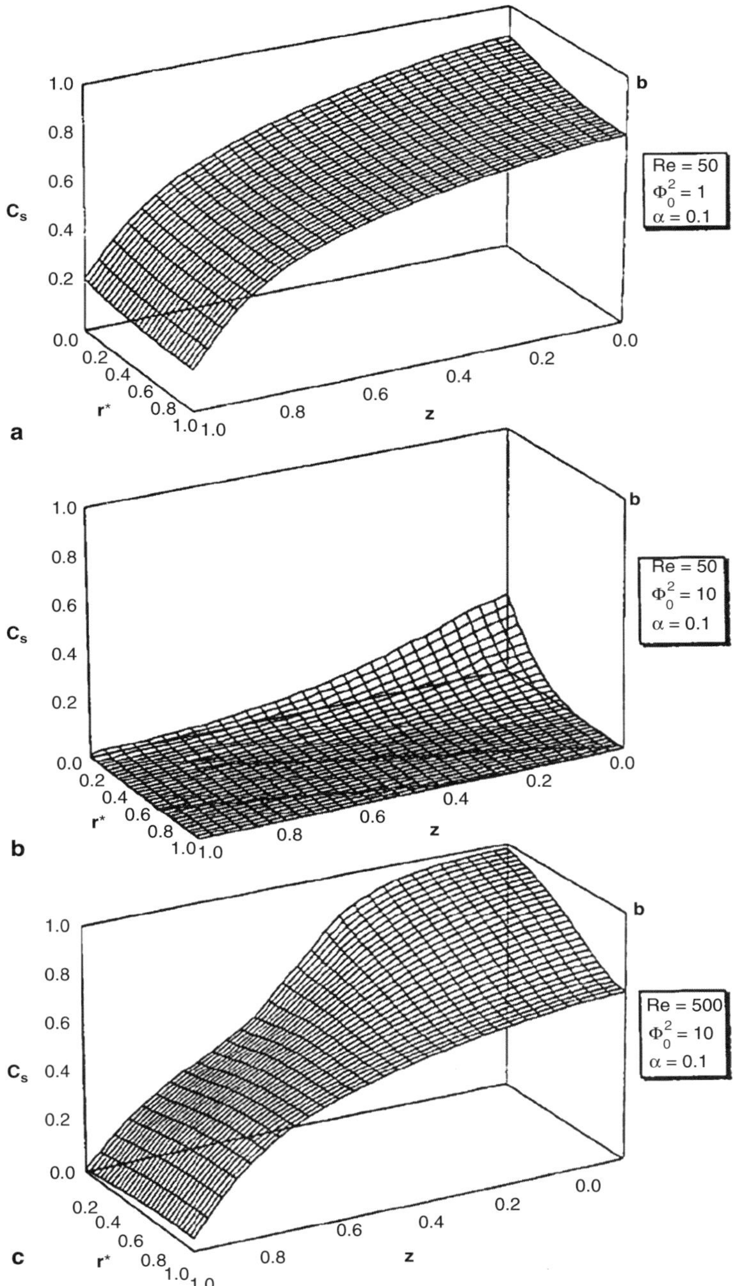

Fig. 5.41 Pseudo-steady-state oxygen profiles in the ECS of a MCCR, with: (**a**) $\alpha = 0.1$, $\mathrm{Re} = 50$ $\Phi_0^2 = 1$; (**b**) $\alpha = 0.1$, $\mathrm{Re} = 50$ $\Phi_0^2 = 10$; (**c**) $\alpha = 0.1$, $\mathrm{Re} = 500$, $\Phi_0^2 = 10$. The coordinate r^* is defined as zero at the fiber outlet wall radius and unity at the Krogh cylinder radius (Reprinted with permission from Brotherton and Chau 1996)

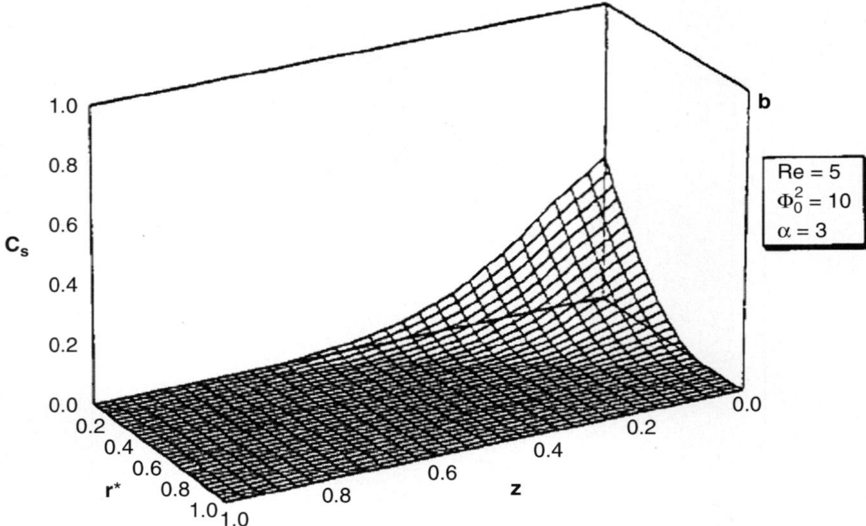

Fig. 5.42 Pseudo-steady-state oxygen profiles in the ECS of a MCCR, with $\alpha = 3$, Re = 5, $\Phi_o^2 = 10$. The coordinate r^* is defined as zero at the fiber outlet wall radius and unity at the Krogh cylinder radius (Reprinted with permission from Brotherton and Chau 1996)

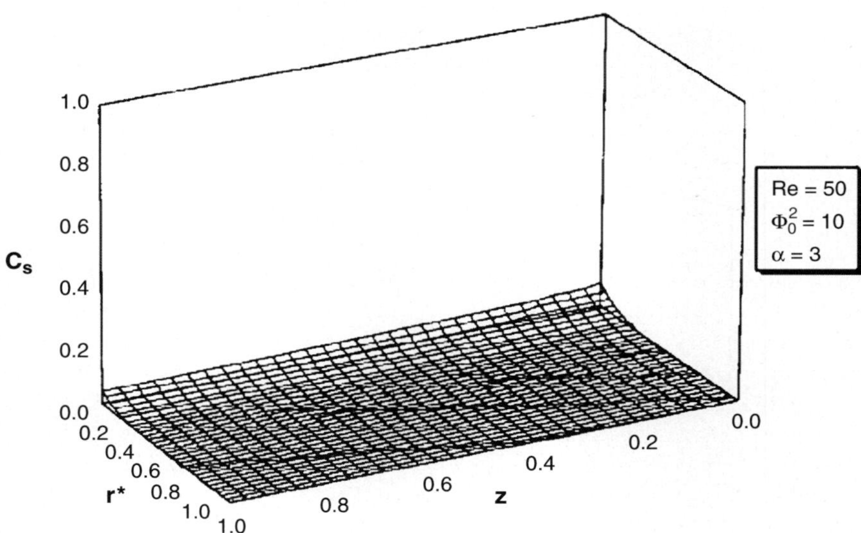

Fig. 5.43 Pseudo-steady-state oxygen profiles in the ECS of a MCCR using Darcy's permeabilities, with: $\alpha = 3$, Re = 50, $\varphi_o^2 = 10$ and $K = 10^{-16}$ m². The coordinate r^* is defined as zero at the fiber outlet wall radius and unity at the Krogh cylinder radius (Reprinted with permission from Brotherton and Chau 1996)

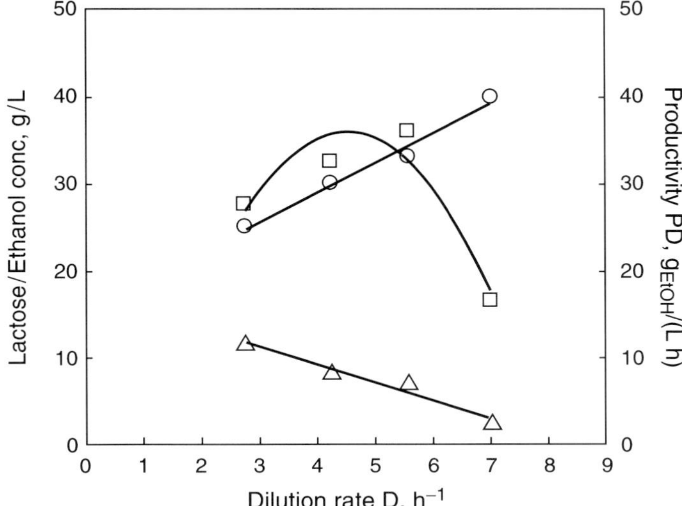

Fig. 5.44 Fermentation performance of a membrane compartmentalized cell reactor with cells of *Kluyveromices* fragilis: cell concentration, X_o = 90 g L^{-1}; lactose concentration, S_f = 150 g L^{-1}: (○) lactose, (△) ethanol, (□) productivity (Adapted from Cheryan and Mehaia 1983)

bioreactors tailored to the specific characteristics of both the cells used and the production goal (e.g., the production of products secreted out of the cells or present only inside the cells). Bioreactors are commercially available featuring volumes up to and exceeding 2000 L coupled to modules equipped with microfiltration or ultra-filtration membranes with surface areas from 4 to 20 m², as in the case of the CellFlo Plus and CellFlo Max concepts by SpectraPor. An interesting application of the CRMR concept is in wastewater treatment. In the Puron® design by Koch, for instance, hollow fiber membranes glued on one end and potted in manifold on the other are loosely assembled in an open-shell module and are submerged into the activated sludge. The barrier properties of the membranes permit retention of the cells and suspended solids in the biological basin: consequently, the bioreactor operates at higher sludge concentrations, and the effluent is much clearer than that with conventional plants. This significantly reduces the bioreactor volume required (with considerable savings of money and time) and permits elimination of the need for a second clari-fier. Figure 5.34c shows also that the loose arrangement permits membranes to fluctuate when the oxygen-rich gas is bubbled in the sludge, thereby getting rid of solids trapped in between neighboring membranes and keeps them clean.

The few MCCRs commercially available are disposable, and are mainly offered for the culture of mammalian cells for the production of proteins for therapeutic or diagnostic applications. The Integra CELLine™ bioreactor is a disposable two-compartment static bioreactor for the production of proteins on a laboratory scale. Cells are cultured in a chamber ca. 1.5 mm thick at the bottom of a vessel shaped like a T-flask compartmentalized between a flat-sheet silicone oxygen-permeable

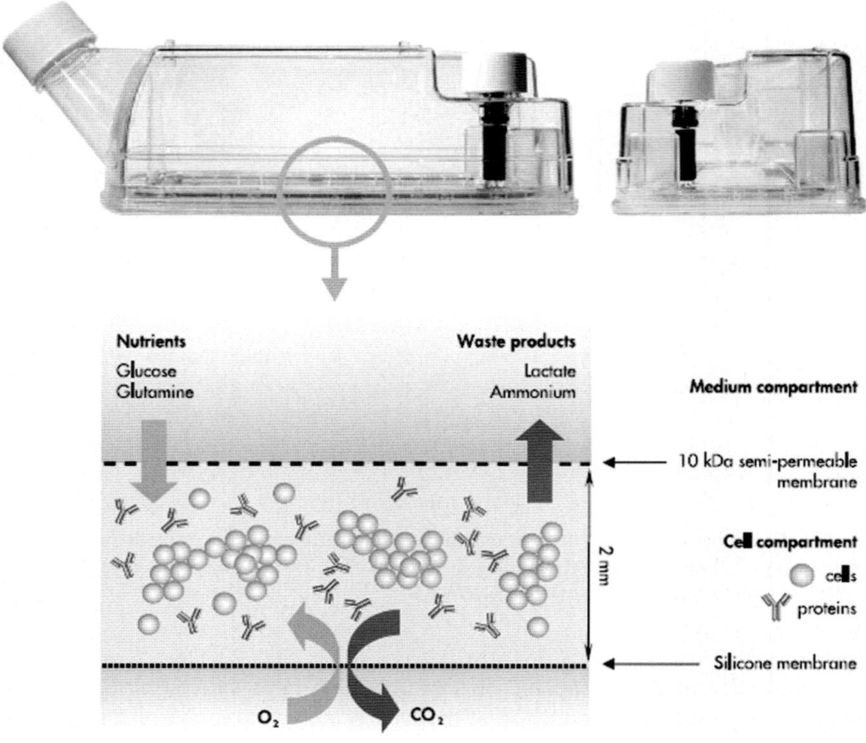

Fig. 5.45 Scheme of the Integra CELLine™ flat sheet membrane bioreactor

membrane (at the bottom) and a flat-sheet 10kD NMWCO semipermeable membrane (at the top), as shown in Fig. 5.45. The bioreactor is easy to operate, and is to be used in an incubator to ensure temperature control and oxygen supply. It is designed to operate with anchorage-independent cells because silicone is highly hydrophobic: a poly(ethyleneterephtalate) (PET) mat is offered for adherent cell culture. The bioreactor is advertised for the production of monoclonal antibodies in hybridoma, recombinant protein expression in transfected cells, and virus production. The CellMax and the FiberCell bioreactors are bench-scale MCCRs for cell culture in the ECS of a bundle of hollow fiber membranes and the small-to-middle-scale production of biologicals. Both bioreactor types are equipped with oxygen-exchanging tubing in the recirculation loop and have to be used in an incubator. The CellMax is a lab-scale bioreactor that may be equipped with hydrophilic ultrafiltration cellulose or polysulfone membranes, with NMWCO from 10 to 30kD, or with microfiltration polyethylene or polypropylene membranes. The latter are also available with a proprietary coating on the lumen surface for the lumen-side culture of endothelial cells under shear stress. FiberCell bioreactors are available from lab-scale to medium-scale production of proteins. Polysulfone membranes, as such or surface-activated to permit the binding of bioactive moieties, are generally used whose NMWCO ranges from ca. 5 kD and 0.1 μm maximal pore diameter and whose surface area ranges from 75 cm^2 to 2.5 m^2. The firm Biovest offers a range of hollow fiber MCCRs for medium-to-large-scale protein production. All bioreactors are equipped with an oxygenator placed in the medium recirculation loop. Starting with the mini-Max, to the Maximizer, to the Xcellerator, the membrane bioreactors are enclosed in a self-contained incubator, and their operation is progressively more and better automated. The number of actual membrane bioreactors in each device increases from two 2.1 m^2 ultrafiltration membrane units with a 35 kD NMWCO in the Maximizer, to 20 units in the Xcellerator.

MCCRs are extremely attractive for the adhesion culture of anchorage-dependent cells. In fact, they make it possible for cells to re-create tissue-like conditions, in terms of cell density and cell–cell and cell–matrix interactions favoring cell polarity and the posttranslational processing of proteins. However, their most widespread use is possibly for the production of monoclonal antibodies (mAbs) in hybridoma. Hybridoma are anchorage-independent cells very sensitive to shear stress. Since 1975, mAbs have been produced by injecting hybridoma cells in the peritoneum of mice and growing them till they produce a fluid containing high mAb concentrations (the ascite method). The ascite method is easy to set up but generates pain and distress in mice. In addition to humane considerations, the mAbs produced may be contaminated with murine viruses, murine immunoglobulins, and other antigenic proteins, among others (Lipman and Jackson 1998). To date, some European countries have imposed a partial or full ban on mAbs production with the ascite method, which is becoming less accepted worldwide. Some pharmaceutical industries (e.g., GlaxoSmithKline Biologicals, GSK Bio) have also switched from mice to the *in vitro* production of mAbs (Dewar et al. 2008). The productivity and economic convenience of bioreactors for mAb production with hybridoma depend on several factors, including cell specific capacity to produce mAbs. Stirred-tank bioreactors are generally preferred for cells with medium to high capacity to produce mAbs. The use of MCCRs has proven itself convenient with hybridoma cells with low to high capacity to produce mAbs. Published reports show that mAb concentration in the medium of MCCRs is 50 to 200 times greater than that of T-flasks, yielding mAb concentrations in the supernatant from 0.7 to 2.1 mg mL^{-1} (Dewar et al. 2008; Gramer and Poeschl 2000). The productivity per run of MCCRs is estimated to be equivalent to that of up to 48 mice, depending on the specific cell line and bioreactor used (Lipman and Jackson 1998). The productivity of commercial bioreactors is advertised to be as high as 420 g$_{mAb}$/month in the case of the Biovest Xcellerator. However, commercial dynamic hollow fiber MCCRs consume more medium than static flat-sheet membrane MCCRs, such as the CELLine™. Capital costs are also higher than the ascite method. However, the mAbs harvested from MCCRs is considerably more pure, especially when a serum-free medium is used. When all these factors are accounted for, it has been estimated that MCCRs such as the CELLine™ bioreactor may produce mAbs at about a half cost per milligram of that of T-flasks (IBS Integra Bioscience 2008).

5.1.4   Disposable Bioreactors

As described in Chapter 3, bioreactors with a presterile cultivation bag made of plastic material are well established in development and manufacturing processes primarily operating with animal and human cells from milliliter to 1000 L scale. While gas-permeable static blood bags are widespread devices for cell expansion, differentiation, and partial freezing in laboratory scale, stirred and wave-mixed bag bioreactors (also called *bioreactors with wave-induced motion*) have become increasingly influential in haematopoietic cell expansion, productions of therapeutic recombinant proteins in BEV/insect cell culture systems or in transient mammalian cells, and monoclonal antibody and virus production processes (Eibl and Eibl 2007, 2008). Therefore, we will subsequently focus on these two bag bioreactor types.

Stirred Bag Bioreactors

Current stirred bag bioreactors mainly differ in impeller type and pumping mode, impeller arrangement, shaft seal type, culture bag shape, and geometry (Card and Smith 2006; Galliher 2007; Thermo Fisher Scientific 2007; Zambaux 2007). Bag bioreactors with rotational impeller (SUB, Single-Use Bioreactor; XDR, Disposable Stirred Tank Bioreactor) have the typical cultivation container geometry and cylindrical form known from standard bioreactors for animal cell cultures. Besides rotational axial flow impeller, they are usually equipped with a microsparger or sparger ring. Whereas SUB (http://www.thermofisher.com) is top driven and has a mechanical seal, the XDR (http://www.xcellerex.com), operating with a magnetically coupled impeller, is bottom driven. In case of the SUB, it is therefore necessary to penetrate its driveshaft through the mixing drive and to lock it into the impeller assembly during the installation procedure. Similar to SUB's driveshaft, the impeller of the Artelis-ATMI Life-Sciences' Pad-Drive disposable bioreactor (http://www.atmi-lifesciences.com or http://www.artelis. be) is protected by a film, which is identical to the bag material, to avoid product contamination.

In contrast to SUB and XDR, where the impeller rotates in a cylindrical bag, the impeller of the Artelis-ATMI Life-Science's Pad-Drive disposable bioreactor tumbles (eccentric motion) in a cubical bag. For all three stirred bag bioreactors, the culture bag, including disposable sparger and impeller assembly, gas filters, and ports for integration of sensor probes and line sets, is shaped and fixed in a customized stainless steel support container with heater or cooler jackets. Trouble-free insertion of standard sensors for pH and dissolved oxygen measurement is accomplished by using aseptic connectors.

Because the SUB, XDR, and the traditional steel stirred bioreactors (unbaffled) for animal cells have substantial similarity in geometry and design, it is assumed that they guarantee similar mixing characteristics, oxygen transfer efficiency, and scale-up, and therefore product quantity and quality. Differences from dominating stirred steel cell culture bioreactors arise from the occurrence of bag irregularities (bag manufacture, folding) and unavailable Newton number for disposable impellers. To date, only hydrodynamic and oxygen transfer efficiency data of a 250 L SUB have been published by Kunas and Keating (2005). Taking achieved cell densities and product titers from batch and perfusion experiments with monoclonal antibody secreting CHO and PER.C6 cells (Brecht 2007; Card 2007; Ozturk 2007; Zijlstra 2007) into account, the reported mixing characterization results and scale-up parameters (Table 5.9) support the conclusion that the SUB is an interesting alternative to stirred steel bioreactors.

Wave-Mixed Bioreactors

It is undoubted that the early version of a bag bioreactor with wave-induced motion, which was introduced in the late 1990s, has promoted the development of disposable bag bioreactors. Various types of such wave-mixed bag bioreactors with rocking or shaking platforms have been tested for growing animal and human cells, but BioWave (http://www.wavebiotech.et) and Wave Reactor (http://www.wavebiotech.com or http://www.wave-europe.com) are the most widely used (Amanullah et al. 2004; Cronin et al. 2007; Fries et al. 2005; Genzel et al. 2006; Hami et al. 2004; Hundt et al. 2007; Levine 2007; Slivac et al. 2006; Weber et al. 2002). BioWave and Wave Reactor are based on the first prototype of a wave-mixed bioreactor. Here, mechanical power input produced by the moving platform facilitates mixing and aeration (Chaps. 3 and 6).

In order to compare the BioWave (1, 10, and 100 L culture volumes) to traditional stirred cell culture bioreactors with surface and membrane aeration, hydrodynamic and oxygen transfer efficiency studies were carried out. It was found that fluid flow, mixing time, residence time distribution, specific power input, and oxygen transfer efficiency were dependent on the rocking angle, rocking rate, bag type and its geometry, and culture volume. In addition, a modified Reynolds number was introduced to determine the fluid flow in the culture bag of the BioWave. The quantum of existing data in our working group and our experience in animal-cell-

Table 5.9 SUB (Single-Use Bioreactor) – mixing characterization results and scale-up parameters (Eibl and Eibl 2008, modified)

Parameter	Data		
Tip speed (m s^{-1})	0.53	1.06	2.13
Power input per volume (W m^{-3})	1.6	13.4	106.6
Reynolds number	34,000	69,000	137,000
Mixing time (s)	90	60	45
Gas–liquid mass transfer coefficient (aeration rate between 0.5–2 L min^{-1}) (h^{-1})	7–11	7–15	No data available

culture-based BioWave cultivations encouraged us to use this modified Reynolds number for successful scale up from T-flask via BioWave 20 SPS with 1 and 10 L working volume to BioWave 200 SPS with 50 and 100 L working volume. In other words, after the optimization of the 1 L culture volume process, we only calculate the required parameters (filling level, rocking rate, rocking angle) for the next scale while guaranteeing constant fluid flow, and have to realize three validation experiments. For more detailed information, the reader is referred to Eibl (2006) and Eibl and Eibl (2006).

Mixing times based on 40 and 50% filling level range between 10 and 1400 s and can be considered as satisfactory values for cell-culture bioreactors. As supposed, the most ineffective mixing was found at the lowest possible rocking rate and rocking angle and maximum bag filling level (50%). A reduction of the mixing time is guaranteed by increasing the rocking rate and/or the rocking angle, which causes a more intensive wave movement. Whereas the most effective mixing (9–264 s) was obtained in the 2 L bag, the most ineffective mixing (40–1402 s) was observed in the 20 L bag (Eibl et al. 2003).

Power input analysis for a BioWave operating with 2 L culture bag demonstrated that a minimum filling level as well as maximum rocking rate and rocking angle result in maximum specific power input (Lisica 2004). Surprisingly, this maximum specific power input was one decimal power higher than operation with the maximum filling level (Fig. 5.45). Furthermore, Fig. 5.45 indicates that a stepwise increase in the rocking rate up to 20 rpm raises the power input. If the rocking rate is increased further, the power input levels out and may even be followed by a slightly decrease. This last observation may be explained by the occurring phase shift of the wave towards the rocking movement. Numerous experiments confirmed our hypothesis that the subsequent increase in the rocking rate results in lower hydrodynamic cell stress but improves nutrient and oxygen transfer efficiency, which in turn give higher cell densities and product titers.

Residence time distribution experiments have shown that a continuously operating BioWave in perfusion mode can be described by the ideally mixed stirred tank model (Eibl and Eibl 2006). Finally, oxygen transfer coefficients provided by the BioWave reach comparable or even higher values than those that have been ensured by stirred cell bioreactors with membrane aeration or surface aeration (Eibl and Eibl 2006). For example, in a 2 L culture bag operating with 1 L culture volume at rocking angles between 6 and 10°, rocking rate of 30 rpm and aeration rate of 0.25 vvm, kLa values between 4 and 10 h^{-1} were found. High oxygen transfer efficiency can be guaranteed by increased rocking rate, rocking angle and aeration rate. Resulting from increased surface area, a decreased filling level in the bag increases kLa at constant parameters. Nevertheless, even small changes in the rocking rate and rocking angle increase the kLa more significantly. In BioWave, oxygen transfer coefficients in ranges above 11 h^{-1} can only be achieved by aeration rates over 0.5 vvm or aeration with pure oxygen.

To conclude, we would like to say that engineering aspects of disposable bioreactors are insufficiently investigated to this day. Computational fluid dynamics (CFD) may be a useful tool to overcome this problem.

5.2   Selection of Bioreactor and Operation Mode

Selection of a suitable type of cell culture bioreactor system and/or an appropriate operation mode (batch, fed-batch, and perfusion) is affected by technical, biological, economical, and regulatory considerations. To some extent, guidelines given in the literature can be helpful, but these are mostly qualitative. Only very few reports compare different reactors system, and in many cases these are intended to promote a certain type of (new) bioreactor (Varley and Birch 1999; Fenge and Lüllau 2006). The main characteristics of cell culture systems and bioreactors have been summarized in Table 5.1 (compare Sect. 5.1).

To start a selection process, it is helpful to address some questions defining the required specifications. The factors summarized in Fig. 5.46 will give some basic ideas in this respect, but is far from complete. It would be beyond the scope of this book to discuss the above-mentioned aspects in detail. For further details refer to Fenge and Lüllau (2006) as well as Varley and Birch (1999). Characteristics of different bioreactor systems applied for mammalian cell culture are discussed extensively in Sect. 5.1 and operation modes in Sect. 4.4. A further aspect might be development time for a process run under GMP (compare Fig. 5.47). The different advantages offered by certain bioreactors or culture systems, the range of applications they can preferably be applied for, and their limitations have been considered. It is quite obvious that there is no single cultivation system suitable for all applications in mammalian cell culture technology. In addition, to compare different bioreactor systems or operation modes experimentally for a certain application is time

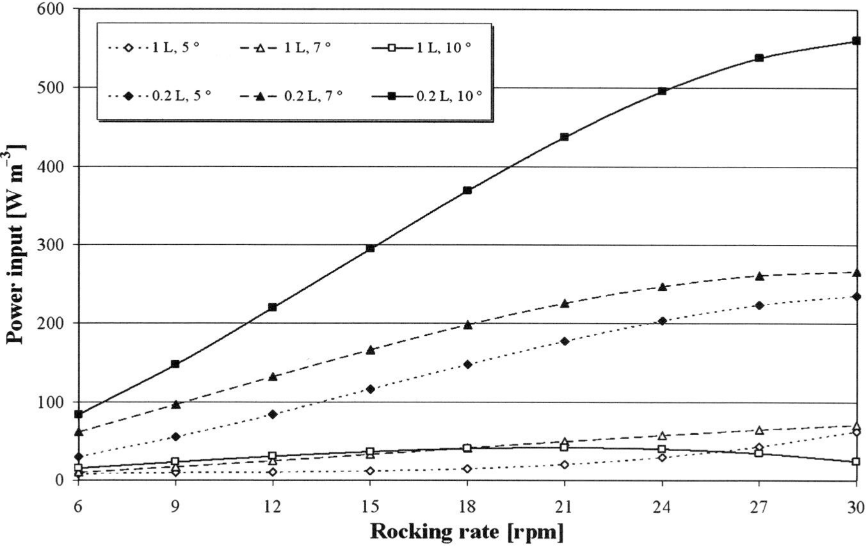

Fig. 5.46 Specific power input course in 2 L bag of BioWave

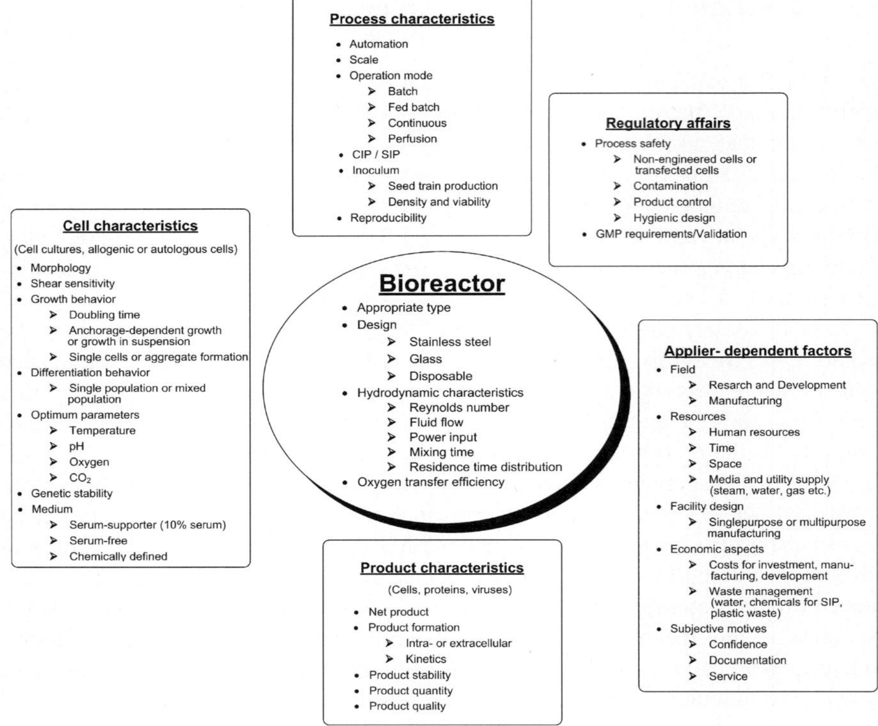

Fig. 5.47 Factors of relevance for selection of a suitable bioreactor system for cultivation of mammalian cells

consuming and expensive. Therefore, selection of an appropriate bioreactor or cultivation system requires an extensive knowledge and expertise of the team involved or professional advice from specialists.

5.3 How to Grow Mammalian Cells from Cryopreserved Vial to Production Bioreactor

A mammalian-cell-based GMP manufacturing process requires a defined robust production process in order to produce highly purified biopharmaceuticals of consistent quality. Both a well-characterized (i.e., in terms of freeze–thaw viability, growth rate, product expression level, absence of mycoplasms) and genetically stable high expressing production cell line are needed. As a matter of course, the production cell line should also be well documented including traceability of any serum or other animal-derived products as to its history. However, it should be borne in mind that serum-free or even chemically defined culture media (compare

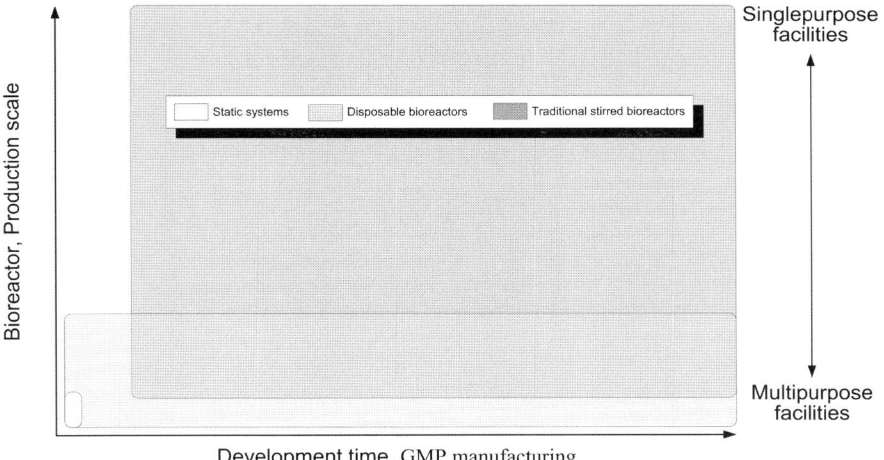

Fig. 5.48 Relationship between type of bioreactor and development time for GMP production

Sect. 2.3) providing lot-to-lot consistency and improved product biosafety are used increasingly today and the addition of antibiotics is restricted to substances keeping the cells under selection pressure. Moreover, phenol red indicating pH changes by color alteration is stringently omitted in commercial manufacturing processes owing to complicated downstream processing.

Once the Master Cell Bank (MCB) has been established and characterized, it is subcultured to provide seed stock that produces Working Cell Banks (WCBs) for future production runs (Sect. 2.2.3). In other words, cells from the WCB represent the real basis material for the manufacturing process depicted in Fig. 5.48, which consists of four main steps: cultivation in milliliter scale, maintenance of the seed train, maintenance of the inoculum train, and fermentation and downstream processing. So far, we have assumed that the production cells coming from cryopreserved vials (cell densities of approximately 1×10^7 cells mL^{-1} per vial), which have been stored at $-196\,°C$ in liquid nitrogen, commonly grow in suspension after their pooling and transfer into disposable T-flasks or shake flasks. Inoculation rates of 2×10^5 cells mL^{-1} and rocking rates between 100 and 150 rpm in the case of the shake flasks are usual. Directly after thawing, cell growth may occur with strong aggregation over the first three to four passages, which are usually carried out every three days. This phenomenon is often seen to be initially independent of the selected cultivation system. With respect to the culture medium used, cultivation in T-flasks and shake flasks are aimed at cell densities between 1 and 2×10^6 cells mL^{-1} and viabilities above 90%. Besides the addition of Pluronic (compare Sect. 2.3) to protect cells against shear stress in shake flasks, the trouble-free growth of very sensitive suspension cells is sometimes facilitated by cultivation in T-flasks during three passages before their transfer to shake flasks (not shown in Fig. 5.48).

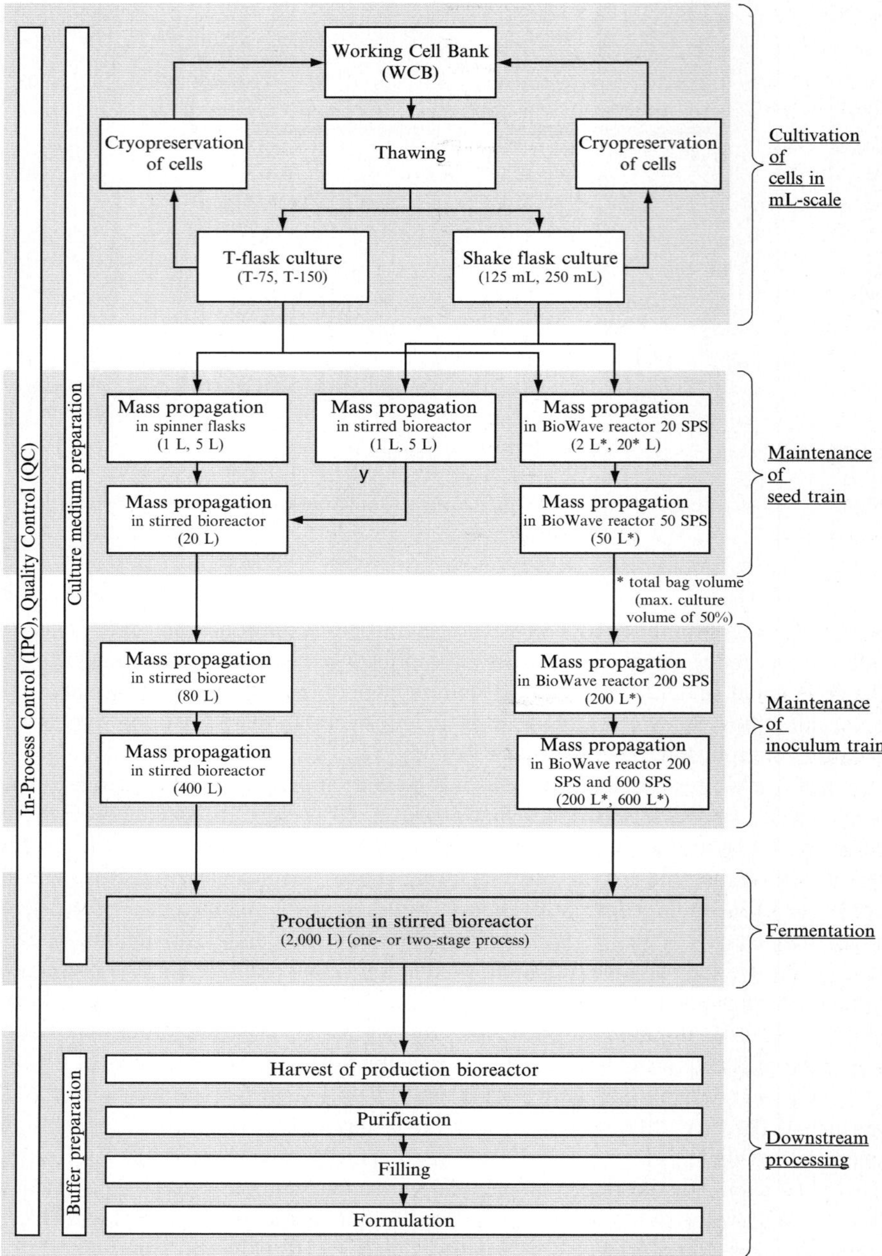

Fig. 5.49 Flowchart of mammalian-suspension-cell-based biopharmaceutical production in a 2000 L stirred bioreactor

The subsequent establishment and maintenance of the seed and the inoculum trains is executed by cell splitting and scale-up of the culture volume in the ratio of 1:5 while keeping the specific power input or fluid flow constant in the bioreactor (Sect. 5.1.1.1). The most frequently used bioreactors for these purposes are spinners, stirred bioreactors, and the BioWave reactor, all of which operate in the feeding mode. Whereas cells in spinners, stirred laboratory reactors and lab-scale Biowave reactors maintained for a long period of between 100 and 120 days are referred to as *seed train*, those from pilot- and industrial-scale bioreactors generating the final inoculum for the production bioreactor within a short time (10–15 days) are referred to as *inoculum train*. Both seed train and inoculum train should achieve cell densities of between 2 and 3×10^6 cells mL^{-1} and viabilities above 95% in order to ensure middle to high cell densities (5–10×10^6 cells mL^{-1}) and product titers between 1 and $3\,\mathrm{gL}^{-1}$ for a production bioreactor operating in the fed batch mode. Although spinners and stirred bioreactors represent the gold standard for seed train and inoculum train to date, many new processes increasingly include disposable bag bioreactors for seed and inoculum production. These include Hyclone's SUB or Xcellerec's XDR in addition to the BioWave (Wave Biotech AG) and the Wave Bioreactor (GE Healthcare), which have already been used in upstream steps in biomanufacturing.

In the particular case we are describing, mammalian-cell-based fermentation is realized in a 2000 L stirred bioreactor as a one- or two-stage process. A further scale-up to 10,000 L in production scale will consequently result in an additional mass propagation stage accomplished in a 2000 L stirred bioreactor or multiple of BioWave reactors on the inoculum train side. In contrast to the one-stage process mentioned, where growth and protein expression occur simultaneously, the two-stage process is characterized by a growth phase and a production phase. The production phase is most frequently induced by a simple temperature shift from 37 °C to values ranging between 34 and 29 °C or a specially designed production medium supporting product secretion. With regard to product titer, cell density, and cell viability influencing desired product quantity and product quality, the production phase is stopped. For example, it is known that cell viabilities below 80% can result in changed protein folding. Normally, production bioreactors are run in fed batch mode for between 6 and 10 days in the commercial manufacture of biopharmaceuticals. Finally, the secreted product of interest is separated from the fermentation broth, purified, filled, and formulated.

5.4 Questions and Problems

- Name the demands important for design of cell culture reactors.
- Name the advantages and disadvantages of suspension reactors.
- What kind of stirrer would you recommend for stirred tanks used for cultivation of mammalian cells? Give reasons.
- Name the techniques for cell retention in a suspension culture.

- Which parameter (criterion) can be used for scale-up of suspension reactors (suspension cells) with respect to cell damage?
- Name the advantages and disadvantages of fixed bed reactors.
- Discuss aspects relevant for selection of bioreactor and operation mode.

List of Symbols

C_{o2}	oxygen concentration
d_R	stirrer diameter
D_R	vessel diameter
n_R	stirrer speed
P/V	mean power input per volume
Re	Reynolds number
θ_m	mixing time
$k_L a$	mass transfer coefficient
U_g	superficial gas velocity
A	active membrane area
Δc_i	transmembrane concentration difference of component i
P_i	permeability coefficient of a single component i
r_p	particle radius
S_i	transport of a certain component i through the membrane
U_p	particle velocity in the gravitational field
η_{fl}	fluid viscosity
ρ_{fl}	fluid density
g	gravitational force
A	cross-sectional area of the fluidized bed vessel
D_{FB}	diameter of the bed of the fluidized bed or fixed bed
H_{FB}	height of the fluidized bed or fixed bed
K	constant depending on particle shape and surface properties
k_{aV}	decay constant of retrovirus vector
q_{O2}	cell specific oxygen consumption rate
q^*_{FB}	volume-specific uptake or production rate related to fixed bed volume
S_v	surface area per unit volume of particles
U_{fl}	fluid velocity
V_{FB}	volume of fluidized bed or fixed bed
X	number of immobilized cells
Δp	pressure drop
ε	void fraction
ρ_{fl}	fluid density

ρ_s	solids apparent density
a	solute equivalent spherical radius (m)
$C = X_2/X_1$	cell concentration factor
D_s	unhindered solute diffusion coefficient (m^2 s^{-1})
D_M	solute diffusion coefficient in the membrane (m^2 s^{-1})
$D_R = Q_F/V$	dilution rate (s^{-1})
$D_{R,crit}$	critical dilution rate at wash-out (s^{-1})
J_s	solute flux (gmol s^{-1} m^{-1})
J_v	solvent flux (m s^{-1})
k	overall cell mass transport coefficient (m^3 m^{-2} s^{-1})
k_c	kinetic constant of substrate utilization (s^{-1})
K	partition coefficient of a solute between the membrane and the neighboring fluid
K_S	Michaelis constant for substrate utilization (gmol m^{-3})
L	active fibre length (m)
L_p	membrane hydraulic permeability (m^2 s kg^{-1})
MW	molecular weight
$N_A = 6.023 \times 10^{23}$	Avogadro's number (molecules/mole)
P	pressure (kPa)
$Pe_{ax} = \dfrac{U_0 L}{D_s}\left(\dfrac{R_i}{L}\right)$	reduced axial Peclet number
P_M	solute diffusive permeability in the membrane (m s^{-1})
$Pe_w(Z) = \dfrac{V_w(Z)R_i}{D_s}$	local wall Peclet number
Q_F	feed flow rate (m^3 s^{-1})
Q_R	recycle flow rate (m^3 s^{-1})
Q_1	flow rate of the stream leaving the bioreactor tank (m^3 s^{-1})
$R = (1 - S)$	membrane rejection coefficient towards the solute
$Re_{in} = \dfrac{U_0 R_i}{\eta \pi}$	inlet Reynolds number
R_i	inner membrane radius (m)
R_K	radius of Krogh cylinder (m)
R_M	membrane resistance to diffusive transport (s m^{-1})
R_o	outer membrane radius (m)
$R_R = Q_R/Q_F$	recycle ratio
$R_u = 8.314 \times 10^7$	universal gas constant, (dynes cm (mole K)$^{-1}$)
S	membrane sieving coefficient towards the solute

S_F	substrate concentration in the feed stream, (gmol m^{-3})
S_1	substrate concentration in the stream leaving the bioreactor tank (gmol m^{-3})
T	temperature (K)
u_o	fiber lumen inlet axial velocity (m s^{-1})
V	bioreactor volume
v_w	transmembrane wall velocity (m s^{-1})
X	cell concentration in the permeate or removal stream (cells/m^3)
X_{gell}	cell concentration in the gel/cake (cells/m^3)
X_o	initial cell concentration in the bioreactor (cells/m^3)
X_1	cell concentration in the stream leaving the bioreactor tank (cells/m^3)
X_2	cell concentration in the stream leaving the membrane module (cells/m^3)
$Y_{X/S}$	cell yield coefficient
z	axial coordinate (m)

Greek Symbols

$\alpha = L\left(\dfrac{16\eta_{fi}L_P}{R_i^3}\right)^{\frac{1}{2}}$	fiber Pressure modules
δ	membrane wall thickness (m)
η_{fl}	bulk solution viscosity (kg m^{-1} s^{-1})
μ	specific cell growth rate (s^{-1})
μ_{max}	maximal cell growth rate (s^{-1})
Π	osmotic pressure (kPa)
σ	Staverman reflection coefficient towards the solute
$\Phi^2 = \dfrac{\pi_{max}x_o}{y_{x/s}S_F D_S}(R_k - R_o)^2$	squared Thiele modulus
$\Psi = \dfrac{S_F Y_{X/S}}{X_o}$	dimensionless yield coefficient

Superscripts and Subscripts

i	refers to device inlet
L	liquid phase
M	membrane phase
o	refers to device outlet

w	refers to the membrane interface with the liquid
BEV	Baculovirus expression vector
	CHO cells Chinese hamster ovary cells
PER.C6	cells human embryogenic retinoblast cells
$k_L a$	gas–liquid mass transfer coefficient

References

Akiyama I, Tomiyama K, Sakaguchi M, Takaishi M, Mori M, Hosokawa M, Nagamori S, Shimizu N, Huh NH, Miyazaki M (2004) Expression of CYP3A4 by an immortalized human hepatocyte line in a three-dimensional culture using a radial-flow bioreactor. Int J Mol Med 14: 663–668

Amanullah A, Burden E, Jug-Dujakovic M, Mikola M, Pearre C, Herber W (2004) Development of a large-scale cell bank in cryobags for the production of biologics. http://www.wavebiotech.com/pdfs/literature/Merck_Cancun_2004.pdf. Cited January 3, 2008

Apelblat A, Katzir-Katchalsky A, Silberberg A (1974) A mathematical analysis of capillary tissue fluid exchange. Biorheol 11: 1–49

Belfort G (1989) Membrane and bioreactors: A technical challenge in biotechnology. Biotechnol Bioeng 33: 1047–1066

Biotechnology by open learning: Bioreactor design and product yield (1992) Butterworth-Heinemann, London

Bohmann A, Pörtner R, Märkl H (1995) Performance of a membrane dialysis bioreactor with radial-flow fixed bed for the cultivation of a hybridoma cell line. Appl Microbiol Biotechnol 43: 772–780

Bohmann A, Pörtner R, Schmieding J, Kasche V, Märkl H (1992) Integrated membrane dialysis bioreactor with radial-flow fixed bed - a new approach for continuous cultivation of animal cells. Cytotechnol 9: 51–57

Born C, Biselli M, Wandrey C (1995) Production of monoclonal antibodies in a pilot scale fluidized bed bioreactor. In: Beuvery EC et al. (eds) Animal Cell Technology: Developments Towards the 21st Century, pp. 683–686. Kluwer, The Netherlands

Brecht R (2007) Disposable bioreactor technology: Challenges and trends in cGMP manufacturing. BioProduction 2007, Berlin, Germany (presentation)

Breman BB, Beenackers AACM, Bouma MJ, Van der Werf MH (1996) Axial liquid mixing and a gas–liquid multi-stage agitated contactor (MAC). Chem Eng Res Des 74: 669–678

Brotherton JD, Chau PC (1996) Modeling of axial-flow hollow fiber cell culture bioreactors. Biotechnol Prog 12: 575–590

Burger C, Carrondo MJ, Cruz P, Cuffe M, Dias E, Griffiths JB, Hayes K, Hauser H, Looby D, Mielke C, Moreira JL, Rieke E, Savage AV, Stacey GN, Welge T (1999) An integrated strategy for the process development of a recombinant antibody-cytokine fusion protein expressed in BHK cells. Appl Microbiol Biotechnol 52: 345–353

Butler M (2004) Animal Cell Culture Technology – The Basics, 2nd ed. Oxford University Press, New York

Card C (2007) Large scale, animal free production of human monoclonal antibody using PERC.6™ cells in a disposable, stirred-tank bioreactor. 20th ESACT meeting 2007, Dresden, Germany (poster)

Card C, Smith T (2006) Application report draft – SUB05601

Celonic (2007) http://www.celonic.de (27.05.2007)

Cheryan M, Mehaia MA (1983) A high performance membrane bioreactor for continuous fermentation of lactose to ethanol. Biotechnol Lett 5: 519–524

Chisti MY, Moo-Young M (1987) Airlift reactors: characteristics, applications and design considerations. Chem Eng Sci 43: 451–457

Chisti Y (2000) Animal-cell damage in sparged bioreactors. Trends Biotechnol 10: 420–432

Chmiel H (2006) Bioprozesstechnik. Spektrum Akademischer Verlag. München

Chu L, Robinson DK (2003) Industrial choices for protein production by large-scale cell culture. Biochemical Eng 10: 180–187

Cong C, Chang Y, Deng J, Xiao C, Su Z (2001) A novel scale-up method for mammalian cell culture in packed-bed bioreactor. Biotechnol Lett 23: 881–885

Cronin CN, Lim KB, Rogers J (2007) Production of selenomethionyl-derivatized proteins in baculovirus-infected insect cells. Protein Sci 16: 2023–2029

Dean RC, Karkare SB, Ray NG, Runstadler PW, Venkatasubramanian K (1987) Large-scale culture of hybridoma and mammalian cells in fluidized bed bioreactors. Ann NY Acad Sci 506: 129–146

Deng J, Yang Q, Cheng X, Li L, Zhou J (1997) Production of rhEPO with a serum-free medium in the packed bed bioreactor. Chin J Biotechnol 13: 247–252

Dewar V, Voet P, Denamur F, Smal J (2008) Industrial implementation of in vitro production of monoclonal antibodies. ILAR J 46

Doran PM (2006) Bioprocess Engineering Principles. Academic Press, London

Ducommun P, Bolzonella I, Rhiel M, Pugeaud P, von Stockar U, Marison IW (2001) On-line determination of animal cell concentration. Biotechnol Bioeng 72: 515–522

Ducommun P, Rueux PA, Kadouri A, von Stockar U, Marison IW (2002) Monitoring of temperature effects on animal cell metabolism in a packed bed process. Biotechnol Bioeng 77: 838–842

Dwyer JL (1984) Scale-up of bioproduct separation with high performance liquid chromatography. Bio/Technology 2: 957

Eibl D (2006) Engineering aspects of the BioWave reactor. BioWave training course at UAS Wädenswil, Switzerland (presentation)

Eibl R, Eibl D (2006) Design and use of the Wave Bioreactor for plant cell culture. In: Dutta Gupta S, Ibaraki Y (eds) Plant Tissue Culture Engineering, Series: Focus on Biotechnology, vol 6, pp 203–227. Springer, Dordrecht

Eibl R, Eibl D (2007) Disposable bioreactors for cell culture based-bioprocessing, Achema Worldwide News: 8–10

Eibl R, Eibl D (2008) Application of disposable bag-bioreactor in tissue engineering and for the production of therapeutic agents. In: Scheper T (ed) Advances in Biochemical Engineering/Biotechnology. Springer, Berlin Heidelberg New York, in press.

Eibl R, Eibl D, Pechmann G, Ducommun C, Lisica L, Lisica S, Blum P, Schär M, Wolfram L, Rhiel M, Emmerling M, Röll M, Lettenbauer C, Rothmaier M, Flükiger M (2003) Produktion pharmazeutischer Wirkstoffe in disposable Systemen bis zum 100 L Massstab, Teil 1. KTI-Projekt 5844.2 FHS, Final Report, University of Applied Sciences Wädenswil, Switzerland

Fassnacht D (2001) Fixed-Bed Bioreactors for the Cultivation of Animal Cells. VDI Verlag GmbH, Düsseldorf

Fassnacht D, Pörtner R (1999) Experimental and theoretical considerations on oxygen supply for animal cell growth in fixed bed bioreactors. J Biotechnol 72: 169–184

Fassnacht D, Reimann I, Pörtner R, Märkl H (2001) Scale-up von Festbettreaktoren zur Kultivierung tierischer Zellen. Chem Ing Tech 73: 1075–1079

Fassnacht D, Rössing S, Singh R, Al-Rubeai M, Pörtner R (1999) Influence of BCL-2 Expression on antibody productivity in high cell density hybridoma culture systems. Cytotechnol 30: 95–105

Fassnacht D, Rössing S, Stange J, Pörtner R (1998) Long-term cultivation of immortalised mouse hepatocytes in a high cell density fixed bed bioreactor. Biotechnol Technol 12: 25–30

Fenge Ch, Lüllau E (2006) Cell culture bioreactors. In: Ozturk SS, Hu WS (eds) Cell Culture Technology for Pharmaceutical and Cell-Based Therapies. Taylor & Francis, New York

Forestell SP, Dando JS, Chen J, de Vries P, Bohnlein E, Rigg RJ (1997) Novel retroviral packaging cell lines: complementary tropisms and improved vector production for efficient gene transfer. Gene Ther 4: 600–610

Frahm B, Lane P, Märkl H, Pörtner R (2003) Improvement of a mammalian cell culture process by adaptive, model-based dialysis fed-batch cultivation and suppression of apoptosis. Bioproc Biosyst Eng 26: 1–10

Fries S, Glazomitsky K, Woods A, Forrest G, Hsu A, Olewinski R, Robinson D, Chartrain M (2005) Evaluation of disposable bioreactors. BioProcess Int 10(Supplement): 36–44

Fussenegger M, Fassnacht D, Schwartz R, Zanghi JA, Graf M, Bailey JE, Pörtner R (2000) Regulated overexpression of the survival factor bcl-2 in CHO cells increases viable cell density in batch culture and decreases DNA release in extended fixed-bed cultivation. Cytotechnol 32: 45–61

Galliher P (2007) Case study: Scale up to 1,000 L perfusion in a disposable stirred tank bioreactor. BioProduction 2007, Berlin, Germany (presentation)

Genzel Y, Olmer RM, Schaefer B, Reichl U (2006) Wave microcarrier cultivation of MDCK cells for influenza virus production in serum containing and serum-free media.Vaccine 24: 6074–6087

Gion T, Shimada M, Shirabe K, Nakazawa K, Ijima H, Matsushita T, Funatsu K, Sugimachi K (1999) Evaluation of a hybrid artificial liver using a polyurethane foam packed-Bed culture system in dogs. J Surg Res 82: 131–136

Glacken MW, Fleishaker RJ, Sinskey AJ (1983) Large-scale production of mammalian cells and their products: engineering principles and barriers to scale-up. Ann NY Acad Sci 413: 355–372

Golmakany N, Rasaee MJ, Furouzandeh M, Shojaosadati SA, Kashanian S, Omidfar K (2005) Continuous production of monoclonal antibody in a packed-bed bioreactor. Biotechnol Appl Biochem 41: 273–278

Gramer MJ, Poeschl DM (2000) Comparison of cell growth in T-flasks, in micro hollow fiber bioreactors, and in an industrial scale hollow fiber bioreactor system. Cytotechnol 34: 111–119

Hambach B, Biselli M, Runstadler PW, Wandrey C (1992) Development of a reactor-integrated aeration system for cultivation of animal cells in fluidized beds. In: Spier RR, Griffiths JB, MacDonald C (eds) Animal Cell Technology: Developments, Process and Products, pp 381–385. Oxford, Butterworth-Heinemann

Hami LS, Green C, Leshinsky N, Markham E, Miller K, Craig S (2004) GMP production and testing of Xcellerated T Cells™ for the treatment of patients with CLL. Cytotherapy 6: 554–562

Heidemann R, Riese U, Lütkemeyer D, Büntemeyer H, Lehmann J (1994) The Super-Spinner: a low cost animal cell culture bioreactor for the CO_2-incubator. Cytotechnol 14: 1–9

Himmelfarb P, Thayer PS, Martin KE (1969) Spin filter culture: The propagation of mammalian cells in suspension. Science 164: 555–557

Hongo T, Kajikawa M, Ishida S, Ozawa S, Ohno Y, Sawada J, Umezawa A, Ishikawa Y, Kobayashi T, Honda H (2005) Three-dimensional high-density culture of HepG2 cells in a 5-ml radial-flow bioreactor for construction of artificial liver. J Biosci Bioeng 99: 237–244

Hu YC, Kaufman J, Cho MW, Golding H, Shiloach J (2000) Production of HIV-1 gp120 in packed-bed bioreactor using the vaccinia virus/T7 expression system. Biotechnol Prog 16: 744–750

Hundt B, Best C, Schlawin N, Kassner H, Genzel Y, Reichl U (2007) Establishment of a mink enteritis vaccine production process in stirred-tank reactor and Wave® Bioreactor microcarrier culture in 1–10L scale. Vaccine 25: 3987–3995 IBS Integra BioScience Website – January 31, 2008

Ijima H, Nakazawa K, Mizumoto H, Matsushita T, Funatsu K (1998) Formation of a spherical multicellular aggregate (spheroid) of animal cells in the pores of polyurethane foam as a cell culture substratum and its application to a hybrid artificial liver. J Biomater Sci Polym 9: 765–778

Iwahori T, Matsuno N, Johjima Y, Konno O, Akashi I, Nakamura Y, Hama K, Iwamoto H, Uchiyama M, Ashizawa T, Nagao T (2005) Radial flow bioreactor for the creation of bioartificial liver and kidney. Transplant Proc 37: 212–214

Jockwer A, Medronho RA, Wagner R, Anspach FB, Deckwer W-D (2001) The use of hydrocyclones for mammalian cell retention in perfusion bioreactors. In: Lindner-Olsson E, Chatzissavidou N, Lüllau E (eds) Animal Cell Technology: From Target to Market, pp 301–305. Kluwer, Dordrecht

Kaiser K (2006) Current Trends in Downstream Processing. 3rd Annual Biologicals Manufacturing, 8–9 March 2006, London, UK

Kataoka K, Nagao Y, Nukui T, Akiyama I, Tsuru K, Hayakawa S, Osaka A, Huh NH (2005) An organic-inorganic hybrid scaffold for the culture of HepG2 cells in a bioreactor. Biomat 26: 2509–2516

Kawada M, Nagamori S, Aizaki H, Fukaya K, Niiya M, Matsuura T, Sujino H, Hasumura S, Yashida H, Mizutani S, Ikenaga H (1998) Massive culture of human liver cancer cells in a newly developed radial flow bioreactor system: ultrafine structure of functionally enhanced hepatocarcinoma cell lines. In Vitro Cell Dev Biol Anim 34: 109–115

Kelsey LJ, Pillarella MR, Zydney AL (1990) Heoretical analysis of convective flow profiles in a hollow-fiber membrane bioreactor. Chem Eng Sci 45: 3211–3220

Knazek RA, Gullino PM, Frankel DS, Dedrick RL (1972) Cell culture on artificial capillaries: an approach to tissue growth in vitro. Science 178: 65–67

Kompala DS, Ozturk SS (2006) Optimization of high cell density perfusion bioreactors. In: Ozturk SS, Hu WS (eds) Cell Culture Technology For Pharmaceutical and Cell-Based Therapies. Taylor & Francis, New York

Kossen NWF (1994) Scale-up. In: Galindo E, Ramirez OT (eds) Advances in Bioprocess Engineering, pp 53–65. Kluwer, Dordrecht

Krahe M (2003) Biochemical Engineering. Ullmann's Encyclopedia of Industrial Chemistry. http://www.mrw.interscience.wiley.com/ueic/articles/b04_381/frame.html

Kunas KT, Keating J (2005) Stirred tank single-use bioreactor: Comparison to traditional stirred bioreactor. bioLOGIC Europe 2005, Geneva, Switzerland (presentation)

Kurosawa H, Yasumoto K, Kimura T (2000) Polyurethane membrane as an efficient immobilization carrier for high-density culture of rat hepatocytes in the fixed-bed bioreactor. Biotechnol Bioeng 70: 160–166

Kwon MS, Kato T, Dojima T, Park EY (2005) Application of a radial-flow bioreactor in the production of beta1,3-N-acetylglucosaminyltransferase-2 fused with GFPuv using stably transformed insect cell lines. Biotechnol Appl Biochem 42: 41–46

Kyung KH, Gerhardt P (1984) Biotechnol Bioeng 26: 252–256

Levine B (2007) Making waves in cell therapy: the wave bioreactor for the generation of adherent and non-adherent cells for clinical use. http://www.wavebiotech.com/pdf/literature/ISCT_2007_Levine_Final.pdf.Cited January 3, 2008

Lipman NS, Jackson LR (1998) Hollow fibre bioreactors: an alternative to murine ascites for small scale (<1 gram) monoclonal antibody production. Research Immunol 149: 571–576

Lisica S (2004) Energieeintrag in Wave-Bioreaktoren. Modelling Approaches, University of Applied Sciences Wädenswil, Switzerland, unpublished

Looby D, Racher AJ, Griffiths JB, Dowsett AB (1990) The immobilization of animal cells in fixed bed and fluidized porous glass sphere bioreactors. In: de Bont JAM, Visser J, Mattiasson B, Tramper J (eds) Physiology of Immobilized Cells. (Elsevier Science B.V., pp 255–264. Amsterdam, Netherlands

Lu JT, Chung YC, Chan ZR, Hu YC (2005) A novel oscillating bioreactor BelloCell: implications for insect cell culture and recombinant protein production. Biotechnol Lett 27: 1059–1065

Lüdemann I, Pörtner R, Schaefer C, Schick K, Šrámková K, Reher K, Neumaier M, Franék F, Märkl H (1996) Improvement of the culture stability of non-anchorage-dependent animal cells grown in serum-free media through immobilization. Cytotechnol 19: 111–124

Lundgren B, Blüml G (1998) Microcarriers in cell culture production. In: Subramanian G (ed) Bioseparation and Bioprocessing – A Handbook. Wiley-VCH, New York

Ma N, Mollet M, Chalmers JJ (2006) Aeration, mixing and hydrodynamics. In: Ozturk SS, Hu WS (eds) Cell Culture Technology for Pharmaceutical and Cell-Based Therapies. Taylor & Francis, New York

Märkl H (1989) Folien und Membranen als neue Elemente im Fermenterbau. Forum Mikrobiologie 5: 234–237

Manual "Cytopilot" (2007) GE Healthcare. http://www1.gelifesciences.com

Matsushita T, Ketayama M, Kamihata K, Funatsu K (1990) Anchorage-dependent mammalian cell culture using polyurethane foam as a new substratum for cell attachment. Appl Microbiol Biotechnol 33: 287–290

McDonogh RM, Bauser H, Stroh N, Chmiel H (1992) Separation efficiency of membranes in biotechnology: an experimental and mathematical study of flux control. Eng Sci 47: 271–279

McTaggart S, Al-Rubeai M (2000) Effects of culture parameters on the production of retroviral vectors by a human packaging cell line. Biotechnol Prog 16: 859–865

Mehaia MA, Cheryan M (1984) Hollow fiber bioreactor for ethanol production: Application to the conversion of lactose by *Kluyveromyces fragilis*. Enzyme Microb Technol 6: 117–120

Merten OW, Cruz PE, Rochette C, Geny-Fiamma C, Bouquet C, Goncalves D, Danos O, Carrondo MJT (2001) Comparison of different bioreactor systems for the production of high titer retroviral vectors. Biotechnol Prog 17: 326–335

Meuwly F, Ruffieux PA, Kadouri A, von Stockar U (2007) Packed-bed bioreactors for mammalian cell culture: Bioprocess and biomedical applications. Biotechnol Adv 25: 45–56

Mishra VP, Joshi JB (1994) Flow generated by a disc turbine. 4. Multiple-impellers. Chem Eng Res Des 73: 657–668

Miyoshi H, Yanagi K, Fukuda H, Ohshima N (1994) Long-term continuous culture of hepatocytes in a packed-bed bioreactor utilizing porous resin. Biotechnol Bioeng 43: 635–644

Moro AM, Rodrigues MT, Gouvea MN, Silvestri ML, Kalil JE, Raw I (1994) Multiparametric analyses of hybridoma growth on glass cylinders in a packed-bed bioreactor system with internal aeration. Serum-supplemented and serum-free media comparison for MAb production. J Immunol Meth 176: 67–77

Moser A (1985) Special cultivation techniques. In: Rehm HJ, Reed G (eds) Biotechnology, vol. 2, pp 311–347. VCH, Weinheim

Mulder M (1990) Basic Principles of Membrane Technology. Kluwer, Dordrecht, The Netherlands

Nagata S (1975) Mixing: Principles and Applications. Kodansha, Tokyo

Nehring D, Gonzalez R, Pörtner R, Czermak P (2006) Experimental and modelling study of different process modes for retroviral production in a fixed bed bioreactor. J Biotechnol 122: 239–253

Nienow AW (1990) Gas dispersion performance in fermeter operation. Chem Eng Prog 86: 61–71

Nienow AW, Langheinrich C, Stevenson NC, Emery AN, Clayton TM, Slater NKH (1996) Homogenisation and oxygen transfer rates in large agitated and sparged animal cell bioreactors: Some implications for growth and production. Cytotechnol 22: 87–94

Noll TBM, Wandrey C (1997) On-line biomass monitoring of immobilised hybridoma cells by dielectrical measurements. In: Carrondo MJT, Griffiths JB, Moreira JLP (eds) Animal Cell Technology: from Vaccine to Genetic Medicine, pp 289–294. Kluwer, Dordrecht

Ong CP, Pörtner R, Märkl H, Yamazaki Y, Yasuda K, Matsumura M (1994) High density cultivation of hybridoma in charged porous carriers. J Biotechnol 34: 259–268

Ozturk SS (2007) Comparison of product quality: Disposable and stainless steel bioreactor. BioProduction 2007, Berlin, Germany (presentation)

Park S, Stephanopoulos G (1993) Packed bed bioreactor with porous ceramic beads for animal cell culture. Biotechnol Bioeng 41: 25–34

Porter MC, Michaels AS (1970) Membrane ultrafiltration. Chem Tech 1: 56–61

Pörtner R (1998) Reaktionstechnik der Kultur tierischer Zellen. Shaker Verlag, Aachen

Pörtner R (ed) (2007) Animal Cell Biotechnology – Methods and Protocols. Humana Press, Clifton, UK

Pörtner R, Lüdemann I, Märkl H (1997) Dialysis cultures with immobilized hybridoma cells for effective production of monoclonal antibodies. Cytotechnol 23: 39–45

Pörtner R, Lüdemann I, Reher K, Neumaier M, Märkl H (1998) Fixed-bed dialysis culture of a transfectoma cell line producing chimeric Fab-fragments with "nutrient-split"-feeding strategy. Biotechnol Technol 12: 501–505

Pörtner R, Märkl H (1998) Dialysis cultures. Appl Microbiol Biotechnol 50: 403–414

Pörtner R, Märkl H, Fassnacht D (1999) Immobilization of Mammalian Cells in Fixed Bed Bioreactors. Bioforum. Bioresearch plus Biotechnology. G.I.T. Verlag, Darmstadt, Germany

Pörtner R, Platas Barradas OBJ (2007) Cultivation of mammalian cells in fixed bed reactors. In: Pörtner R (ed) Animal Cell Biotechnology – Methods and Protocols. Humana Press, Clifton, UK

Pörtner R, Platas OB, Fassnacht D, Nehring D, Czermak P, Märkl H (2007) Fixed bed reactors for the cultivation of mammalian cells: design, performance and scale-up. Open Biotechnol J 1: 41–46

Ray NG, Karkare SB, Runstadler PW (1989) Cultivation of hybridoma cells in continuous cultures: Kinetics of growth and product formation. Biotechnol Bioeng 33: 724–730

Ray NG, Tung AS, Hayman EG, Vorunakis JN, Rundstadler PW (1990) Continuous cell cultures in fluidized-bed bioreactors – cultivation of hybridomas and recombinant chines hamster ovary cells immobilized in collagen micropheres. Ann NY Sci 589: 443–457

Reiter M, Blüml G, Gaida T, Zach N, Doblhoff-Dier O, Unterluggauer F, Noe M, Plail R, Huss S, Katinger H (1991) Modular Integrated Fluidized Bed Bioreactor Technology. Bio/Technology 9: 1100–1102

Schlegel HG, Doelle HW, Fiechter A, Yamada H, Shimizu S, Becker Th, Breithaupt D, Kasper C, Liese A, Lütz St, Pörtner R, Sell D, Stahl F, Suck K, Ulber R, Wegener J, Würges K, Zorn H (2007) Biotechnology. In: Ullmann's Encyclopedia of Industrial Chemistry

Schonberg JA, Belfort G (1987) Enhanced nutrient transport in hollow fiber perfusion bioreactors: a theoretical analysis. Biotechnol Prog 3: 80–89

Sendresen C, Fassnacht D, Benati C, Pörtner R (2001) Possible strategies for the production of viral vector: the role of the engineering design. In: Lindner-Olsson E, Chatzissavidou N, Lüllau E (eds) Animal Cell Technology: From Target to Market, pp 538–540. Kluwer, Dordrecht, Netherlands

Simpson N, Milner A, Al-Rubeai M (1997) Prevention of hybridoma cell death by Bcl-2 during suboptimal culture conditions. Biotechnol Bioeng 54: 1–16

Slivac I, Sráek VG, Radoševic K, Kmetiá I, Kniewald Z (2006) Aujeszky's disease virus production in disposable bioreactor. J Biosci 3: 363–368

Spier RE, Griffiths B (1984) An examination of the data and concepts germane to the oxygenation of cultured animal cells. Dev Biol Stand 55: 81–92

Starling EH (1896) On the absorption of fluid from the convective tissue space. J Physiol 19: 312–326

Strathmann H (1979) Trennung von molekularen Mischungen mit Hilfe synthetischer Membranen. Dr. Dietrich Steinkopf Verlag, Darmstadt

Thelwall PE, Anthony ML, Fassnacht D, Pörtner R, Brindle KM (1998) Analysis of cell growth in a fixed bed bioreactor using magnetic resonance spectroscopy and imaging. In: Merten OW, Perrin P, Griffiths JB (eds) New Developments and New Applications in Animal Cell Technology, pp 627–633. Kluwer, The Netherlands

Thermo Fisher Scientific (2007) Application note: AN003 Rev 1

Varley J, Birch J (1999) Reactor design for large scale suspension animal cell culture. Cytotechnol 29: 177–205

Vick Roy TB, Mandel DK, Dea DK, Blanch HW, Wilke CR (1983) The application of recycle to continuous fermentative lactic acid production. Biotechnol Lett 5: 665–670

Voisard D, Meuwly F, Ruffieux PA, Baer G, Kadouri A (2003) Potential of cell retention techniques for large-scale high-density perfusion culture of suspended mammalian cells. Biotechnol Bioeng 80: 751–765

Wang G, Zhang W, Jacklin C, Freedman D, Eppstein L, Kadouri A (1992) Modified CelliGen-packed bed bioreactors for hybridoma cell cultures. Cytotechnol 9: 41–49

Warnock JN, Bratch K, Al-Rubeai M (2005) Packed bed bioreactors. In: Chauduri J, Al-Rubeai M (eds) Bioreactors for Tissue Engineering. Springer, Dordrecht, The Netherlands

Weber W, Weber E, Geisse S, Memmert K (2002) Optimization of protein expression and establishment of the wave bioreactor for baculovirus/insect cell culture. Cytotechnol 38: 77–85

Wei J, Russ MB (1977) Convection and diffusion in tissues and tissue cultures. J Theor Biol 66: 775–787

Werther J (2007) Fluidized-bed bioreactors. In: Ullmann's Encyclopedia of Industrial Chemistry, Wiley-VCH Verlag GmbH & Co. KGaA, Weinheim

Yamashita Y, Shimada M, Tsujita E, Tanaka S, Ijima H, Nakazawa K, Sakiyama R, Fukuda J, Ueda T, Funatsu K, Sugimachi K (2001) Polyurethane foam/spheroid culture system using human hepatoblastoma cell line (Hep G2) as a possible new hybrid artificial liver. Cell Transplant 10: 717–722

Yang ST, Luo J, Chen C (2004) A fibrous-bed bioreactor for continuous production of monoclonal antibody by hybridoma. Adv Biochem Eng Biotechnol 87: 61–96

Yoshida H, Mizutani S, Ikenaga H (1993) Production of monoclonal antibodies with a radial-flow bioreactor. In: Kaminogawa S et al. (eds) Animal Cell Technology: Basic and Applied Aspects. Kluwer, Dordrecht, Boston, London, 5, 347–353

Zambaux JP (2007) How synergy answers the biotech industry needs. BioProduction 2007, Berlin, Germany (presentation)

Zijlstra G (2007) Scale-up of a PER.C6® fed-batch process in 50 and 250 L Hyclone single bioreactors compared to 50 L and 250 L stainless steel bioreactors. 20th ESACT Meeting 2007, Dresden, Germany (poster)

Complementary Reading

Biotechnology by Open Learning: Bioreactor Design and Product Yield (1992). Butterworth-Heinemann, London

Butler M (2004) Animal Cell Culture Technology – The Basics, 2nd ed. Oxford University Press, New York

Chisti Y (2000) Animal-cell damage in sparged bioreactors. Trends Biotechnol 10: 420–432

Doran PM (2006) Bioprocess Engineering Principles. Academic Press, London

Eibl R, Eibl D (2006) Design and use of the wave bioreactor for plant cell culture. In: Dutta Gupta S, Ibaraki Y (eds) Plant Tissue Culture Engineering, Series: Focus on Biotechnology, vol 6. Springer, Dordrecht, pp 203–227

Eibl R, Eibl D (2008) Application of disposable bag-bioreactor in tissue engineering and for the production of therapeutic agents. In: Scheper T (ed) Advances in biochemical engineering/biotechnology. Springer, Berlin Heidelberg, in press.

Krahe M (2003) Biochemical Engineering. Ullmann's Encyclopedia of Industrial Chemistry. http://www.mrw.interscience.wiley.com/ueic/articles/b04_381/frame.html

Lundgren B, Blüml G (1998) Microcarriers in cell culture production. In: G. Subramanian: Bioseparation and Bioprocessing – A Handbook. Wiley-VCH, New York

Meuwly F, Ruffieux PA, Kadouri A, von Stockar U (2007) Packed-bed bioreactors for mammalian cell culture: Bioprocess and biomedical applications. Biotechnol Adv 25: 45–56

Mulder M (1991) Basic principles of membrane technology. Kluwer, Dordrecht, The Netherlands

Ozturk SS, Hu WS (eds) (2006) Cell Culture Technology For Pharmaceutical and Cell-Based Therapies. Taylor & Francis, New York

Pörtner R (ed) (2007) Animal Cell Biotechnology – Methods and Protocols. Humana Press, Clifton, UK

Shuler ML, Kargi F (2002) Bioprocess Engineering – Basic Concepts. Prentice Hall PTR, Englewood Cliffs, NJ

Varley J, Birch J (1999) Reactor design for large scale suspension animal cell culture. Cytotechnol 29: 177–205

Voisard D, Meuwly F, Ruffieux PA, Baer G, Kadouri A (2003) Potential of cell retention techniques for large-scale high-density perfusion culture of suspended mammalian cells. Biotechnol Bioeng 80: 751–765

Werther J (2007) Fluidized-bed bioreactors. In: Ullmann's Encyclopedia of Industrial Chemistry. Wiley-VCH Verlag GmbH & Co. KGaA, Weinheim

Part II
Special Applications

Chapter 6
Insect Cell-Based Recombinant Protein Production

W. Weber and M. Fussenegger

Abstract Insect cells in conjunction with the baculovirus expression vector (BEV) system represent a fast and efficient system for the generation of recombinant tool proteins (mg- to g-range). This is quite evident from scores of application reports which have been published since 1970. Moreover, the European approval of the vaccine Cervarix (for cervical cancer) from Glaxo SmithKline in autumn 2007 and the development of further potential product candidates (e.g. influenza vaccine FluBIok from Protein Sciences Inc.) indicate the growing industrial interest in commercial use of the BEV/insect cell culture system for manufacture of biopharmaceuticals.

Starting with a short description of main characteristics of insect cells, this chapter delineates generation methods of recombinant virus, the scale-up to protein production (suitable bioreactors and optimum process design), application examples and current trends in insect cell-based protein production. Finally, pros and cons of the BEV/insect cell culture system are summarized.

6.1 Insect Cell Culture

Insect cell culture is a mature technology and is being applied in the routine production of recombinant proteins in processes reminiscent of an assembly line. The main advances in insect cell technology are the development of insect cell lines able to grow in suspension mode in chemically defined, serum- and protein-free culture media (Ikonomou et al. 2003; Schlaeger 1996; Weber and Fussenegger 2005). Today's workhorse cell line is the *Spodoptera frugiperda*-derived ovarian cell line Sf-9, which is routinely used to produce intracellular or membrane proteins, whereas the *Trichoplusia ni* egg cell homogenate-derived cell line BI-TN-5B1-4

W. Weber, M. Fussenegger
Department of Biosystems Science and Engineering, ETH Zurich, Mattenstrasse 26,
CH-4058 Basel, Switzerland
Fussenegger@bsse.ethz.ch

R. Eibl et al., *Cell and Tissue Reaction Engineering: Principles and Practice,*   263
© Springer-Verlag Berlin Heidelberg 2009

(also known as High Five™) was reported to increase the specific and volumetric yield of secreted proteins (Palomares et al. 2003). For isolation and propagation of baculoviruses, Sf-21 (parental cell line of Sf-9) remains the preferred cell line. All cell lines grow rapidly (doubling time approximately 24h), thereby enabling fast expansion and short overall processes.

Insect cells have similar requirements as mammalian cells except that they grow at 28 °C and tolerate higher levels of free amino acids and glucose without switching to overflow metabolism, which enables the design of more nutrient-rich media. In contrast to mammalian cells, the pH is lower (6.2–6.9) and commonly maintained by a phosphate buffer, thus obviating the need for a culture containing carbon dioxide, as required for the open bicarbonate buffer system in mammalian cell culture media.

While first-generation media required the addition of serum or insect hydrolysates, today several (proprietary) media types are commercially available, which are devoid of serum and protein and enable growth in suspension mode, a prerequisite for large-scale cultivation in bioreactors. The most important media include the SF900-II and SF900-III medium for Sf-9 and Sf-21 cells. For cell growth in adherent mode, as required for plaque purification or titration of baculoviruses (see below), the media can be supplemented with 10% fetal calf serum.

Insect cells have been reported to be more sensitive to shear forces than mammalian cells, which indicates the need for medium supplementation with shear force protecting agents such as block polymers [e.g. Pluronic F68 (Palomares et al. 2000)], which adhere to the cell surface like a barrel hoop and thus stabilize the cell membrane. Protection from shear forces is particularly required in bioreactor operation since insect cells show higher rates of specific oxygen uptake than mammalian cells and thus need either higher rates of stirring or increased gas sparging.

6.2 Special Aspects: Engineering Baculoviruses as Vectors

Introduction of genetic material for expression of desired recombinant proteins in insect cells can be performed by transfection of plasmid DNA, according to protocols similar to mammalian cells (e.g. lipofection); however, the overwhelming majority of insect cell-based protein production is performed by transducing the product gene into the target cell by infection with a recombinant baculovirus.

Today's commonly used baculoviruses belong to the *Eubaculoviridae* (King and Possee 1992), a subfamily of double-stranded DNA viruses, which infect and propagate in host arthropods like insect larvae. Following infection, wild-type baculoviruses propagate and produce proteinaceous structures known as occlusion bodies, in which the viruses can persist for extended periods, even after the death of the infected host and until the next insect is infected following ingestion of virus-containing occlusion bodies. High-level expression of occlusion body-forming proteins (representing up to 50% of total cellular protein) is driven by the very late and extremely strong polyhedrin and p10 promoters (King and Possee 1992).

For application in biotechnology, baculoviruses have been engineered to express recombinant target genes under the control of the p10 or polyhedrin promoters while eliminating the genes that form occlusion bodies. The biotechnologically most important baculoviruses are derived from the *Autographa californica* nuclear polyhedrosis virus (AcNPV), which has been engineered to enable straightforward target gene integration into the viral genome (total size: 129 kbp). Since such large genomes are not amenable to standard molecular biology cloning, a wide variety of strategies has been developed to incorporate target gene sequences into the viral backbone (Vialard et al. 1995) (Fig. 6.1):

(a) Insect cell-based homologous recombination: The target gene is inserted into a transfer vector between sequences that are homologous to the viral genome. Following transfection of the viral genome along with the transfer vector into insect cells, recombination occurs and thus results in intact viral genomes harboring the target gene sequence (BacPAK, BD Biosciences, Bac-N-Blue, Invitrogen, Fig. 6.1a).
(b) Homologous recombination in *Escherichia coli*: The transfer vector encoding the target gene between homologous sequences is transformed into *E. coli* along with the baculoviral genome for recombination. Isolated recombinant viral genomes can be rescued from *E. coli* and used subsequently for transfection of insect cells. For the rapid identification of correctly recombined *E. coli* clones, the 3'-terminal *lacZ* fragment was placed on the viral genome, whereas the 5'*lacZ* terminus was encoded on the transfer vector, thus resulting in reconstitution of functional β-galactosidase expression upon successful recombination, which can be monitored easily by plating the *E. coli* on X-Gal-containing agar (Bac-to-Bac, Invitrogen, Fig. 6.1b).
(c) In vitro recombination: The target gene can be transferred into the viral backbone by phage lambda integrase-mediated recombination of corresponding *attL* and *attR* sites. Recombined plasmids can be used directly for transfection of insect cells (BaculoDirect, Invitrogen, Fig. 6.1c).

Transfection of viral genomes, harboring target genes, into insect cells results in expression of viral genes and production of intact viral particles, budding from the cell into the medium and final cell lysis. Freshly budded viruses infect neighboring cells and trigger formation of progeny virus (approximately 500 viruses per infected cell (Power et al. 1994)) as well as production of the target recombinant protein. The entire process, from infection to cell lysis, takes approximately three days and is accompanied by a 50% increase in the rate of specific oxygen uptake (Kamen et al. 1996) and a substantially larger cell volume. Viral stocks for infection of the production batch can thus be produced by several cycles of virus propagation in insect cells followed by removal of cell debris by centrifugation. Viral stocks can be stored at + 4 °C for several weeks, whereas long-term storage should be performed at −80 °C. Storage at −20 °C was associated with a rapid decrease in virus titer.

Viral stocks are commonly titrated by plaque assays, where a lawn of adherently growing insect cells is infected by serial dilutions of the viral stock and subsequently

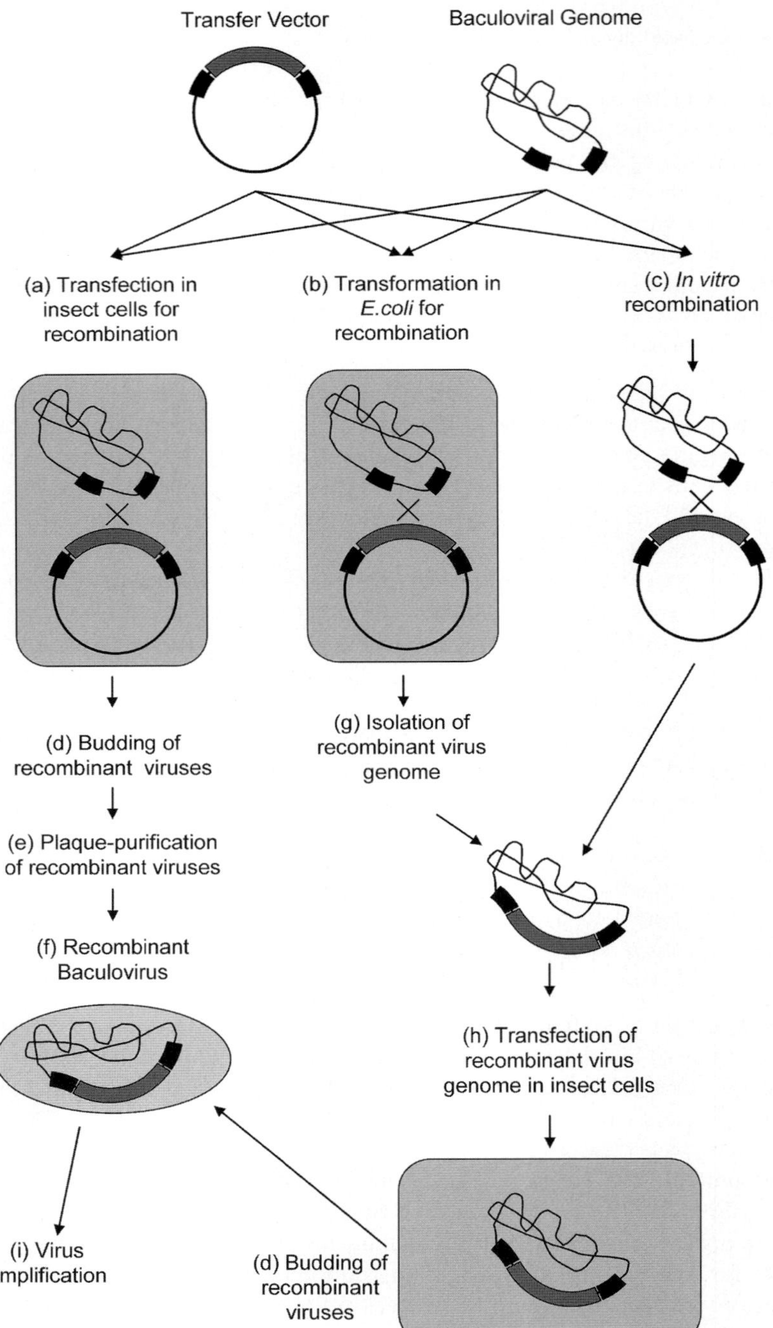

Fig. 6.1 Engineering of recombinant baculoviruses. The target gene is cloned into a transfer vector, which is subsequently recombined into the baculoviral genome either by homologous recombination in insect cells (**a**) or *E. coli* (**b**) or by addition of isolated recombinase in vitro (**c**). While the virus, which budded after recombination in insect cells, requires plaque purification (**e**) to eliminate incorrectly recombined viruses, *E. coli* or in vitro-recombined viral genomes can be analyzed by standard molecular biology tools (**g**). Transfection of recombinant baculoviral genome in insect cells (**h**) results in budding (**d**) of recombinant viruses (**f**), which can subsequently be amplified (**i**) by infection of insect cells resulting in approx. 500 progeny viruses per infected cell

overlaid with agarose to limit virus diffusion. The initial number of infection-competent viral particles can be determined after five days by counting the plaques in the cell layer formed by virus-mediated cell lysis. The titer of a virus stock is thus indicated in pfu (plaque-forming units). To accelerate virus quantification, immunology-based (Kwon et al. 2002) or quantitative real-time PCR-based (Lo and Chao 2004) approaches have been developed, which, however, continue to require empirical correction factors to reach the same accuracy as the gold-standard plaque assay titration method.

6.3 Bioreactor Concepts

Scale-up of insect cell-based protein production is performed in insect cells adapted for suspension growth in serum-free media and bioreactors, which resemble those used for mammalian cells such as stirred tank, BioWave, perfusion and rotating wall vessel bioreactors or lab-scale roller bottles and spinner flasks. Since typical baculovirus-based production processes are conducted within four to seven days, the desired bioreactor types have to enable simple assembly, sterilization, inoculation, cleaning and disassembly. This limits the suitability of rotating wall vessels or perfusion bioreactors due to their operational complexity, although it has been reported that those bioreactor types provide superior cell densities. The bioreactors of choice are thus simple, conventional, stirred tank bioreactors or, preferably, the recently developed BioWave (Ikonomou et al. 2003; Weber et al. 2001) reactors, where all the parts in direct contact with the culture medium are sterilized before purchase and are disposable, thus eliminating the need for sterilization, validation and cleaning (http://www.wavebiotech.net). The BioWave reactor consists of an inflated plastic bag containing the medium (WaveBag). The entire bag is gently agitated, thereby inducing a wave on the surface of the medium to ensure high rates of oxygen transfer from the headspace to the culture medium. The low shear forces (free of stirrer and air bubbles), together with low manpower requirements and the possibility of scaling-up to 500 L, result in 40% lower operation costs compared to the stirred tank alternative. As a result, the BioWave reactor is the preferred option for insect cell-based protein production in industry (Weber and Fussenegger 2005).

During cultivation in the bioreactor there must be an adequate oxygen supply, especially after infection when the rates of specific oxygen uptake increase by 50%. Despite some controversy (Ikonomou et al. 2001), there seems to be agreement that dissolved oxygen concentrations should be above 30% air saturation (Weber and Fussenegger 2005). This can be ensured by increasing the rate of oxygen transfer accordingly, preferably by aeration with pure oxygen and not by increasing the gas flow or stirrer speed in order to minimize shear forces. However, the gas flow must be sufficiently high in order to strip the produced carbon dioxide from the flow, because it is suggested to inhibit growth at elevated concentrations (Ikonomou et al. 2001).

6.4 Process Design

At the level of the single cell, recombinant protein production starts approximately 12 h after infection and continues until cell lysis, approximately 48 h later, thus limiting the maximum number of recombinant protein molecules produced by a single cell. To maximize volumetric protein yields it is therefore crucial to time infection so that a maximum number of cells can terminate recombinant protein production before nutrients become limiting, implying that the nutrient supply and infection kinetics must be optimized in order to reach the theoretically possible protein yield (Licari and Bailey 1992).

6.4.1 *Optimization of Nutrient Supply*

In nutrient optimization experiments peak cell density under non-infected conditions is considered to be the best lumped value for reflecting nutritional parameters as it integrates all the factors modulating growth and viability. Nutrient supply can best be maximized by a fed-batch design, where concentrated nutrients are added at elevated cell densities to sustain further growth and recombinant protein production. Integration of several optimization studies revealed that the addition of concentrated amino acids, glucose, yeastolate and lipids is most effective in increasing the nutrient supply, thus allowing cell densities of up to 5.2×10^7 cells mL^{-1} in fed-batch mode (Elias et al. 2000).

6.4.2 *Optimization of Production Kinetics*

Optimum infection timing would result in an infection kinetic, where the last cell terminates protein synthesis prior to lysis, at the same moment as the last nutrient molecule is consumed. Premature infection and protein production would result in premature cell lysis, before all the nutrients could be converted to product protein, biomass, progeny virus and byproducts, whereas late infection would result in nutrient limitation during target protein biosynthesis, thus decreasing recombinant protein yield (Licari and Bailey 1992).

To achieve well-timed infection, two main strategies are:

– Addition of sufficient infectious viral particles to synchronously infect all the cells results in a well-determined infection time point, but it requires substantial amounts of virus stock. Mathematical modeling and experimental evidence has shown that at least five infectious particles per cell (multiplicity of infection, MOI = 5) are required to achieve infection of all the cells due to the Poisson-like distribution of viral entry into the target cells.

– Substantially lower amounts of virus stocks are required when in situ virus amplification is performed. Therefore, small amounts of virus are added (MOI < 0.1), which initially infect a small fraction of insect cells and thus trigger the production of a progeny virus, which subsequently infects the rest of the culture. It has been shown that successful recombinant protein production is possible with MOIs as low as 3×10^{-5} (Liebmann et al. 1999). However, small deviations in virus quantification or time of infection are amplified during in situ virus production and, thus, may lead to irreproducible processes.

Since both strategies, high MOI for synchronous infection and low MOI for in situ virus amplification, have been shown to yield comparable product protein titers (Licari and Bailey 1992), the infection strategy can be chosen according to the characteristics of the target protein and factors that limit the technology. Criteria for selecting the strategy are highlighted in Table 6.1.

To achieve overall optimized recombinant protein titers it is therefore crucial to optimize both the nutrient conditions and the time of infection. To limit the number of factorial experiments while synchronously optimizing nutrient supply and infection kinetics, a four-step process optimization is usually performed (Table 6.2) (Weber and Fussenegger 2005).

Table 6.1 Criteria for selecting a high or low multiplicity of infection (MOI) infection strategy

High MOI infection	In situ virus amplification following low MOI infection
Unstable product protein. High MOI infection shortens the overall process and thus limits protein degradation or denaturation	Stable protein
Toxic or growth-retarding product protein	Non-toxic protein. Premature toxic protein production during virus amplification would compromise cell viability and delay production kinetics thus resulting in uncontrollable variations in the overall process
Unknown infection kinetics and cell growth rate	Well-known infection kinetics and cell growth rate are the key to developing strategies for low MOI, where subsequent rounds of in situ virus amplification are required
High-titer virus stocks available	Only low titer virus stocks available
Expression at low to moderate cell densities. Infection at high cell densities $(1–2 \times 10^6$ cells mL$^{-1})$ would require addition of substantial amounts of virus, which might be difficult to produce or to concentrate	Expression at very high cell densities, at which to add enough virus is technically demanding

Table 6.2 Optimization scheme for recombinant protein production

Step	Action
Step 1	Optimization of overall process. Optimize fed-batch design by means of different nutrient mixes, oxygen supply and bioreactor settings. The peak cell density can be taken as the response parameter, since it represents the integration of all growth promoting and growth-retarding parameters. When using common cell lines like Sf-9, established protocols can be applied (Weber and Fussenegger 2005; Weber et al. 2001; Weber et al. 2002)
Step 2	Deciding on the infection strategy (low or high MOI, see Table 6.1)
Step 3	Infection of the culture at different times. This step can also be performed on a small scale (Bahia et al. 2005) under conditions imitating the process in Step 1
Step 4	Sampling of the culture at different points in time post infection and determination of titer and quality of product protein (degradation products). A premature harvest might result in low titers, since protein synthesis has not been completed, whereas a late harvest can impact protein quality and yield due to liberation of proteases from lysed cells or loss of intracellularly produced protein by cell lysis

6.5 Applications

Baculoviruses and insect cells have been applied as a tool to express thousands of different proteins and are today almost as common in protein production labs as *E. coli* or yeast. Insect cells are used mainly to produce "tool proteins" (from mg to g) for subsequent structure determination, screening assays or evaluation of biochemical parameters like enzyme kinetics. The use of insect cells to express therapeutic proteins such as hormones or antibodies is limited due to a non-mammalian glycosylation pattern, which would result in impaired protein activity or even immunogenicity (see below, Paragraph 6.6.1). Table 6.3 lists selected studies on production of proteins belonging to the most important classes and serves as an overview and reference when designing the production of similar proteins.

6.6 Current Trends in Insect Cell-Based Protein Production

Parameters for baculovirus-based protein production in insect cells are well characterized with regard to process design, medium requirements and scaling up of facilities, thus leading to a high, reproducible production yield in a minimum of time (Bahia et al. 2005). The research focus has therefore shifted from maximizing yields to optimizing product protein quality, mainly by humanizing post-translational modifications by engineering the glycosylation pathway. Wild-type insect cells produce potentially allergenic glyco-motifs of the Fucα(1,3)Glc-NAc-Asn-type and are incapable of producing mammalian cell-specific complex *N*-glycans

Table 6.3 Selected case studies for the production of proteins belonging to different protein classes

Protein class	Protein	Feature	References
Antibodies	Diverse	Review of expression of antibodies in insect cells	Guttieri and Liang (2004)
Enzyme	Cytosine-C5 methyltransferase	Optimization of product yield and purification	Brank et al. (2002)
Enzyme	beta-1,4-N-acetylglucosaminyltransferase III (GnTIII)	Enzyme production and application for humanization of antibodies	Hodoniczky et al. (2005)
Hormone	Eryhtropoietin	Glycosylation analysis of erythropoietin expressed in Drosophila S2 cells	Kim et al. (2005)
Hormone-like factors	Coagulation factor VIII	Expression of mutated coagulation factor VIII	Sarafanov et al. (2006)
Kinase	BTK, Brutons tyrosine kinase	Optimization of BTK expression parameters with regard to nutrient and infection parameters in a BioWave reactor	Weber and Fussenegger (2005)
Kinase	IRAK4	Process optimization by co-expression of GFP	Philipps et al. (2005)
Kinase	PDK1/PKBbeta/Akt2	Optimization of kinase expression using different MOIs and time of infection	Gao et al. (2005)
Membrane transporter	H^+K^+-ATPase	Synchronous expression of both the a and b subunits from one baculovirus result in active, membrane-localized H^+K^+-ATPase. Functional protein titers were increased by ethanol addition, thus impacting membrane characteristics	Klaassen et al. (1995)

(continued)

Table 6.3 (continued)

Protein class	Protein	Feature	References
Multiprotein complexes	Human transcription factor IID complex	Construction of baculoviruses for simultaneous expression of several proteins for assembly of multiprotein complexes	Berger et al. (2004a)
Virus like particles	HIV-1 Pr55gag	Impact of MOI and time of infection on proteolytic activities and resulting impact on Pr55gag integrity	Cruz et al. (1999)
Virus like particles	HIV-1 Pr55gag	Optimization of virus-like particle titers and quality under different agitation and aeration conditions	Cruz et al. (1998)
Virus like particles	Porcine parvovirus like particle	Optimization of product yield at low MOI infection strategy under fed-batch conditions	Maranga et al. (2003)
Virus structural proteins	Infectious bursal disease virus proteins	Optimization of product yield and quality using different media and protease inhibitors	Hu and Bentley (1999)

thus limiting the use of insect cell-derived proteins for therapeutic in vivo applications (Tomiya et al. 2003). Therefore, the baculovirus community is focusing on improving the glycosylation pattern by means of different approaches with the aim of competing for industrial application:

- Screening for new insect cell lines that produce mammalian-like glycosylation patterns. For example, the *Danau plexippis*-based DpN1 line showed up to 26% complex glycoforms compared to standard High Five™ cells (<5%). However, standard protocols and media have not yet been adapted for this newly introduced cell line, resulting in less competitive protein titers (Palomares et al. 2003).
- Supplementation of medium with glycosylation precursors, such as *N*-acetylmannosamine (ManNAc), or addition of specific glycosyltransferase inhibitors

to High Five™ cells were reported to result in N-glycan sialyalation of recombinant interferon-g.

- One of the most promising approaches relies on co-expression of glycosyltransferases as well as enzymes producing the glycan precursors. Expression of β-1,2-*N*-acetylglucosaminyltransferase II, under control of a baculoviral immediate an early promoter, resulted in galactosylated, biantennary, complex-type glycans on the model protein human transferrin (Tomiya et al. 2003). Glycosylation could be improved further by biosynthesis of the glycan precursor N-acetylneuraminic acid (Neu5Ac) through expression of the human sialic acid 9-phosphate synthase in combination with UDP-GlcNAc 2-epimerase/ManNAc kinase, the bifunctional enzyme initiating sialic acid biosynthesis in mammals (Viswanathan et al. 2003). Finally, human-like biantennary, terminally sialylated *N*-glycans could be produced in Sf-9 insect cells by upregulation of five glycosyltransferases with the model human β-trace protein, a prostaglandin-*D* synthase (Mimic™ Sf-9 cells, Invitrogen, Hollister et al. 2002). However, despite overexpression of those five glycosyltransferases, mammalian-like glycosylation is still deficient. Equine chorionic gonadotropin (eCG), synthesized in Mimic™ cells, showed differential glycosylation compared to Sf-9 cells but failed to produce significant in vivo bioactivity, probably because of insufficient terminal sialylation of its carbohydrate chains, leading to its rapid removal from blood (Legardinier et al. 2005).

6.6.1 Elimination of Product Protein Proteolysis

Besides glycoengineering, attention has been paid to minimizing protease activities in order to increase product protein quality. Namely, the baculoviral protease *v-cath*, a mammalian cathepsin L homologue, has been correlated with increased proteolysis. Product quality improved either by adding cysteine protease-specific inhibitors, such as peptin or pepstatin (Hu and Bentley 1999; Pyle et al. 1995), or by genetic inactivation of *v-cath* and *chiA*, a chitinase involved in *v-cath* activation (Berger et al. 2004a), thus significantly increasing product protein integrity for high-end applications like crystallographic studies or NMR analysis.

6.7 Limitations

Insect cells, in combination with baculovirus infection, face competition from alternative expression systems, namely *E. coli*, yeast, plant and mammalian cells, for the production of tool proteins as well as for biopharmaceuticals. The pros and cons of using insect cells in conjunction with the baculovirus expression system are highlighted in Table 6.4.

Table 6.4 Pros and cons of using insect cells for recombinant protein production

Parameter	Pros and cons
Medium requirements	Serum-free, chemically defined media, which are similar to the complexity and price of mammalian cell culture media, are commonly used
Process design	Fed-batch and batch processes are typically used since the production period is limited to 48–72 h, after which cell lysis occurs. Thus, protein yield cannot be increased in the same way as mammalian cells by prolonging the production phase or by using continuous cultures. The short overall process favors bioreactor designs (e.g. the BioWave reactor), requiring minimal manpower for installation and de-installation
Introduction of target genes	Target gene transfer is mediated by baculovirus infection with very high efficiency and eliminates the need for transfection or selection of stable clones as required for mammalian cells. However, construction and amplification of viral stocks is necessary and takes up to 2–3 weeks
Reactor types	Similar to mammalian cells but without CO_2 or pH control. Cultivation temperature is 28 °C
Posttranslational modifications	Insect cells are superior to bacteria and yeast with regard to posttranslational modifications. Human cell-like glycosylation has been described for specially engineered cell lines; however, terminal sialylation still seems to be insufficient
Yield	Recombinant protein yields are commonly high and superior to mammalian cells without gene amplification. The very late viral polyhedrin and p10 promoters are extremely strong. In wild-type baculoviruses these promoters produce inclusion bodies that make up approx. 50% of the total cell protein
Ease of handling	Ready-to-use kits for virus construction and protein expression are available, thus making protein expression in insect cells almost as straightforward as using *E. coli*-based protocols
Range of target proteins	Very broad range. Numerous reports exist for intracellular, secreted or membrane proteins. Baculovirus-based expression is especially suitable for toxic or growth-retarding proteins, since gene expression is induced only at optimum cell density by infection, thus eliminating the need for stable cell clones as in mammalian cells
Safety	Baculoviruses and insect cells are considered to be safe no special precautions must be taken, in contrast to mammalian cell-specific lentiviral or adenoviral vectors

References

Ailor E, Betenbaugh MJ (1999) Modifying secretion and post-translational processing in insect cells. Curr Opin Biotechnol 10:142–145

Bahia D, Cheung R, Buchs M, Geisse S, Hunt I (2005) Optimisation of insect cell growth in deep-well blocks: development of a high-throughput insect cell expression screen. Protein Expr Purif 39:61–70

Bedard C, Kamen A, Tom R, Massie B (1994) Maximization of recombinant protein yield in the insect cell/baculovirus system by one-time addition of nutrients to high-density batch cultures. Cytotechnology 15:129–138

Berger I, Fitzgerald DJ, Richmond TJ (2004a) Baculovirus expression system for heterologous multiprotein complexes. Nat Biotechnol 22:1583–1587

Berger I, Fitzgerald DJ, Richmond TJAG (2004b) Baculovirus expression system for heterologous multiprotein complexes. Nat Biotechnol 22:1583–1587

Brank AS, Van Bemmel DM, Christman JK (2002) Optimization of baculovirus-mediated expression and purification of hexahistidine-tagged murine DNA (cytosine-C5)-methyltransferase-1 in Spodoptera frugiperda 9 cells. Protein Expr Purif 25:31–40

Cruz PE, Cunha A, Peixoto CC, Clemente J, Moreira JL, Carrondo MJ (1998) Optimization of the production of virus-like particles in insect cells. Biotechnol Bioeng 60:408–418

Cruz PE, Martins PC, Alves PM, Peixoto CC, Santos H, Moreira JL, Carrondo MJ (1999) Proteolytic activity in infected and noninfected insect cells: degradation of HIV-1 Pr55gag particles. Biotechnol Bioeng 65:133–143

Elias CB, Zeiser A, Bedard C, Kamen AA (2000) Enhanced growth of Sf-9 cells to a maximum density of 5.2 × 10(7) cells per mL and production of beta-galactosidase at high cell density by fed batch culture. Biotechnol Bioeng 68:381–388

Gao X, Yo P, Harris TK (2005) Improved yields for baculovirus-mediated expression of human His(6)-PDK1 and His(6)-PKBbeta/Akt2 and characterization of phospho-specific isoforms for design of inhibitors that stabilize inactive conformations. Protein Expr Purif 43:44–56

Geisse S, Gram H, Kleuser B, Kocher HP (1996) Eukaryotic expression systems: a comparison. Protein Expr Purif 8:271–282

Guttieri MC, Liang M (2004) Human antibody production using insect-cell expression systems. Methods Mol Biol 248:269–299

Hodoniczky J, Zheng YZ, James DC (2005) Control of recombinant monoclonal antibody effector functions by Fc N-glycan remodeling in vitro. Biotechnol Prog 21:1644–1652

Hollister J, Grabenhorst E, Nimtz M, Conradt H, Jarvis DL (2002) Engineering the protein N-glycosylation pathway in insect cells for production of biantennary, complex N-glycans. Biochemistry 41:15093–15104

Hu YC, Bentley WE (1999) Enhancing yield of infectious Bursal disease virus structural proteins in baculovirus expression systems: focus on media, protease inhibitors, and dissolved oxygen. Biotechnol Prog 15:1065–1071

Ikonomou L, Bastin G, Schneider YJ, Agathos SN (2001) Design of an efficient medium for insect cell growth and recombinant protein production. In Vitro Cell Dev Biol Anim 37:549–559

Ikonomou L, Schneider YJ, Agathos SN (2003) Insect cell culture for industrial production of recombinant proteins. Appl Microbiol Biotechnol 62:1–20

Jarvis DL (2003) Developing baculovirus-insect cell expression systems for humanized recombinant glycoprotein production. Virology 310:1–7

Kamen A, Bedard C, Tom R, Perret S, Jardin B (1996) On-line monitoring of respiration in recombinant baculovirus-infected and uninfected cell bioreactor cultures. Biotechnol Bioeng 50:36–48

Kim YK, Shin HS, Tomiya N, Lee YC, Betenbaugh MJ, Cha HJ (2005) Production and N-glycan analysis of secreted human erythropoietin glycoprotein in stably transfected Drosophila S2 cells. Biotechnol Bioeng 92:452–461

King LA, Possee RD (1992) The baculovirus expression system. Chapman & Hall, London

Klaassen CH, Swarts HG, De Pont JJ (1995) Ethanol stimulates expression of functional H+,K(+)-ATPase in SF9 cells. Biochem Biophys Res Commun 210:907–913

Kost TA, Condreay JP, Jarvis DL (2005) Baculovirus as versatile vectors for protein expression in insect and mammalian cells. Nat Biotechnol 23:567–575

Kwon MS, Dojima T, Toriyama M, Park EY (2002) Development of an antibody-based assay for determination of baculovirus titers in 10 hours. Biotechnol Prog 18:647–651

Legardinier S, Duonor-Cerutti M, Devauchelle G, Combarnous Y, Cahoreau C (2005) Biological activities of recombinant equine luteinizing hormone/chorionic gonadotropin (eLH/CG) expressed in Sf9 and Mimic insect cell lines. J Mol Endocrinol 34:47–60

Licari P, Bailey JE (1992) Modeling the population dynamics of baculovirus-infected insect cells: optimizing infection strategies for enhanced recombinant protein yields. Biotechnol Bioeng 39:432–441

Liebmann JM, LaScala D, Wang W, Steed PM (1999) When less is more: enhanced baculovirus production of recombinant proteins at very low multiplicities of infection. BioTechniques 26:36–42

Lo HR, Chao YC (2004) Rapid titer determination of baculovirus by quantitative real-time polymerase chain reaction. Biotechnol Prog 20:354–360

Maranga L, Brazao TF, Carrondo MJ (2003) Virus-like particle production at low multiplicities of infection with the baculovirus insect cell system. Biotechnol Bioeng 84:245–253

Palomares LA, Gonzalez M, Ramirez OT (2000) Evidence of Pluronic F-68 direct interaction with insect cells: impact on shear protection, recombinant protein, and baculovirus production. Enzyme Microb Technol 26:324–331

Palomares LA, Joosten CE, Hughes PR, Granados RR, Shuler ML (2003) Novel insect cell line capable of complex N-glycosylation and sialylation of recombinant proteins. Biotechnol Prog 19:185–192

Philipps B, Rotmann D, Wicki M, Mayr LM, Forstner M (2005) Time reduction and process optimization of the baculovirus expression system for more efficient recombinant protein production in insect cells. Protein Expr Purif 42:211–218

Power JF, Reid S, Radford KM, Greenfield PF, Nielsen LK (1994) Modeling and optimization of the baculovirus expression vector system in batch suspension culture. Biotechnol Bioeng 44:710–719

Pyle LE, Barton P, Fujiwara Y, Mitchell A, Fidge N (1995) Secretion of biologically active human proapolipoprotein A-I in a baculovirus-insect cell system: protection from degradation by protease inhibitors. J Lipid Res 36:2355–2361

Sarafanov AG, Makogonenko EM, Pechik IV, Radtke KP, Khrenov AV, Ananyeva NM, Strickland DK, Saenko EL (2006) Identification of coagulation factor VIII A2 domain residues forming the binding epitope for low-density lipoprotein receptor-related protein. Biochemistry 45:1829–1840

Schlaeger EJ (1996) Medium design for insect cell culture. Cytotechnology 20:57–70

Tomiya N, Betenbaugh MJ, Lee YC (2003) Humanization of lepidopteran insect-cell-produced glycoproteins. Acc Chem Res 36:613–620

Vialard JE, Arif BM, Richardson CD (1995) Introduction to the molecular biology of baculoviruses. Methods Mol Biol 39:1–24

Viswanathan K, Lawrence S, Hinderlich S, Yarema KJ, Lee YC, Betenbaugh MJ (2003) Engineering sialic acid synthetic ability into insect cells: identifying metabolic bottlenecks and devising strategies to overcome them. Biochemistry 42:15215–15225

Weber W, Fussenegger M (2005) Baculovirus-based Production of Biopharmaceuticals using Insect Cell Culture Processes. In "Modern Biopharmaceuticals", J. Knäblein (ed). Wiley-VCH, Weinheim, Germany, 1045–1062

Weber W, Weber E, Geisse S, Memmert K (2001) Catching the Wave: the BEVS and the Biowave. In "From target to market", Lindner-Olsson, E., Chatzissavidou, N., and Lüllau, E. (ed.) Kluwer academic publishers, 335–337

Weber W, Weber E, Geisse S, Memmert K (2002) Optimization of protein expression and establishment of the Wave bioreactor for baculovirus/insect cell culture. Cytotechnology 38:77–85

Complementary Reading

This section highlights original and review articles for in-depth information on the baculovirus expression vector system in conjunction with insect cell culture (Table 6.5).

Table 6.5 Selected reading for in-depth information on different aspects of insect cell-based protein production

Scope	Summary	References
Baculoviral vector optimization	Engineering of baculoviruses for expression of multiple proteins and deletion of viral proteases	Berger et al. (2004b)
Case study	Detailed review of protein production optimization in a case study for expression of a secreted protein. Establishment of the BioWave reactor for insect cells	Weber et al. (2002)
Comparison of expression systems	A detailed comparison of mammalian and insect cell-based production processes	Geisse et al. (1996)
General overview	This article gives a comprehensive overview of the baculovirus system as well as detailed protocols for recombinant protein expression, process design and scaling up together with a cost-based choice of the bioreactor type	Weber and Fussenegger (2005)
General overview	Exhaustive overview of the baculovirus-insect cell system with the focus on medium design, protease activities, feeding strategies and bioreactor design together with an extensive list of case studies	Ikonomou et al. (2003)
Glycoengineering	Review article on baculoviruses as vectors for production of recombinant proteins in insect cells together with the latest development in glycoengineering and the use of baculoviruses as vectors for mammalian cells	Kost et al. (2005)
Glycoengineering	Development of baculovirus-insect cell systems for production of humanized glycoproteins	Jarvis (2003)
Post-translational modifications	This article gives a comprehensive overview of the engineering of post-translational modification in insect cells	Ailor and Betenbaugh (1999)
Process design	Multiparallel optimization of protein production parameters for high-throughput protein expression	Bahia et al. (2005)
Process design	Development of a model describing baculovirus infection kinetics and optimization of overall yield. The model developed here is the basis of all the successive work on optimizing multiplicity of infection and time of infection	Licari and Bailey (1992)
Process optimization	Development of a factorial experimental design for multi-parameter optimization	Bedard et al. (1994)

Chapter 7
Bioreactors for Bioartificial Organs

G. Catapano

Abstract In this chapter, the design of bioreactors constituting the core of some bioartificial organs is discussed. Initially, the problem of cell sourcing is shortly addressed. Then, criteria and limitations to current bioreactor design for bioartificial organs are presented. Design equations are separately obtained and discussed for bioreactors implanted in extravascular body compartments or connected to the blood circulation of the recipient. In the design, attention is focused mainly on oxygen delivery to the cells cultured in the bioreactor to prevent oxygen starvation and cell death, and to ensure long-term bioreactor function. The properties of the membranes used to protect the cell graft and the bioreactor as a whole from the host immune response are also discussed briefly. In the end, the bioreactors proposed for bioartificial pancreas, liver and the delivery of bioactive molecules are briefly presented and discussed.

7.1   Introduction

Artificial (i.e., made by man) instruments or devices have long been used to replace or augment the functions of malfunctioning or failing human tissues. Egyptian mummies have shown that wooden stumps were used to replace parts of legs and fingers already in 3000 B.C. Wearable eye glasses have been used since the XIV century to augment the eyesight. In time, the development of prostheses for the substitution of bodily physical functions, such as mechanical support or sight, by artificial means has become more rigorous, and has given rise to a discipline termed "substitutive medicine" that paralleled and complemented the pharmaceutical approach of traditional medicine to healing patients or improving the quality of their lives. In the twentieth century, artificial means were developed and introduced in the clinics to substitute for some mass and heat exchange function of complex human organs. In the forties, for the first time low molecular weight toxins were

G. Catapano
Department of Chemical Engineering and Materials, University of Calabria, I-87030
Rende (CS), Italy
catapano@unical.it

R. Eibl et al., *Cell and Tissue Reaction Engineering: Principles and Practice,*   279
© Springer-Verlag Berlin Heidelberg 2009

selectively and successfully removed from the blood of kidney failure patients with a membrane device termed artificial kidney (Kolff and Berk 1943). In the fifties, the first artificial gas exchange devices were used to replace the lung functions in an extracorporeal cardio-pulmonary by-pass and this permitted open heart surgery (Gibbon 1954). Today, orthopaedic and eye prostheses bear but a pale resemblance to the first ones used in humans. Artificial kidneys and lungs have also become established clinical treatments. Yet, artificial prostheses may not substitute for the complex regulatory and synthetic metabolic functions of malfunctioning or failing complex organs such as liver, pancreas or kidney.

In the last 20 years, research efforts have been concentrated on the development of bioartificial organs to overcome the limitations of artificial prostheses. In bioartificial organs, artificial devices are coupled with living cells capable of performing most or all of the biochemical metabolic functions of a complex organ. In therapeutic treatments, bioartificial organs are connected to, or implanted into, the body of diseased patients and are expected to replace, or assist, a failing or malfunctioning organ temporarily until a compatible transplantation organ is available, or the natural organ (and the patient) heals. The bioreactor (i.e., the vessel where cells are cultured), its configuration, scale and operation affects the success of bioartificial organs-based treatments to an extent that depends on the importance of the organ's synthetic and regulatory functions to the healing process, or to the stabilization of the patient's state till transplantation. However, in the treatment of severe organ failure, separation (i.e., physical) processes are often needed to: remove endogenous or exogenous toxins (the latter is typically the case of intoxicated acute liver failure patients) thus fostering the healing process; reduce the burden of cytotoxic species on the cells in the bioreactor, and prevent their premature death; protect the cellular graft in the bioreactor from rejection of the host (i.e., for immunoprotection purpose); or to simply enable the use of particular bioreactor configurations, by separating plasma from the blood and treating it in complex bioreactor designs in which blood is likely to clot. Success of treatments based on a bioartificial organ depends on the optimization of all the devices making up the bioartificial organ (the bioreactor being one of them) as a whole, as well as on the patient's state at the beginning of treatment.

In this chapter, information is provided on the bioreactors proposed for some developmental bioartificial organs and the criteria currently used to design and operate them.

7.2 Cells for Bioartificial Organs

An effective bioartificial organ should perform either all the functions of the parenchymal cells of the organ in vivo (i.e., the cells that characterize the organ's metabolic functions), or only those lacking in the failing or malfunctioning organ and most needed for patient recovery. In the absence of clear information on the pathophysiology of the diseased organ (or tissue), mammalian cells expressing the adult differentiated organotypical phenotype (Chap. 2) are generally used. The problem is that, except for tumour cells, differentiation generally limits cell proliferation. To overcome this limitation, investigations have recently been performed on the use of some kind of progenitor cells.

Working with animal or human-derived cells requires observation of ethical and safety caution measurements. Obtainment of tissue for primary cell isolation from animals is strictly regulated by rules set by governmental agencies (e.g., the American Food and Drug Administration (FDA), the European International Standard Office (ISO), etc.) to minimize the number of animals sacrificed, and to ensure proper animal handling and minimize sufferings. Work with adult or foetal human tissue requires the consent of the local ethical committee, and of the patient or her/his relatives. On safety grounds, human tissue should also be handled only in a Class II laminar flow hood at Containment Level 2 (Freshney 2005).

Several types of cells have been proposed for bioartificial organs. Embryonic or neonatal animal cells, and human progenitor or stem cells have been proposed to exploit their proliferation capacity. Differentiation is generally induced during culture when the required cell mass has been obtained. Transfected or transgenic cells are selected for the expression of a single organ function: this restricts their use to those cases when healing requires provision only of that specific metabolic function. Tumour cells, preferably non-malignant, have also been proposed for their capacity to proliferate and for not requiring anchorage to a support. However, their use is limited by the expression of only a subset of the differentiated functions of the organ parenchymal cells, and by the risks of malignancy generally associated with the leak of cells or tumorigenic soluble factors from the bioreactor into the patient's body. Primary cells and immortalized cell lines are the cell types mostly used in bioartificial organs that have been (or are still being) tested in the clinics.

Primary cells. The most straightforward approach to bioartificial organ design capable of fostering the healing process is to use parenchymal cells isolated from the same organ of similar mammals. In fact, the scarce availability of tissue-compatible donor organs and rejection of allografts (i.e., organs from different individuals of the same species) or xenografts (i.e., organs from an individual of a different species) still limit transplantation to a minor fraction of patients on the national waiting lists. Parts of, or whole, xenogeneic organs cultured in bioartificial organs exhibit differentiated organotypic functions *ex vivo* or *in vitro* for just a few hours. Thus, primary cells are preferably isolated from a tissue (or the whole organ) by disaggregation with mechanical, chemical or enzymatic means to free them from the extracellular matrix (ECM), from undesired non-parenchymal cells and from other unwanted cells. Mechanical disaggregation causes more damages to cells than the other two techniques. However, hard tissue specimens (e.g., bone) are generally mechanically reduced to minute fragments prior to chemical and/or enzymatic disaggregation. Organs or parts thereof, are often mechanically disaggregated prior to, or after, chemical or enzymatic treatment to facilitate harvesting of the isolated cells. Chemicals capable of chelating calcium ions, Ca^{+2}, (e.g., EDTA, citrate) are used to disaggregate a tissue or only to aid cell dissociation. In fact, the function of many molecules mediating cell adhesion to other cells or a substratum (e.g., cadherins, cell adhesion molecules (CAMs)) is mediated by Ca^{+2} ions. Ca^{+2} ion depletion loosens up cell junctions and facilitates cell dissociation, albeit with a lower isolation yield than enzymatic methods. Enzymes (e.g., collagenase, trypsin) are used to digest the collagen and protein-rich ECM that envelopes and scaffolds the cells in the natural tissues, thus making it possible to harvest isolated cells. All disaggregation techniques remove some cell surface receptors or may damage permanently, or even digest, the cells themselves. For instance, parenchymal liver cells (i.e., the hepatocytes) appear to lose some of their oligosaccharide-lectin binding capacity after isolation (Jaregui 1994). In some cases (e.g., bone tissue), after mechanical disaggregation cells migrate out of minute tissue fragments adhering to the surface of the culture vessel and are trypsinized and harvested when they grow to confluence. After tissue disaggregation by any means, the parenchymal cells are generally purified from cell debris, unwanted cells and remnants of the ECM by repeated centrifugation and rinsing with isotonic medium and by filtration through molecular sieves. Centrifugation in medium with a density gradient is occasionally used to sort out a given subset of cell population. In fact, contamination of epithelial cells with unwanted cells may be a major problem when cells are isolated to be cultured in a bioartificial organ. In fact, unwanted cells such as the fibroblasts may have less stringent metabolic requirements than the epithelial or parenchymal cells (e.g., the hepatocytes) and may outgrow them, ending up with a bioreactor that is no longer capable of performing the required metabolic functions. A specialized cell type may be selected during isolation with proper molecular sieves, when cells differ in size, or during the first days of culture by using selective, serum-free media supplemented with specific mitogenic factors (e.g., promoting proliferation of the epithelial cells and inhibiting the growth of unwanted cells). However, cell identity should be carefully characterized prior to using them in a bioartificial organ. This may be accomplished by checking many cell phenotypic features, such as their morphology, DNA, the expression of specific enzymatic activities, or the presence of cell-specific surface antigens, among others. Excellent textbooks and reviews are available in the specialized literature where optimized cell isolation techniques are described for various tissues and processes that maximize the isolation yield (i.e., the number of cells per unit mass of tissue) and cell viability, and minimize cell damage (e.g., Freshney 2005).

Strictly speaking, only the cells seeded in a bioreactor and used before their first sub-culture (often termed *passage*) ought to be termed *primary cells* (Freshney 2005). After the first sub-culture, cells retaining a limited life span are often referred to as *finite cell lines*. Although they might exhibit some differences, both cell types generally grow slowly (some human cell types do not proliferate at all in the typical therapeutic treatment times), are contact-inhibited, have important metabolic requirements (as compared to continuous cell lines) but, under proper *in vitro* culture conditions, may be highly differentiated and may express most or all the functions of the organ from which they were isolated (Freshney 2005).

Immortalized cell lines. A large number of continuous cell lines have been developed and patented that have been mostly obtained from tumour cells modified to be immortal and to express some organotypic differentiated functions. These cells generally grow faster and have less important stringent metabolic requirements than primary cells (e.g., in terms of oxygen consumption rate), yet they are often as contact-inhibited as primary cells (e.g., HepG2 or C3a liver cell lines) (Sussman et al. 1992). Their features make a virtually unlimited cell mass available for the bioreactor, and permit easy bioreactor operation at cell densities typical of natural tissue. However, these cells generally express only a few differentiated functions typical of the natural organ that the bioartificial organ should replace or assist. Concerns do also exist for the risks associated with the leak of cells into the patient or soluble factors suspected of being carcinogenic. Immortalized cell lines expressing one single metabolic function have been proposed for the treatment of pathological states in which the presence of that specific metabolite is thought to relieve the symptoms or heal the disease. This is the case of the dopamine-secreting PC12 cells that have been proposed for the treatment of Parkinson's disease (Lindner and Emerich 1998) or of the catecholamine-secreting bovine chromaffin cells that have been proposed for relieving pain in terminal cancer patients.

7.2.1 *Expression of the Adult Cell Phenotype*

In vivo, epithelial (or parenchymal) cells are generally organized with other cell types (e.g., in the liver the hepatocytes co-exist with Kuppfer, Ito and endothelial cells) in hierarchical architectural structures surrounded and mechanically supported by the extracellular matrix (ECM). Epithelial (or parenchymal) cells establish intercellular junctions with homologous cells that hold them close together (i.e., the desmosomal and tight junctions) or allow transmission of biochemical signals between adjacent cells (the gap junctions). The endothelial cells line the fine capillary network that vascularize an organ and permit the necessary supply of oxygen and nutrients to, and removal of waste metabolites from, the cells. The ECM generally consists of different types of collagen, laminin, fibronectin, hyaluronan, proteoglycans, and bind growth factors with an actual composition that is determined by the cell types present in the tissue/organ. The ECM provides mechanical support to the cells and affects their differentiation. In fact, it provides immobilized moieties (e.g., the arginine- glycine-aspartic acid (RGD) amino acid sequence) and growth factors that promote cell adhesion, direct cell motion and foster cell differentiation. In many organs, this complex network of interactions:

- Causes epithelial cells to polarize, so that each cell side performs a specialized function, e.g., in the hepatocytes in the natural liver, nutrients are transported through the apical surface facing the vascular endothelial cell lining, and bile is eliminated through the basal surface;
- Induces the expression of different enzymatic activities in cells located in different zones of the organ, a phenomenon termed as 'zonation', e.g., in the natural liver, hepatocytes farther away from the blood capillaries exhibit a greater capacity to remove exogenous drugs;

284	G. Catapano

- Optimizes nutrient and waste metabolite transport to and away from cells;
- Makes complex, multicellular physiological feedback control mechanisms possible, e.g., in the pancreas, in the islets of Langerhans three cells types (i.e., β-cells, α-cells and δ-cells) closely interact to provide for glucose homeostasis;
- Makes repair and regeneration possible when tissue is somehow damaged;
- Ultimately, makes it possible also for large organs to live and perform the specialized metabolic functions needed for the homeostasis of the organism.

Owing to the great complexity of tissue/organ structure and physiology, it is highly recommended to refer to available textbooks of anatomy and physiology and the specialized literature for more exhaustive information. Such a complex network of effects leads to "the expression of constitutive (i.e., stably expressed) and adaptive (i.e., subject to regulation) phenotypic properties characteristic of the functionally mature cell in vivo", in other words to cell differentiation (Freshney 2005).

When epithelial cells are isolated from an organ or tissue, they are deprived of this complex network of relations. This causes cell depolarization, hence their morphology changes from polygonal or elongated (depending on the cell type) to round, as shown in Fig. 7.1; cell de-differentiation, hence cells are no longer able to perform their specialized metabolic functions; cell death often within a few hours, if the cells are cultured as singlets in suspension. Thus, prior to using cells for therapeutic purposes in a bioreactor strategies have to be devised and put in practice to induce the cells to regain their polarity and differentiate, and to prevent their premature death, as well. Fortunately, cells may be induced to differentiate *in vitro*, partly or even completely, provided that media with suitable biochemical stimuli are used, the appropriate cell types are co-cultured at high cell density, and proper

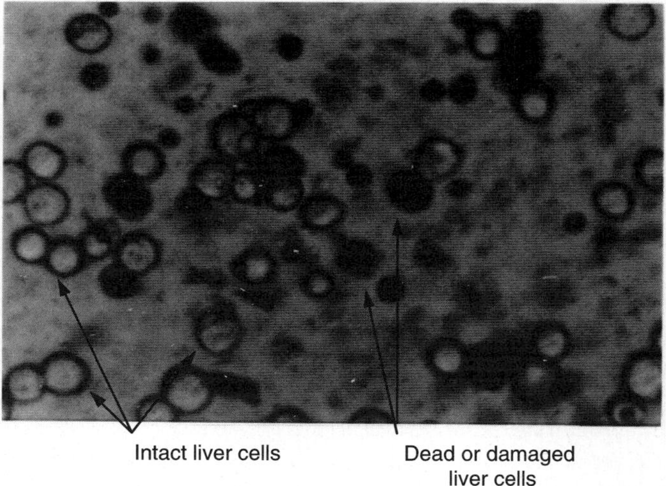

Fig. 7.1 Optical microscopy of rat liver cells immediately after isolation and staining with trypan blue. Dead cells do not exclude trypan and are stained blue

support for cell attachment and migration is offered that permits cells reorganization as in the natural tissue *in vivo*. In fact, culture at high cell density (i.e., greater than 10^5 cells cm^{-2} in monolayer adhesion culture) fosters homotypic cell-cell adhesion and signal transmission among contacting cells. Co-culture of the different cell types present in the natural tissue fosters heterotypic cell-cell interactions that have been shown to initiate and promote differentiation. For instance, the hepatocyte growth factor (HGF) produced by fibroblasts was shown to induce tubule formation by cells of the MDCK continuous line from dog kidney (Orellana et al. 1996; Montesano et al. 1997). Cell differentiation may also be promoted by supplementing the culture medium with physiological inducers, such as hormones (e.g., insulin and glucagon), steroids (e.g., hydrocortisone), vitamins (e.g., vitamin D3 and retinoids), cytokines (e.g., HGF and bone morphogenetic protein, BMP), minerals and proper dissolved oxygen tension, or/and non-physiological inducers, such as dimethylsulfoxide (DMSO) or dexamethasone. Chemical composition, morphology and topography of the artificial scaffolds on which cells adhere and in which they migrate have been shown to regulate gene expression and the functional expression of many cells. Scaffolds resembling the ECM have been produced by mixing purified constituents of the natural ECM (e.g., collagen, fibronectin and laminin), by using ECM generated by cells that are subsequently destroyed and washed off (e.g., acellularized intestinal submucosa), or by using commercial matrices such as that generated by the Engelberth-Holm-Swarm mouse sarcoma and sold under the trade name Matrigel®. Cell culture in three-dimensional (3D) scaffolds made of polymeric or ceramic foams, thin fibres arranged in non-woven fabrics, and hollow fibre membranes, or embedded in shrunk gels made of natural proteins (e.g., collagen) or polysaccharides (e.g., alginate) has also been shown to help isolated cells regain their polarity and to foster cell differentiation and survival better than conventional monolayer two-dimensional (2D) culture.

7.3 Bioreactor Design for Bioartificial Organs

Bioreactors are used at any stage of the preparation of bioartificial organs to culture cells, in adhesion on two-dimensional (2D) flat substrata or seeded in three-dimensional (3D) scaffolds, with the aim of subjecting the cells to controlled and closely monitored biochemical, mechanical, electrical or magnetic cues. Primary cells are often cultured in Petri dishes or T-flasks prior to seeding them in the bioreactor used for therapeutic purposes to let them recover from the isolation stress and re-differentiate, or to expand them to the desired mass if tissue was obtained from a biopsy. Petri dishes and T-flasks are also used to expand primary progenitor or stem cells in an undifferentiated state, to maximize their proliferation capacity and to minimize the time to implant. In fact, both bioreactors permit cell culture in the absence of mechanical disturbance (e.g., shear stresses) and minimize the amount of medium needed at this phase, which is made expensive by the factors supplemented to either make primary cells polarize and re-differentiate, or to keep progenitor cells in an undifferentiated state. These bioreactors are dealt with in details in Chap. 3, and will not be discussed here.

The rest of this chapter is devoted only to those bioreactors that constitute the core of a bioartificial organ. The bioreactor designs proposed are rather original and often differ from those typically used for technological applications. For the sake of brevity, only those bioreactors used in bioartificial organs that have been tested in the clinics or in animal models are considered. The bioreactors proposed have volumes ranging from a few microlitres to about a litre, and host from a few cells (as in the treatment of diabetes, Parkinson's or Alzheimer's disease) to more than 600 g cells (as in the support to acute liver failure (ALF) patients) generally arranged in a fixed bed. Hence, these bioreactors are operated batch-wise with respect to the cells used independent of how nutrients are supplied to, and waste metabolites removed from, the cells (e.g., continuously or batch-wise). Most bioreactors used for bioartificial organs are *membrane bioreactors* (Sect. 5.1.3). The reason for this is that, when connected to, or implanted into, the recipient's body the bioreactor components are sensed by the body as "foreign" which activates the body's humoral (i.e., soluble) and cellular immune defence systems to reject the bioreactor (i.e., eliminate or expel it). Polymeric permselective membranes have been effectively used to protect the cells in the bioreactor from rejection. These membranes act as selective barriers that compartmentalize the cells in an immuno-privileged compartment (i.e., the cell compartment) by preventing immune-competent cells and proteins from crossing their wall, yet they are freely permeable to the nutrients and metabolites necessary for cell survival and metabolism. Those proposed are mostly *membrane compartmentalized cell reactors* (MCCRs), and share the advantageous features of this type of bioreactors (Sect. 5.1.3).

The design of bioreactors for bioartificial organs is made more complex than those for industrial applications by a number of concomitant factors. In fact, the cells in the bioreactor have to perform multiple metabolic tasks (i.e., excretory, synthetic and regulatory) at the same time and are operated under conditions far more critical than in healthy patients for the accumulation of endogenous toxins or immune proteins caused by multi-organ failure and by the activation of the patient's immune response. For example, this is the case of acute liver failure patients in whose blood hepatic toxins accumulate and who often develop a concomitant multiorgan failure before treatment begins. Ideally, the cells should remain viable and should express differentiated organotypical functions till the patient heals. The bioreactor should also adjust its performance to the changing metabolic state (and needs) of the patient, as treatment progresses in time. Bioreactor connection to, or implantation into, the patient's body restricts the possible external control of the operator on the bioreactor performance, and makes it difficult even to monitor the actual bioreactor performance during treatment. The contact of bioreactor constituents (i.e., cells and materials) with physiological fluid and tissues is likely to change their properties, an actual *host vs. graft response*. In fact, the xeno- or allogeneic cells or the materials used may be recognized as *foreign* by the recipient's immune system and be rejected, e.g., by enclosing the bioreactor in a fibrotic capsule. Plasma or blood proteins may adsorb on the membranes used in the bioreactor to protect the cells from rejection and worsen their separation and transport properties. Activation of the coagulation system may cause blood coagulation resulting in the bioreactor occlusion and require its replacement. Likewise, the contact of physiological fluids and tissues with bioreactor constituents is likely to change their properties, an actual *graft vs. host response*. In fact, xenogeneic cells may shed antigens that may cross the immune barrier and trigger hypersensitivity or anaphylactic reactions. Cytokines may accumulate in the patient's blood as a result of the cellular system activation caused by the contact with the bioreactor materials. Cells of tumoural origin in the bioreactor might leak or might release soluble species that might put the patient at the risk of developing a tumour.

The definition of the therapeutic performance that the bioreactor should deliver is also at times elusive. In fact, the pathophysiology of the organ failure and the cell metabolism in the bioreactor are often only partially known. This has certainly to do with: the strongly case-to-case varying causes of the organ failure; the broadly varying patient's metabolic state at the time that permission is given to start the treatment; and the often purely qualitative knowledge of cell metabolic reactions and their kinetics *in vitro*. This causes a great uncertainty in the definition of the performance that the bioartificial organ as a whole, and the bioreactor in particular, has to deliver. In fact, depending on the extent of the organ damage (e.g., an organ with most parenchymal cells still living has a greater chance of healing than one whose cells are mostly or all dead) and of the damages caused to other organs, both the metabolic functions that should be replaced and the rate at which these metabolic processes should be provided by the bioartificial organ might change.

Today, design objective functions for bioartificial organs and the bioreactors used are generally defined in terms of the cell mass necessary for the organ self-recovery, the rate of metabolic product synthesis or nutrients consumption, and/or the response time for hormone excretion and delivery following a metabolic challenge. These parameters are generally required to match the corresponding physiological values of the healthy natural organ. Bioreactor configuration, operation and scale, as well as its required operating lifetime, greatly vary with the application. The bioreactor may be designed to be operated intermittently for a few hours at a time (as in the support to ALF patients), or continuously (and ideally) for patient's lifetime (as in the treatment of diabetic or Parkinson's patients). Independent of the specific application, all bioreactors have to fulfil a few common requirements: cells should be maintained in a viable and differentiated state for as long as the treatment is needed; cells should be protected from rejection of the recipient's immune system; bioreactors and the treatment as a whole have to be biocompatible: this means that they should not do any short nor long-term harm to the patient; bioreactors should be easily scaled-up and sterilized; bioreactors should be ready for use on demand (this requirement is particularly important for the treatment of acute patients); the bioreactor should be easily replaceable in case of failure with minimal hassle or harm to the patient. Effective therapeutic treatments based on a bioartificial organ have to meet a few additional necessary (albeit not sufficient) requirements: the organ-specific metabolites synthesized by the cells in the bioreactor in response to metabolic signals (e.g., insulin in the case of the bioartificial pancreas) have to return unhindered into the patient's blood circulation in physiological times; blood-borne toxins have to be cleared quickly to stabilize the patient's state; the treatment as a whole (and not only the bioreactor) has to be biocompatible (i.e., it must not cause any harm to, nor must it expose the patient to risks).

Bioreactors for bioartificial organs are better classified based on how they are connected to, and are located with respect to, the patient, each connection posing specific constraints to bioreactor design. Bioreactors implanted into the patient's body are termed *"intra-corporeal"* (IC), whereas those located external to the patient are termed *"extra-corporeal"* (EC). Bioreactors connected to the patient via the blood circulation, generally as an artery-to-vein shunt, are termed *"intra-vascular"* (IV), whereas those in direct contact with the tissues or other body fluids are termed *"extra-vascular"* (EV). Independent of the actual design, most proposed bioreactors for bioartificial organs consist of three compartments (Fig. 7.2): *the body fluid compartment; the membrane compartment; the cell compartment.* The body fluid compartment is filled with the blood or plasma flowing from the patient's circulation in IV bioreactors, or with stagnant physiological fluids or tissue in EV bioreactors whose nature and volume depends on the site where the bioreactor is implanted. The membrane generally acts as a molecular sieve selecting the species exchanged between the liquid and the cell compartment. The organ-specific cells are located and cultured in the cell compartment, in adhesion, free in suspen-

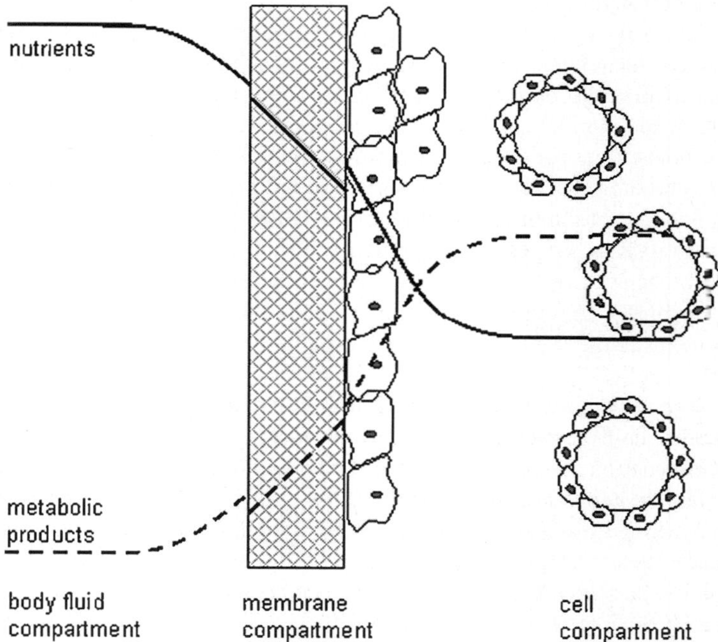

Fig. 7.2 Scheme of the compartments and concentration profiles in a typical bioreactor for bioartificial organs

sion or attached to microcarriers (e.g., beads or fibres). Nutrients, metabolic cues (i.e., oxygen, hormones, growth factors, etc.) and toxins are generally supplied to the cells in the bioreactor from the body fluid compartment. To reach the cells, nutrients and toxins (generally present in the body fluid at the highest concentration) must be transported: from the body fluid bulk to the fluid-membrane interface; across the membrane wall; from the membrane/cell interface into the cell mass. Organ-specific metabolic products and waste metabolites are transported in the opposite direction from the cell compartment (where they are produced and are present at the highest concentration) towards the body fluid compartment. Whether the bioreactor design and operation meet the cell metabolic requirements and the set therapeutic objectives depends on the rate at which essential nutrients and metabolites are exchanged from the body fluid to the cell compartment and vice versa. The actual rate at which metabolic species are transported across compartments, hence the mass per unit time cleared from the body fluid or released into the body fluid (as well as the time that the transfer takes), depends on the mass transport resistance of each compartment. Hence, it depends not only on the solute physical-chemical properties and its mobility in each compartment, but also on the fluid dynamics of each compartment.

In the following section, the bioreactor design problem for bioartificial organs is discussed separately for EV and IV bioreactors for the different problems posed by the different connection to the patient's body. For that said above, in all bioreactors cells are cultured (or co-cultured) at high densities nearing those in natural tissues. *In vivo* experiments with the proposed bioartificial organs show that a necrotic core often forms in the cell compartment regions of a bioreactor farther away from the nutrient and oxygen source, and is likely to be caused by oxygen and nutrient deprivation (Sect. 5.1.3). Thus, currently bioreactor design is also required to ensure that nowhere in the cell compartment cells are cultured at sub-physiological (i.e., hypoxic) or nul (i.e., anoxic) dissolved oxygen concentrations. For this reason, in the following design equations are presented focussing on oxygen transfer in different bioreactor types.

7.3.1 Extravascular (EV) Bioreactors

In EV bioreactors, cells or cell clusters are compartmentalized in a close environment either in the lumen of tubular or hollow fibre membranes sealed at the ends, or between two flat sheet membranes sealed at their periphery, or inside hollow spherical microcapsules coated with a perm-selective outer membrane, as shown in Fig. 7.3. These bioreactors are also referred to as macro- or microencapsulated cell bioreactors depending on their size.

The site of implantation defines the conditions of the body fluid compartment, at which rate species are transported from the bulk body fluid (i.e., tissue or physiological fluid) towards the membrane, and the type of immune activation elicited by the bioreactor. Depending on the specific application, EV bioreactors have been implanted in the peritoneal cavity, the kidney capsule, the thigh muscles, the spleen,

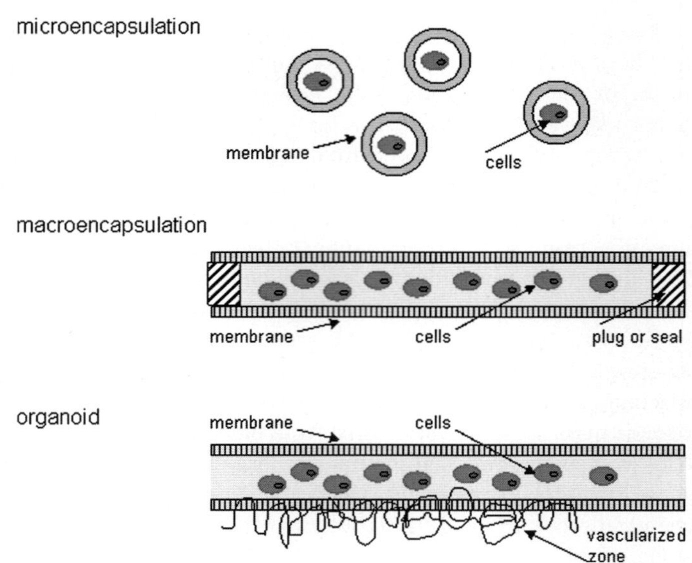

Fig. 7.3 Scheme of EV bioreactors for bioartificial organs

the external wall of the stomach, the back wall of abdominal muscles, subcutaneously, in the cerebrospinal fluid, in the striatum, among others. The peritoneal cavity is often preferred for its vascularization and the lower fibrotic response (see below). In some cases, prior to cell seeding, the bioreactor is implanted subcutaneously for a few days to pre-vascularize the membrane wall and to minimize external diffusion limitations to oxygen and nutrients transfer from the capillaries at the implantation site to the cells (De Vos et al. 1997). These bioreactors are often referred to as *organoids*. Cells have been compartmentalized in spherical microcapsules with diameter ranging from ca. 250 to 1,200 μm, in hollow fibre membranes a few hundred microns in diameter and up to ca. 40 cm long, and between flat sheet membranes in patches up to 4 × 8 cm large and ca. 250 μm thick.

Design of these bioreactors is generally approached with mass transfer models developed under the same assumptions introduced for MBR design (Sect. 5.1.3). Additionally, the body fluid compartment is generally supposed to be completely stirred and the resistance to mass transport in this compartment is often neglected on account of the good vascularization of the implantation site. Based on the fact that bioreactors with larger external surface areas activate to a greater extent the immune system of the recipient, cells are generally cultured at high density in the bioreactor and their concentration is not supposed to vary in time (i.e., the bioreactor is operated at steady state). As soluble nutrients and metabolic cues are transported across the cell mass they are also consumed by the cells. Their concentration decreases more than linearly away from the body fluid compartment (i.e., their source) yielding a spatial concentration profile that locally matches the net solute flux to its consumption rate. A suboptimal bioreactor design might cause nutrients and oxygen to become insufficient to ensure cell survival in the farthest regions from the nutrient source, thus causing cell death. This makes oxygen and nutrients delivery to the cells the actual current design criterion to ensure cell survival in a differentiated state throughout the bioreactor cell compartment.

If transport of relevant solutes is not considered limited, or affected, by membrane resistance, design equations for EV bioreactors of different geometry may be readily obtained by writing the mass balance equation for the nutrient only in the cell compartment. In the following section, reference is made only to the transport of the cell-permeable oxygen (as was done for MBR design) for its importance to cell viability and metabolic functions. As said above, cells are cultured at high densities because this favours cell polarization, re-organization in *in-vivo*-like structures and, eventually differentiation. Collagen, ECM or alginate are also purposely used to provide a substratum for cell attachment or to hold the cells in given spatial positions (and prevent their fall to the bottom of the bioreactor), or ECM is produced by the cells. Thus, in the cell compartment little room is left for a convective liquid flow through the cell mass and solute transport can be assumed to occur mainly by a diffusive mechanism. The value of the effective diffusion coefficient in a tissue-like cell mass has been reported to be about 50% of that in water for the presence of cells and highly concentrated structural proteins (Chresand et al. 1988). This yields a typical D_{eff} for oxygen of ca. 1.6×10^{-5} cm^2 s^{-1} at 37 °C (Wilke and Chang 1955). The rate dependence of oxygen or nutrients consumption on oxygen or nutrients concentration is generally expressed according to a Michaelian equation (Chap. 2). However, oxygen or nutrient concentration is often presumed to exceed the Michaelis constant so that the metabolic reaction rate is often considered independent of nutrient or oxygen concentration (i.e., zeroth order kinetics). Cells inside a micro- or macrocapsular bioreactor implanted extravascularly are generally presumed to be cultured under uniform environmental conditions. Neglecting transport across the ends or the periphery of the capsule, the dissolved oxygen concentration may be assumed to change only along a direction perpendicular to the membrane surface. For a flat sheet macrocapsule, if membrane resistance to oxygen transport is negligible, the dissolved oxygen concentration profile in the cell compartment is obtained from the one-dimensional steady state mass balance equation for dissolved oxygen about the control volume comprised between two parallel planar surfaces at a differential distance from one another (as shown in Table 7.1) yielding:

$$D_{eff}\left(\frac{\mathrm{d}^2 C_{\mathrm{C}}}{\mathrm{d}x^2}\right) = V_{max} \tag{7.1}$$

with the following boundary conditions:

$$\text{B.C.1 } x = 0 \quad \mathrm{d}C_{\mathrm{C}} / \mathrm{d}x = 0$$
$$\text{B.C.2 } x = \delta \quad C_{\mathrm{C}} = C_{\mathrm{CM}} = C_{\mathrm{L}}.$$

B.C.1 states the symmetry of the concentration profile. B.C.2 states that the dissolved oxygen concentration at the cell/membrane interface equals that in the body fluid bulk for negligible membrane and body fluid resistance to oxygen transport. Solving Equation (7.1) for C_{C} yields the non-dimensional spatial solute concentration profile through the cell mass:

Table 7.1 Scheme, mass balance equation, and concentration profile of a consumable solute (e.g., dissolved oxygen) in EV capsular bioreactors of different geometry inside which organotypical cells are cultured, for negligible resistance to oxygen transport in the membrane and the body fluid. See Nomenclature for symbols

Capsular bioreactor geometry	Mass balance equation for a consumable solute	Concentration profile of a consumable solute
a. Flat slab <inline-fig/>	$D_{\mathit{eff}}\left(\dfrac{d}{dx}\left(\dfrac{dC_C}{dx}\right)\right)=V_{max}$	$C_c(x)=C_L-\dfrac{V_{max}}{D_{\mathit{eff}}}\dfrac{\delta^2}{2}\left(1-\left(\dfrac{x}{\delta}\right)^2\right)$ with $\phi^2=\dfrac{V_{max}}{D_{\mathit{eff}}\,C_L}R^2$
b. Cylindrical <inline-fig/>	$D_{\mathit{eff}}\left(\dfrac{1}{r}\dfrac{d}{dr}\left(r\,\dfrac{dC_C}{dr}\right)\right)=V_{max}$	$C_c(r)=C_L\dfrac{V_{max}R^2}{D_{\mathit{eff}}}\dfrac{1}{4}\left(1-\left(\dfrac{dy}{dx}\right)^2\right)$ with $\phi^2=\dfrac{V_{max}}{D_{\mathit{eff}}\,C_L}R^2$
c. Spherical <inline-fig/>	$D_{\mathit{eff}}\left(\dfrac{1}{r^2}\dfrac{d}{dr}\left(r^2\,\dfrac{dC_C}{dr}\right)\right)=V_{max}$	$C_c(r)=C_L+\dfrac{V_{max}R^2}{D_{\mathit{eff}}}\dfrac{1}{6}\left(1-\left(\dfrac{r}{R}\right)^2\right)$ with $\phi^2=\dfrac{V_{max}}{D_{\mathit{eff}}\,C_L}R^2$

$$\frac{C_C(x)}{C_L}=1-\phi^2\frac{1}{2}\left[1-\left(\frac{x}{\delta}\right)^2\right],\qquad\qquad(7.2)$$

where

$$\phi^2=\frac{V_{max}/C_L}{D_{\mathit{eff}}}\delta^2\qquad\qquad(7.3)$$

is the squared Thiele modulus for zeroth order kinetics. ϕ^2 compares the maximal metabolic oxygen consumption rate to the maximal solute transport rate in the cell compartment. Mass balance equations for cylindrical and spherical capsular bioreactors are reported in Table 7.1 together with suitable boundary conditions. Equation (7.2) shows that the dissolved oxygen concentration in the cell mass decreases with the distance from the outer membrane and depends only on the Thiele modulus. Figure 7.4 shows that, when metabolically active primary cells are cultured in the ICS at cell densities nearing that *in vivo*, oxygen concentration, C_C, steeply decays from its body fluid bulk value into the cell mass, and cells farther away are severely deprived of oxygen also at lower cell densities. Figure 7.4 shows

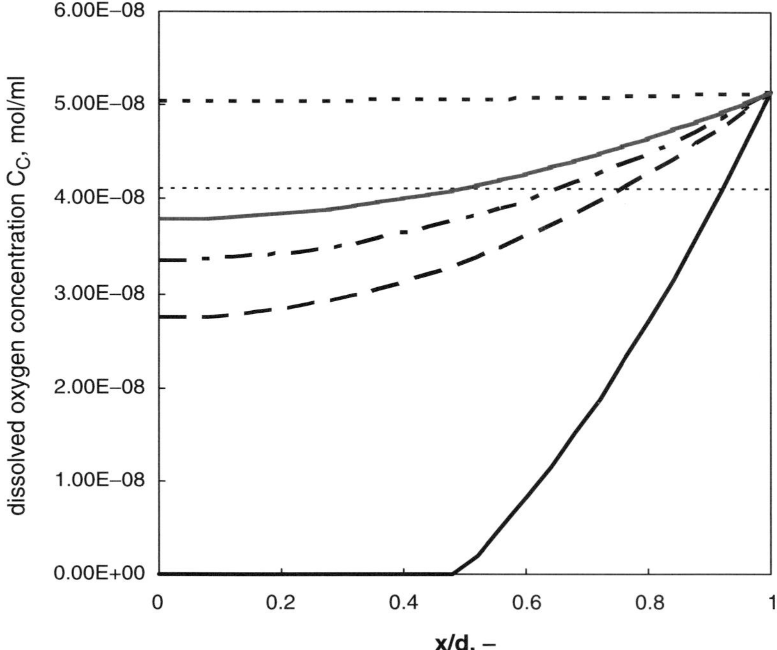

Fig. 7.4 Spatial oxygen concentration profiles for various cell metabolic activities or cell density (i.e., Thiele modulus, ϕ) of bone cells (i.e., osteoblasts) compartmentalized inside a flat slab membrane EV bioreactor 150 μm thick, implanted in peripheral tissue: (——) neonatal pig calvarial cells, $X_c = 5 \times 10^8$ cells mL^{-1}, $\phi = 2.63$; (– – –) rat calvarial cells, $X_c = 5 \times 10^8$ cells mL^{-1}, $\phi = 0.93$; (– · – ·) guinea pigs calvarial cells, $X_c = 5 \times 10^8$ cells mL^{-1}, $\phi = 0.69$; (.....) MC3T3 immortalized cell line, $X_c = 5 \times 10^8$ cells mL^{-1}, $\phi = 0.036$; (——) neonatal pig calvarial cells, $X_c = 1 \times 10^8$ cells mL^{-1}, $\phi = 0.53$; (. . .) minimal dissolved oxygen concentration in peripheral tissue.

also that oxygen deprivation is less of a problem for immortalized cell lines that generally feature lower OCRs. Immobilization of primary cells in shrunk gels has often been used to control the spatial cell distribution and hinder cell aggregation as a means to minimize local oxygen depletion and starvation.

If membrane resistance to oxygen transport is not negligible, the dissolved oxygen concentration at the cell/membrane interface, C_{CM}, is no longer equal to the body fluid bulk value, C_L. Its value may be estimated from an oxygen balance at the cell/membrane interface, where the oxygen molar flow rate across the whole membrane area matches the rate at which oxygen is metabolically consumed by the cells in the cell compartment, to give:

$$2P_m A\left(C_L - C_{CM}\right) = V_{max}\, 2A\delta \tag{7.4}$$

and

$$C_{CM} = C_L - \frac{V_{max}}{P_m}\delta. \tag{7.4a}$$

Integrating (7.1) with the following B.C.2

$$\text{B.C.2} \quad x = \delta \quad C_C = C_{CM} \tag{7.1a}$$

yields the actual dissolved oxygen concentration profile in the cell compartment, as follows:

$$C_C(x) = C_{CM} - \frac{V_{max}}{D_{eff}}\frac{\delta^2}{2}\left[1 - \left(\frac{x}{\delta}\right)^2\right]. \tag{7.2a}$$

Substituting (7.4a) in (7.2a) yields:

$$\frac{C_C(x)}{C_L} = \left[1 - \frac{V_{max}}{P_m C_L}\delta^2 - \frac{V_{max}}{D_{eff}C_L}\delta^2\frac{1}{2}\left[1 - \left(\frac{x}{\delta}\right)^2\right]\right]. \tag{7.5}$$

or

$$\frac{C_C(x)}{C_L} = \left[1 - \frac{\phi^2}{Bi_{mem}} - \phi^2\frac{1}{2}\left[1 - \left(\frac{x}{\delta}\right)^2\right]\right], \tag{7.5a}$$

where

$$Bi_{mem} = \frac{P_m\delta}{D_{eff}} \tag{7.5b}$$

is an analogue of the dimensionless membrane Biot number, which compares the diffusive oxygen transport across the membrane and into the cell compartment. Equation (7.5a) shows that the dissolved oxygen concentration in the cell compartment decreases away from the cell/membrane interface into the cell mass more steeply than for negligible membrane resistance to oxygen transport. The actual concentration profile depends not only on the Thiele modulus but also on the dimensionless ϕ^2/Bi_{mem} group that accounts for such membrane resistance.

7.3.2   Intravascular (IV) Bioreactors

The proposed IV bioreactors may be classified either as membrane compartmentalized cell reactors (MCCRs) or as perfused cell reactors (PCRs). In MCCR-type IV bioreactors, the cells are compartmentalized in a closed shell outside hollow fibre membranes arranged in a shell-and-tube configuration, or outside a single large diameter tubular membrane, as a suspension of single or clustered cells, in adhesion on the membrane surface, or attached to microbeads, as shown in Fig. 7.5. Blood or plasma flows in the membrane lumen, supplies the cells with nutrients, oxygen and metabolic cues, and is enriched in cell-synthesized metabolic products that are brought back to the host blood circulation. In a few bioreactor types, cells have been compartmentalized in the lumen of hollow fibre membranes arranged in a shell-and-tube configuration and whole blood has been perfused in cross-flow through the bioreactor shell. In PCR-type IV bioreactors, the cells are attached to a 3D scaffolding matrix (e.g., non-woven fabrics, porous foams, or a membrane network), in clusters at low density or at tissue-like density, packed in a bioreactor housing and are continuously perfused with plasma separated from the patient's blood with plasmapheresis membranes or continuous centrifuges. Flowing whole blood in these bioreactors is considered inconvenient for the possible blood clotting caused by the activation of the coagulation system.

Design of MCCR-type bioreactors. The design of MCCR-type IV bioreactors may be approached as was done for the MCCRs used for technical applications (Sect. 5.1.3). The only difference lies in the properties of the liquid flowing along or across the membranes, in the ICS or the ECS, respectively. The plasma is Newtonian (i.e., the shear stress is proportional to the shear rate) and features a viscosity of $\eta_p = 1.2\,cP$ slightly higher than water for the high protein concentration. When blood flows in the liquid compartment, description of its fluid dynamics, particularly in complex channel geometry, is made awkward by its non-Newtonian behaviour caused by the concomitant presence of cells and blood proteins (in particular fibrinogen). In fact, when subjected to a shear field, blood features a yield stress (i.e., blood does not flow below a threshold shear rate) largely dependent on the blood cell volume fraction (i.e., the hematocrit), and, as it flows, the square root of the shear stress increases proportional to the square root of the shear rate, as adequately described by the Casson equation (Middleman 1972). When the shear rate exceeds about $100\,s^{-1}$ the blood exhibits a Newtonian behaviour (Whitmore 1968). Moreover, when blood flows in small bore membranes, the shear rate field exerts a spin on the larger cells in the flow channel (i.e., the red blood cells, RBCs) that pulls them away from the liquid-membrane interface. This causes a cell-free layer to form near the membrane surface where transport takes place. On the account of all this, blood is often assumed Newtonian with a viscosity $\eta_B = 3.0\,cP$ (Fournier 1999). The diffusion coefficient of the soluble species in the plasma, D_P, or blood, D_B, is also different than in water for the presence of blood proteins and cells. Estimates of D_P can be obtained by correcting the solute diffusion coefficient in water for the different viscosity according to Equation (5.11). This gives an approximate value of

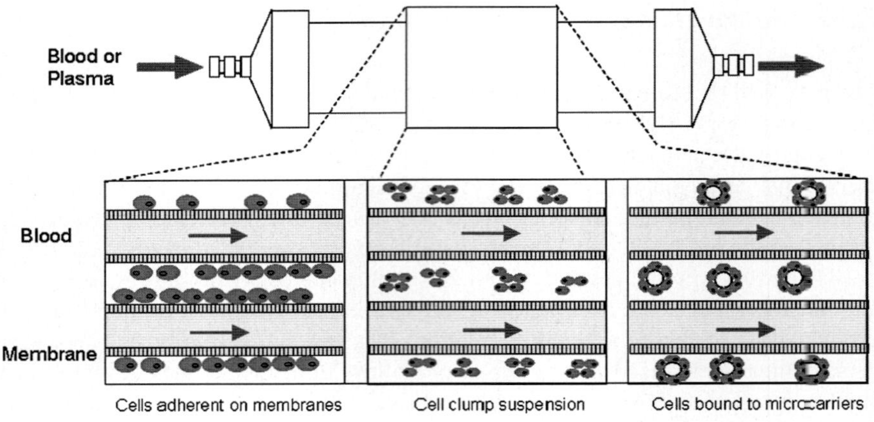

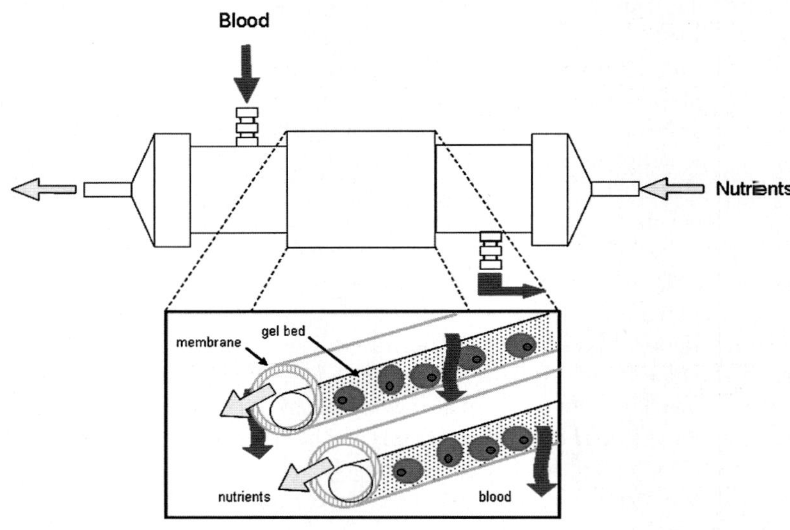

Fig. 7.5 Scheme of different IV bioreactors for bioartificial organs: (**a**)–(**d**) MCCR-type bioreactors, of which (**a**)–(**c**) proposed for bioartificial livers, (**d**) proposed for bioartificial pancreas; (**e**)–(**f**) PCR-type bioreactors proposed for bioartificial livers. (Published with permission from Catapano and Gerlach 2007)

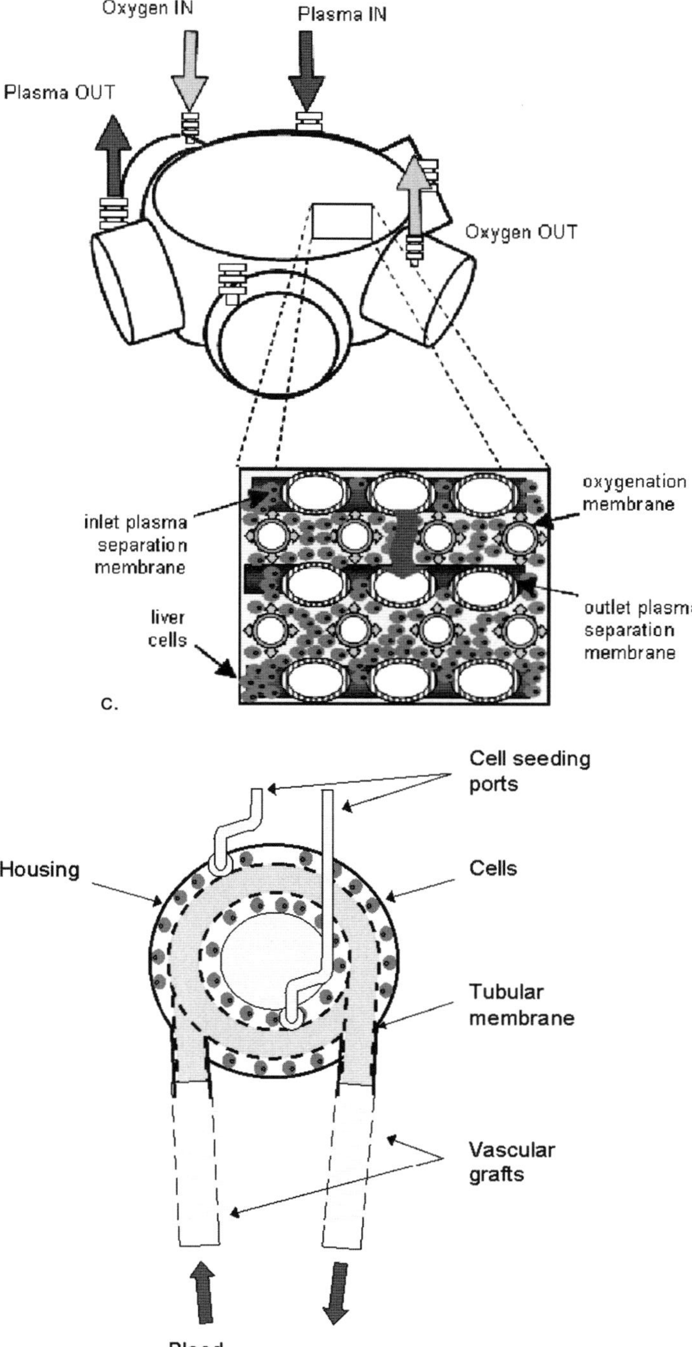

Fig. 7.5 (continued)

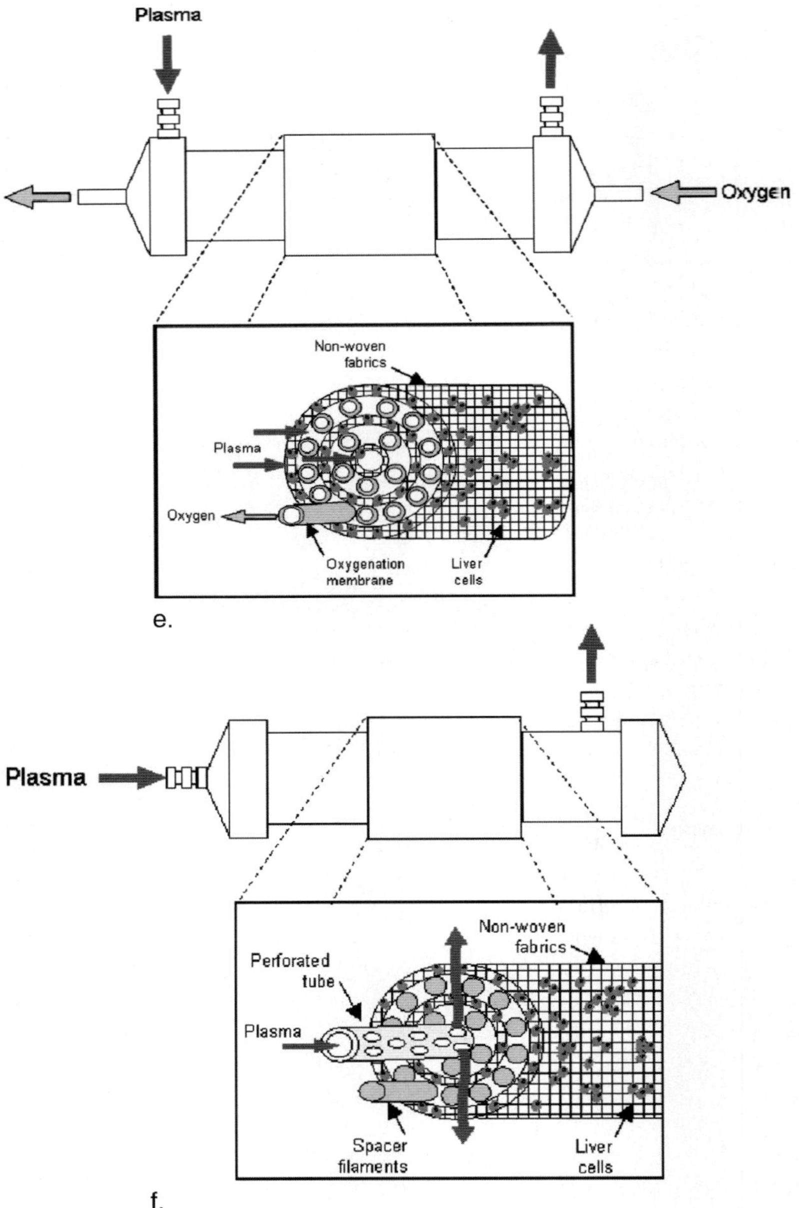

Fig. 7.5 (continued)

$$D_P = \frac{\eta_w}{\eta_P} D_S \approx 0.63\, D_S$$ at 37 °C. Estimates of D_B can be obtained with models

accounting for the cell volume fraction, H (i.e., the hematocrit), such as that by Maxwell (1873):

$$D_{\mathrm{B}} = D_{\mathrm{p}} \frac{2D_{\mathrm{P}} + D_{\mathrm{cell}} - 2H(D_{\mathrm{P}} - D_{\mathrm{cell}})}{2D_{\mathrm{P}} + D_{\mathrm{cell}} + H(D_{\mathrm{P}} - D_{\mathrm{cell}})}, \tag{7.6}$$

where D_{cell} is the solute diffusivity in the cells. Hence, considerations and conclusions presented for MCCRs for technical applications may be extended also to those used for bioartificial organs provided that proper viscosities and diffusion coefficients are used for the estimation of the relevant dimensionless groups determining the bioreactor behaviour.

Design of PCR-type bioreactors. The design of PCR-type IV bioreactors is addressed differently depending on the actual cell density. At tissue-like density, a similar approach may be used to that done to model transport in the ECS of MCCRs (Sect. 5.1.3). The plasma flux across the cell mass promoted by the applied hydrostatic pressure difference is estimated according to Darcy's law, as in Equation (5.26), and provision is made for the rates of the metabolic production or consumption reactions. At low cell density, plasma flow is relatively unobstructed through highly porous scaffolds and around clusters of cells hanging from the scaffold surface. The concentration of oxygen and nutrients in the plasma bulk decreases along the bioreactor length but also from the cluster periphery bathed by the plasma into the cell cluster, at any given bioreactor section. Once more, let us refer to oxygen for its importance to the cells. The dissolved oxygen concentration anywhere in the bioreactor or the cell cluster requires solution of coupled oxygen balance equations both in the plasma and the cell cluster. However, a pseudo-homogeneous approach may still be used to obtain the dissolved oxygen concentration profile along the bioreactor length provided that the metabolic reaction rate term accounts for the resistance to oxygen transport from the bulk plasma to the cluster periphery and inside the cell cluster, to account for oxygen transport and reaction into it. In the clusters, cells are generally aggregated at such a high density that oxygen transport inside the cluster may be assumed to occur by a purely diffusive mechanism. In this case, the oxygen concentration decay inside the cell cluster may be accounted for by introducing an efficiency factor, η, as in the traditional chemical reaction engineering approach to transport in porous catalysts. The effect of the mass transport resistance inside and outside the cell cluster on the bioreactor behaviour may be more easily understood when the rate of oxygen consumption linearly depends on the dissolved oxygen concentration. This condition is likely to be established in sections close to the bioreactor outlet where the plasma is oxygen depleted and its concentration is likely *to be* $C_{L} << K_{\mathrm{m}}$. In this case, the external and internal resistance to oxygen transport add up and an overall efficiency factor, η_{ov}, may be defined accounting for all occurring phenomena, which is constant throughout the bioreactor. At steady state, at any section of the bioreactor the oxygen flux from the bulk plasma to the cell cluster periphery matches the rate at which oxygen is consumed inside the cell cluster, as follows:

$$k_{L}\left(C_{L} - C_{LC}\right) A_{ext} = \eta k_{c} C_{LC} A_{\mathrm{int}}, \tag{7.7}$$

where: A_{ext} is the cluster external surface area (e.g., $A_{ext} = \pi d_{c}^{2}$, for a spherical cluster of diameter d_{c}); A_{int} is the outer cell surface area in the cluster (e.g., $A_{\mathrm{int}} = a_{\mathrm{v}} \pi d_{c}^{3}/6$, for a spherical cluster); k_{c} is the kinetic constant for oxygen consumption per unit cell surface area in the cluster (i.e., $k_c \approx V_{max}''/K_m$); k_L is the oxygen mass transport coefficient in the plasma; and a_v is the external cell surface area per unit cluster volume, which can be assumed proportional to the actual cell concentration per unit cluster volume, X_c. Equation (7.7) may be solved for oxygen concentration at the plasma–cluster interface, C_{LC}, to give:

$$C_{LC} = \frac{k_L A_{ext}}{k_L A_{ext} + \eta k_c A_{int}} C_L.$$

(7.8)

The overall efficiency factor, η_{ov}, is obtained by setting the right hand side of Equation (7.7) equal to $\eta_{ov} \, kc \, A_{int} \, C_L$ (i.e., by relating the reaction rate to the dissolved oxygen concentration in the bulk plasma as the driving force) and by substituting Equation (7.8) for C_{LC} to give:

$$\eta_{ov} = \frac{\eta}{1+\eta\left(\dfrac{k_c A_{int}}{k_L A_{ext}}\right)} = \frac{\eta}{1+\eta\left(\dfrac{1}{6}\dfrac{k_c a_v}{k_L}d_c\right)} = \frac{\eta}{1+\dfrac{2}{3}\eta\dfrac{\phi^2}{Bi}},$$

(7.9)

where $\phi^2 = k_c \, a_v \, d_c^2/(4 \, D_{eff})$ is the squared Thiele modulus for first order kinetics, and Bi is the Biot number, $Bi = k_L d_c/D_{eff}$. Equation (7.8) shows that the external resistance to oxygen transport reduces the dissolved oxygen concentration at the periphery of the cell cluster as compared to its bulk value. The resistance to oxygen transport external to the membrane encapsulating the cells may be expected to have similar effects on the dissolved oxygen concentration at the membrane–tissue interface in EV bioreactors when the tissue/physiological fluid at the implantation site is poorly vascularized or poorly stirred. The actual external resistance to oxygen (or other metabolites) transport towards the cell cluster may be estimated from semi-empirical correlations in literature, such as those for flow through porous beds of particles of different geometry (Bird et al. 1960). Semi-empirical correlations in the technical literature show that the solute mass transport coefficient may be maximized (hence mass transport resistance may be minimized) by optimizing the channel geometry and maximizing the plasma velocity, to an extent permitted by the limits on the shear rate tolerated by proteins and cells and by the sustainable pressure drops.

Under the following assumptions: solutes are uniformly distributed at a bioreactor section normal to its axis; the kinetics of oxygen consumption in the cell cluster is first order in its concentration; cell density is constant; and the axial Peclet number is high (i.e., solute axial diffusion is negligible with respect to axial convection), taking the limit for $\Delta x \rightarrow 0$ of the oxygen mass balance about a differential control volume of plasma comprised between two bioreactor cross-sections at distance (x) and $(x + \Delta x)$ from the bioreactor entrance (Fig. 7.6) yields:

$$U\frac{dC_L}{dx} = -\eta_{ov} k_c C_L.$$

(7.10)

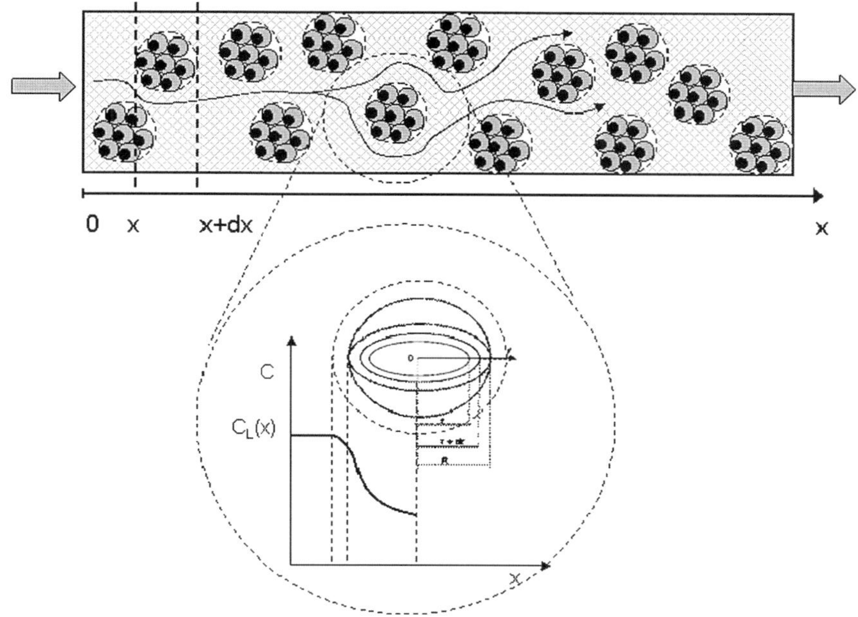

Fig. 7.6 Scheme of a Perfused Cell Reactor (PCR) with cell clusters adherent to a 3D porous scaffold showing the qualitative dissolved oxygen concentration profile around the cell cluster

In (7.10), U is the superficial velocity in the plasma flow channel (i.e., the velocity estimated as if the whole bioreactor cross-sectional area were available for flow). Equation (7.10) may be integrated with the following boundary condition:

$$\text{B.C. } x=0 \quad C_L = C_{L,in},$$

where $C_{L,in}$ is the dissolved oxygen concentration in the plasma stream entering the bioreactor, to give the dissolved oxygen concentration at any section along the bioreactor length, as follows:

$$C_L(x) = C_{L,in} \exp^{-\frac{\eta_{ov} k_c}{U} x} \text{ (Fogler 1999).} \tag{7.11}$$

The dissolved oxygen concentration profile across spherical cell clusters positioned at a distance x from the plasma entrance may be obtained from a mass balance on oxygen in the cluster, as follows:

$$C_c(r/R_c, x) = C_L(x) + \frac{R_c}{r} \frac{\sinh\left(\phi \frac{r}{R_c}\right)}{\sinh(\phi)} \text{ (Bird 1960).} \tag{7.12}$$

Similar considerations apply to cell clusters of different geometry (e.g., flat slab or cylinders) provided that the A_{int}/A_{ext} ratio in Equation (7.9) and the concentration profile in Equation (7.12) used apply to the specific geometry. The latter can be found in textbooks of chemical reaction engineering (e.g., Bird et al. 1960; Levenspiel 1972; Fogler 1999). Different consumption kinetics yield a Thiele modulus, ϕ^2, and an overall efficiency, η_{ov}, depending on the local plasma concentration and require the numerical solution of the resulting equations.

7.3.3 Membrane Immunoprotection in Bioartificial Organs

Semi-permeable membranes are used in bioartificial organs to protect the bioreactor constituents (particularly the allo- or xenogeneic cells cultured in the bioreactor) from the host immune response. The membranes are generally directly located in the bioreactor, as in EV bioreactors. For some applications, they are also used in the extracorporeal circulation loop to continuously separate plasma from the patient's blood which is continuously fed to the bioreactor, as in many bioartificial livers for the support to ALF patients.

The membrane separation properties (Sect. 5.1.3) must prevent the passage of immune-competent cells (e.g., macrophages) and proteins (e.g., antibodies and complement proteins) across the membrane wall, while permitting the unhindered exchange of essential nutrients, growth factors, toxins to be cleared, and metabolic products of therapeutic relevance between the body fluid and the cell compartment. Examples of metabolic products of therapeutic relevance are insulin in the treatment of diabetes (i.e., for a bioartificial pancreas), and coagulation proteins in the treatment of ALF patients (i.e., for a bioartificial liver). Membranes with pores a fraction of a micron in diameter have been shown to reject the cells of the immune system and successfully protect xeno- or allogeneic cells from rejection (Colton 1995). However, these membranes do not prevent the passage of antibodies and complement proteins that are produced by the recipient's immune system in response to the antigens shed by the xenogeneic cells in the bioreactor. Actually, antibody binding to the cell surface antigens in the bioreactor in the absence of immune cells is not lethal to the cells. However, in the presence of the complement proteins the cells may be killed by complement activation and formation of the membrane attack complex (Iwata et al. 1999). Discussion of the complement system and its activation mechanisms is beyond the scope of this chapter and reference should be made to specialized textbooks (e.g., Fournier 1999). It suffices to say here that the first protein to be activated along the classical pathway of complement activation is the protein C1. In particular, its subunit C1q is the first subunit activated that must bind to an antigen bound IgM, or several IgGs, to start the cascade of complement reactions leading to cell destruction. Hence, activation of the complement cascade may be prevented by using membranes totally rejecting the subunit C1q, a protein with a MW of ca. 400,000 and a maximal width of ca. 30 nm (Colton and Avgoustiniatos 1991). Discussion of the separation properties of synthetic membranes in Sect. 5.1.3 has made it clear that they are but ideal separators and that, molecules may cross the membranes with a higher MW than the membrane NMWCO. An effective rule-of-the-thumb is that, for a successful separation, membranes ought to be used whose NMWCO is about ten times lower than the MW of the molecule that must be rejected. Indeed, commercial and tailor-made membranes for UF and MF have been successfully used for bioartificial organs with a NMWCO ranging from ca. 50,000 to 0.1 μm to protect the cells cultured in the bioreactor. However, these membranes generally cannot reject the antigens shed by the xenogeneic cells cultured in the bioreactor. As they enter the host's organism, antigens might trigger an immune reaction of the recipient leading to an anaphylactic reaction or antibody-mediated hypersensitivity reaction that may even be deadly. Transplantation of membrane immuno-protected xenogeneic islets of Langerhans has shown that antigen shedding indeed causes the production of antibodies against the antigens, but without any apparent pathological effect (Scharp et al. 1994).

An important aspect of the membranes used in bioreactors for bioartificial organs is their biocompatibility. The biocompatibility concept encompasses a large number of requirements that have to be met. Membranes have to be made of materials that do not start a life-threatening immune response, that are not cytotoxic, or release toxicants during the treatment, among others. National regulatory agencies issue a series of tests that the materials must pass before they can be used in a therapeutic treatment. In addition to these requirements, the way membranes are made and used may trigger an immune response that may cripple the bioreactor performance and require its replacement. For IV bioreactors, the most serious issue is the activation of the coagulation system. In fact, membranes with a rough, platelet-activating blood-contacting surface, and the presence of stagnant regions or sharp edges in the blood flow channel may activate the coagulation system and cause the formation of blood clots. These may threaten the patient's life if released in the blood circulation, or may block the blood flow channels and occlude the bioreactor. In many bioartificial livers using IV bioreactors, plasmapheresis MF membranes are used to continuously separate plasma out of blood which is fed to the bioreactor. Removal of blood cells from the feed stream prevents the formation of blood clots in the bioreactor, and its occlusion, caused by the activation of the host coagulation system. For EV bioreactors, the most serious issue is the activation of the foreign body reaction, a complex *sequelae* of events triggered by the surface chemical composition, morphology, topography and shape of the membrane encasing the implanted bioreactor. The foreign body reaction starts with an acute inflammatory response and leads to the development of granulation tissue that ultimately causes the formation of a thick fibrotic capsule around the bioreactor. Fibrotic capsules are mainly composed of collagen, macrophages and fibroblasts, and are generally poorly permeable to essential solutes (e.g., oxygen and nutrients). The cells in the fibrotic capsule wall further reduce oxygen and nutrients transport to the organotypic cells in the membrane bioreactor because they use up these substrates for their own metabolism. This ultimately leads to the loss of cell function and the death of the organotypic cells.

Membranes made of regenerated cellulose, polyamide, polyetheretherketone, polysulphone and its derivatives, polypropylene and polyvinylchloride-acrylic copolymers have been used for IV bioreactors (Silva et al. 2006). Hollow fibre membranes made of polyethersulphone or polyvinylchloride-acrylic copolymers (NMWCO ranging from ca. 50,000 to 80,000), and flat sheet acrylonitrile ultrafiltration (NMWCO of ca. 65,000) or polycarbonate microfiltration membranes (maximal pore size of ca. 0.1–0.6 μm) have been used for EV macro encapsulated bioreactors (Silva et al. 2006). A sandwich of cross-linked acellular alginate flat sheets sealed at its periphery has also been used to protect cells embedded in an alginate gel, and implanted EV (Storrs et al. 2001). Alginate or agarose are among the most used materials to prepare spherical microcapsular EV bioreactors. Cell-containing microcapsules of alginate are often formed by dropping into a calcium chloride solution droplets of a cell suspension in sodium-alginate. The exchange of ions makes alginate precipitate, and promotes the gel bead formation. The separation properties of the alginate beads are often improved either by coating their surface with a layer of poly-L-lysine or by cross-linking alginate with barium ions. In the former case, the bead biocompatibility have been improved by coating the microcapsule with an additional layer of alginate. Alginate composition, purity, MW and the conditions under which the beads are produced are reported to have an important effect on the properties of the microcapsular bioreactor obtained (Silva et al. 2006).

7.4 Commercial Bioreactors and Applications

In the last 20 years, several bioreactor types have been developed for bioartificial organs designed to assist patients with failing liver (Catapano 1996; Allen and Bathia 2002; Catapano and Gerlach 2007), pancreas (Silva et al. 2006) and kidneys (Humes et al. 2002), or to deliver bioactive molecules to patients that were thought to alleviate their pathological state (Aebischer et al. 1991). The bioreactor configuration and operation, the membranes used and their geometry, vary widely depending on the specific requirements that the bioreactor, and the bioartificial organ as a whole, is to meet. For this reason, in the following section the set of requirements for each application is discussed together with some proposed bioreactors. The Reader should be warned that what is reported below is not intended to provide an exhaustive picture of what has been proposed over the years. In fact, reviewing the state-of-the-art of bioreactors proposed for various bioartificial organs is beyond the scope of this chapter. For this purpose, reference should be made to the excellent reviews published over the years on the topic of interest.

Bioartificial pancreas. Insulin dependent diabetes mellitus (IDDM) is a chronic autoimmune disease in which the patient's own immune system attacks and destroys the cells in the pancreas (located in structured cell clusters termed islets of Langerhans), that secrete insulin in response to increased glucose concentrations in the blood (i.e., the β-cells). This causes higher-than-normal blood glucose concentrations, and in the long-term leads to angiopathic lesions causing blindness, kidney failure, gangrene of the limbs, and neurological complications. IDDM patients take daily injections of exogenous insulin (generally after a meal) and have to frequently monitor their blood glucose level, for their entire life, to delay the occurrence of long-term complications. A minor fraction of them is transplanted for the scarcity of donor organs. The administration of insulin in a way poorly related to the current blood glucose concentration causes wild fluctuations of insulin and glucose concentrations in the blood that are held responsible for the early onset of angiopathic complications (American Diabetes Association 1993). A bioartificial pancreas (BAP) is a bioreactor in which xenogeneic or allogeneic isolated islets of Langerhans are used to maintain normoglycemia of the recipient even during hyperglycemic challenges for the physiological feed-back inherent in the islets. Administration of immune suppressants is not necessary because the islets are protected by an immunoprotective synthetic membrane against the host response. It has been estimated that, to maintain a diabetic patient normoglycemic, a number of islets should be loaded in a BAP ranging all the way from 5,000 to 20,000 $EIN/kg_{body\ weight}$ (Suzuki et al. 1998). A BAP should also ensure a fast transport of stimulatory solutes to the β-cells, a fast secretion and transport of insulin into the blood circulation to minimize the long-term complications caused by the wild fluctuation of secretagogue concentrations in time, and the expression of the pancreas differentiated functions for as long as possible. In fact, the treatment is purely supportive (e.g., it does not heal the organ) and IDDM patients would need life long treatment. EV bioreactors have been proposed for BAP where islets of Langerhans are implanted in hollow fibres sealed at the ends, empty macrospheres, and flat sheet pouches, or in alginate microspheres (Silva et al. 2006). These bioreactors have been mainly implanted in the peritoneal cavity for its low fibrotic overgrowth, although other implantation sites have been explored. Islet aggregation was often prevented by immobilization in a gel bed made of collagen, hyaluronic acid, or alginate. Islet loading ranged from 16,000 to 20,000 $EIN/kg_{body\ weight}$ in the *in vivo* tests where normoglycemia was successfully maintained in animal models of IDDM. The favourable surface area-to-volume ratio of microcaspules permits to use much smaller bioreactor volumes also when one islet is implanted in each microcapsule. The flat sheet EV bioreactors designed according to the Isletsheet™ concept (Fig. 7.7) are among the few commercially available today. Implanted in the peritoneal cavity they were reported to maintain fasting normoglycemia in dogs for about 84 days (Storrs et al. 2001). The formation of a fibrotic capsule around the membrane and central necrosis for poor islet oxygenation are typical findings when these bioreactors are used *in vivo*. In the proposed IV bioreactors, islets of Langerhans have been immobilized outside hollow fibre or tubular membranes (the latter featuring a lumen diameter of ca. 4–6 mm), in the space between concentric tubular membranes, or inside narrow hollow fibre membranes coiled to form the outer surface of a vessel with blood flowing in the membrane lumen. IV bioreactors are advantageous with respect to EV bioreactors from the mass transport point of view. In fact, the cells are located much nearer to the oxygen and nutrient source (i.e., the flowing blood), blood flow enhances transport of stimulants and secretagogues from the blood to the cells and vice versa, and permits to deliver glucose to or remove insulin from the islets much faster. This causes a significant decrease of the insulin response time after a glucose challenge. A IV bioreactor made of a coiled tubular membrane 6 mm inner diameter and loaded with 42,000 canine islets equivalent was reported to maintain normoglycemia in a canine recipient for about 140

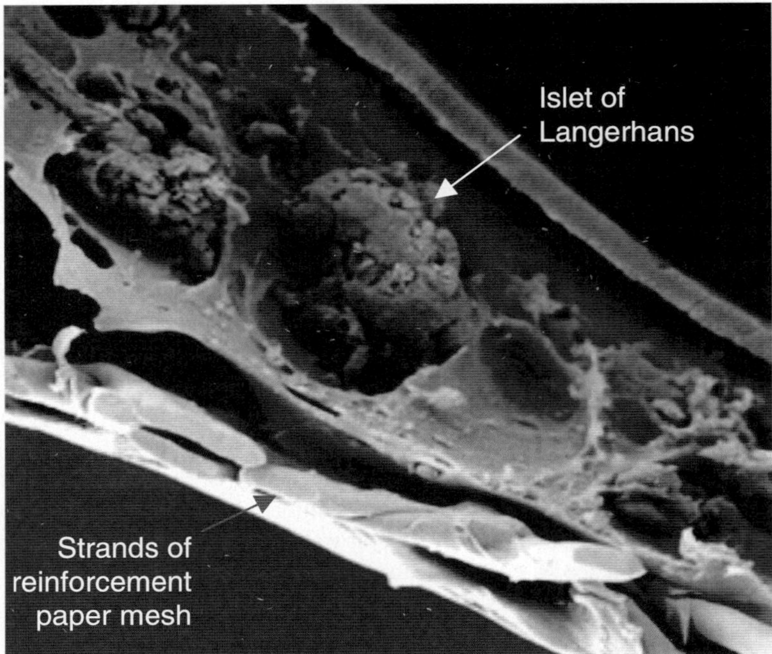

Fig. 7.7 Cross section of the Isletsheet concept proposed by Cerco Medical. Islets of Langerhans are immobilized in alginate gel between two selective layers of cross-linked alginate. The fibers are the strands of a paper mesh added to reinforce the pouch and permit an easier suture to the omentum (Published with permission of CercoMedical XXXX)

days (Maki et al. 1996). However, occlusion for blood clot formation and vascular thrombosis are typical drawbacks of these bioreactors for the difficulty of anti-coagulating the recipient long-term.

Bioartificial liver. The liver is a complex highly vascularized organ that performs about 1,200 known life-sustaining functions. The cells characterizing the liver functions are the hepatocytes. They are about 25–30 μm in diameter, and there are about 250×10^9 of them in a human liver. They play an important role in carbohydrate, fat and protein metabolism. Most blood proteins are made in the liver, in particular albumin and most blood clotting proteins. The liver has a storage function for vitamins and iron. It plays also an important role in the detoxification and conjugation of exogenous toxins (e.g., drugs, poisons, etc.), that are made water soluble and excreted by the kidneys. Different from other organs, the liver has the capacity to regenerate. Acute liver failure (ALF) is caused by drug poisoning, viral infections or decompensated cirrhosis and evolves within a few weeks from the first symptoms (e.g., elevated blood ammonia concentration or jaundice) into multiorgan failure and death (Stockman and JN 2002). The more rapidly evolving form of ALF is often referred to as fulminant hepatic failure (FHF). With the most advanced supportive therapies, the mortality rate of FHF patients still is of about 90% (Schiodt et al. 1999). The elective treatment of ALF is liver transplantation.

Unfortunately, only a minor number of donor organs is available as compared to those required. ALF is associated with elevated levels of endogenous neural and hepatic toxins (e.g., ammonia, mercaptans, bilirubin, etc.). This has led to the development of artificial organs aimed at removing either soluble or protein-bound plasmatic toxins to support ALF patients till recovery. Up to now, those proposed have not yet led to statistically significant improvements over conventional pharmacological supportive treatments. Liver specific products (e.g., blood clotting proteins, hepatopoietin A and hepatic growth factors) seem necessary to relieve some of the symptoms of ALF (e.g., bleeding and cerebral oedema) and promote liver self-regeneration, but only a few of them are known. In the last two decades, research efforts have focussed on the development of a bioartificial liver (BAL) where isolated hepatocytes are used to provide the patient with the liver biosynthetic, regulatory and biotransformation functions (Galletti and Jaregui 1995). The scarce knowledge of the pathophysiology of ALF makes it difficult to define objective functions to optimize the BAL design. Thus, bioreactors for BALs have generally been designed to host a mass of cells necessary to maintain the patient alive, promote the patient's own liver regeneration, and to allow good oxygen and nutrient delivery to the cells to keep them viable and functional. Liver resection experiments suggest that a mass of about 30% of the total liver mass is sufficient for complete liver regeneration (Higgins and Anderson 1931). More recent reports suggest that as little as 10% of the total liver mass is compatible with life. For a standard 70 kg man, this would correspond to 2.5–7.5×10^{10} cells, or ca. 250–750 g of cells. EV bioreactors similar to those proposed for BAPs have also been proposed for BALs. Implanting such a high cell mass with reasonable bioreactor volumes is a major problem for these bioreactor types. In fact, it has been estimated that about 2,500 mL of microencapsulated cells or 1,300 mL of cells macroencapsulated in hollow fibre membranes would be needed (Fournier 1999). For this reason, research efforts have mostly concentrated on the development of extra-corporeal IV bioreactors. In the first bioreactors proposed for BALs, liver cells were cultured outside MF hollow fibre membranes arranged in a shell-and-tube configuration. In the Hepatassist® 2000 concept proposed by Demetriou et al. (1986), ca. 6×10^9 primary porcine hepatocytes are cultured attached to microcarriers outside the membranes in the bioreactor shell. The bioreactor is fed with a stream of plasma that is continuously separated from the patient's blood with plasmapheresis membrane module. Prior to entering the bioreactor, the plasma flows through an oxygenator where it is enriched in oxygen and an adsorption column to reduce the toxin load on the hepatocytes in the bioreactor. ALF patients have been treated with this BAL intermittently and for no longer than 4–6 h per treatment. In the concept proposed by Millis et al. (1999), the plasma produced by plasmapheresis of the patient's blood is fed to four MCCR-type bioreactors in parallel each loaded with ca. 200 g of the immortalized HepG2 hepatocyte cell line. Cells are cultured in the bioreactor shell in adhesion on the outer surface of hollow fibre membranes. Actually, each bioreactor is seeded with a few grams of cells that rapidly grow to the final mass. Since the bioreactors use tumour-derived cells, the plasma is filtered across MF membranes prior to returning it to the patient, to prevent the leak of cells into the patient's circulation. In the Algenix™ concept first proposed by Nyberg et al. (1993), ca. 5×10^7 primary porcine hepatocytes are embedded in a collagen gel in the lumen of hollow fibre membranes arranged in shell-and-tube configuration. Anti-coagulated blood continuously flows in cross-flow outside and around the membranes in the bioreactor. While blood carries much more oxygen than plasma thus improving oxygen transport to the cells (as compared to bioreactors fed with plasma), blood clotting may be a problem in this bioreactor in the presence of irregular or suboptimal realization of the blood flow path. In the bioreactors described above, a common finding was the formation of a necrotic core in the innermost regions of the bioreactor far away from the oxygen source. When the importance of the presence of different cells, of providing the cells with a scaffolding structure, and with oxygen for cell survival and the expression of differentiated liver functions became clear, new bioreactor concepts were developed in which liver cells were co-cultured in 3D scaffolds (e.g., porous foams, non-woven fabrics, membrane networks, etc.), and distributed oxygen sources were built into the bioreactor.

In the MELS™ concept proposed by Gerlach et al. (1994), up to ca. 600 g of primary liver cells are cultured in the interfibre spaces outside three (or even four) mats of cross-woven hollow fibre membranes organized in a 3D network. The basic repeating unit in the network consists of two mats of hydrophilic high-flux dialysis membranes in between which a mat of blood oxygenation hydrophobic hollow fibre membranes is interposed to supply oxygen and remove CO_2. Membranes in the same mat are bundled together and fitted with a header. Plasma continuously produced from the patient's blood with plasmapheresis membranes is fed to the first hydrophilic membrane mat, a part filters across the membrane wall, bathes the cells, flows around the hydrophobic membranes and is enriched in oxygen (and depleted in CO_2), and is re-absorbed into the second hydrophilic membrane mat from which it is returned to the patient. Today, a BAL based on this bioreactor type is commercially offered by HepaLife™ (Fig. 7.8) with the only exception that it uses cells of the proprietary immortalized cell line PICM-19 (Talbot et al. 2002) In the AMC™ bioreactor concept proposed by Flendrig et al. (1997), up to ca. 230 g of primary liver cells are cultured attached to the fibres of a polyester non-woven fabric, wrapped around a solid core and packed in a cylindrical housing through which plasma flows along the bioreactor length. Later on, it was proposed to arrange the fabric in an annular packed-bed bioreactor and to flow plasma radically across the fabric to reduce the bioreactor pressure drop (Naruse et al. 1996; Morsiani et al. 2001). In the MELS bioreactor, the design equations provided above are not directly applicable because oxygen is also delivered in the cell compartment. Provision for oxygen supply might be introduced by adding a term to the mass balance equation in the cell compartment, accounting for the moles of oxygen delivered across the hydrophobic membrane surface area in the control volume. It is interesting to notice that the actual number of cells used in the various bioreactors varies from ca. 0.5×10^8 (or 0.5 g) to 2.0×10^{10} cells. The assumption on which the bioreactors using lower number of cells have been designed is that the cells in the bioreactor have to complement the residual liver function of the patient and foster the 'own' liver regeneration, rather than replacing it.

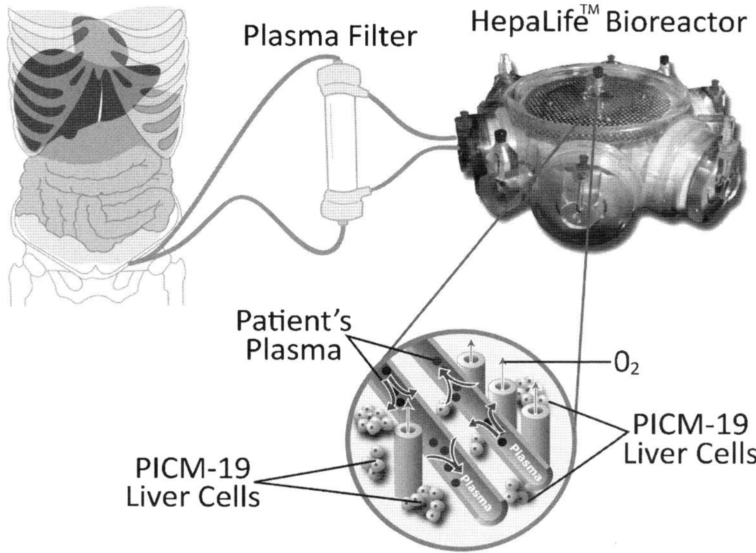

Fig. 7.8 Scheme and cross-section of the MELS bioreactor concept marketed by Hepalife © Copyright Hepalife 2007 (Published with permission of Hepalife XXXX)

Delivery of bioactive molecules. Cell encapsulation in an immuno-privileged compartment represents an alternative to gene therapy for the continuous and sustained site-specific delivery of biologically active molecules in the treatment of pathological states in which the lack of such molecules is thought to play an important role. Parkinson's disease is characterized by poor motor control (e.g., akinesia/ bradykinesia, rigidity, tremor) primarily mediated by the progressive degeneration of nigrostriatal dopaminergic neurons (Lindner and Emerich 1998). Systemic administration of dopamine or L-DOPA produces some behavioural improvements. However, it is difficult to maintain constant blood levels of the biomolecule and systemic delivery may lead to side-effects. Cells of the PC12 cell line derived from tumours derived from adrenal chromaffin cells have the advantage over primary adrenal chromaffin cells of being easy to culture and to produce dopamine and L-DOPA. However, when implanted in primates, they are completely rejected within 4 weeks. To solve these problems, PC12 cells were macroencapsulated in semipermeable hollow fibre membranes made of polyacrilonitrile–polyvinylchloride copolymers, embedded in a chitosan layer, and were implanted in the striatum as a treatment for Parkinson's disease (Aebischer et al. 1991). Membrane separation ensured cell survival and that after implantation cells would not produce tumours. Treatment of animal models of Parkinson's disease yielded some behavioural improvement (Lindner and Emerich 1998). However, no clinical trials were performed because

 G. Catapano

the bioreactor volume required was too great. A similar approach was also proposed to reduce pain of terminal cancer patients. In this case, bovine chromaffin cells were macroencapsulated in semipermeable hollow fibre membranes embedded in an alginate gel and were implanted in the cerebrospinal fluid of the patients. The cells produced painkillers (e.g., enkephalines, catecholamines, somatostatin) that were expected to reduce pain more effectively than the opioids administered systemically.

In conclusion, it may be said that *in vivo* tests on animal models or on patients in the clinics show that in many bioreactors proposed for BALs or BAPs the physiological transfer of oxygen to the cells is still a problem. In fact, oxygen starvation often promotes the formation of a necrotic core in the cell compartment far away from the oxygen source. In the case of the treatment of ALF, many of the proposed bioreactors were proven to be safe, but no statistical evidence gathered in multi-center blind trials exists to show that they are more effective in treating ALF than conventional pharmacological supportive treatments. Research on the use of macroencapsulated cells to deliver bioactive molecules in site-specific fashion is still active but is more and more focussed on stopping or slowing the progression of the disease, or more simply to decrease the amount of drugs that are administered systematically to the patient.

List of Symbols

A_{ext}	cell cluster external surface area, m^2
A_{int}	outer cell surface area in the cell cluster, m^2
a_v	external cell surface area per unit cluster volume, m^2/m^3
$Bi = k_L d_c/D_{eff}$	Biot number, –
$Bi_{mem} = P_M \delta/D_{eff}$	analogue of the Biot number for diffusive transport in the membrane and inside the cell compartment, –
C_C	solute concentration in the cell compartment, moles m^{-3}
C_L	solute concentration in the bulk body fluid, moles m^{-3}
C_{LC}	solute concentration at the body fluid/cell interface, moles m^{-3}
C_{CM}	solute concentration at the cell/membrane interface, moles m^{-3}
D_B	solute diffusion coefficient in the blood, $m^2\ s^{-1}$
D_{cell}	solute diffusion coefficient in blood cells, $m^2\ s^{-1}$
D_P	solute diffusion coefficient in plasma, $m^2\ s^{-1}$
D_S	unhindered solute diffusion coefficient, $m^2\ s^{-1}$
D_{eff}	solute diffusion coefficient in the cell mass, $m^2\ s^{-1}$
d_c	cell cluster diameter, m
EIN	Equivalent islet number, i.e., the number of islets $150\,\mu m$ in diameter that are equivalent in volume to a given sample of islets
H	hematocrit, –
k_c	first order kinetic constant of substrate utilization, $1 s^{-1}$
k_L	liquid phase mass transport coefficient, $m^3/(m^2\ s)$

K_M Michaelis constant for oxygen utilization, gmol m^{-3}

MW molecular weight, $-$

P_m solute diffusive permeability in the membrane, m s^{-1}

R_c cell cluster radius, m

r radius, m

U liquid superficial velocity, m s^{-1}

V_{max} solute consumption rate at saturating concentrations per unit bioreactor volume, mole/(s m^3)

V''_{max} solute consumption rate at saturating concentrations per unit cell surface, mole/(s m^2)

X_c cell concentration, cell m^{-3}

x axial coordinate, m

Greek Symbols

δ slab half width, m

η effectiveness factor, $-$

η_B bulk blood viscosity at high shear rates, kg/(m s)

η_{ov} overall effectiveness factor, $-$

η_p bulk plasma viscosity, kg/(m s)

ϕ Thiele modulus, $-$

Superscripts and Subscripts

C cell compartment

in refers to device inlet

L liquid compartment

M membrane compartment

References

Aebischer P, Tresco PA, Winn SR, Greene LA, Jaeger CB (1991) Long-term cross-species brain transplantation of a polymer-encapsulated dopamine-secreting cell line. Exp Neurol 111: 269–275.

Allen JW, Bathia SN (2002) Engineering liver therapies for the future. Tissue Eng 8(5): 725–737.

Bird RB, Stewart WE, Lightfoot EN (1960) Transport Phenomena. Wiley, New York, USA.

Catapano G (1996) Mass transfer limitations to the performance of membrane bioartificial liver support devices. Int J Artif Organs 19(1): 18–35.

Catapano G, Gerlach JC (2007) Bioreactors for liver tissue engineering. In: Ashammakhi N, Reis N, Chiellini E (eds) Topics Tissue Eng 3.

Chresand TJ, Gillies RJ, Dale BE (1988) Optimum fiber spacing in a hollow fiber bioreactor. Biotechnol Bioeng 32: 983–992.

Colton CK (1995) Implantable biohybrid artificial organs. Cell Transplant 4: 415–436.

Colton CK, Avgoustiniatos ES (1991) Bioengineering in development of the hybrid artificial pancreas. J Biomech Eng 113: 152–170.

Demetriou AA, Whiting JF, Levenson SM, Chowdury NR, Schechner R, Michalski S, Feldman D, Chowdury JR. (1986) New method of hepatocyte transplantation and extracorporeal liver support. Ann Surg 204(3): 259–270.

De Vos P, Hillebrands JL, De Haan BJ, Struppe JH, Van Schilfgaarde R. (1997) Efficacy of a prevascularized expanded polytetrafluoroethylene solid support system as a transplantation site for pancreatic islets. Transplant 63: 824–830.

Flendrig LM, La Soe JW, Joerning GGA, Steenbeck A, Karlsen OT, Bovèe WMMJ, Ladiges NCJJ, te Velde AA, Chamuleau AFM. (1997) In vitro evaluation of a novel bioreactor based on an integral oxygenator and spirally wound nonwoven polyester matrix for hepatocyte culture as small aggregates. J Hepatol 26: 1379–1392.

Fogler HS (1999) Elements of Chemical Reaction Engineering, 2nd Ed. Prentice Hall, Englewood Cliffs NJ, USA.

Fournier RL (1999) Basic Transport Phenomena in Biomedical Engineering. Taylor and Francis, Philadelphia PA, USA.

Freshney RI (2005) Culture of Animal Cells – A Manual of Basic Technique, 5th Ed. Wiley-liss.

Galletti PM, Jaregui HO (1995) Liver support systems. In Bronzino JD (ed) The Biomedical Engineering Handbook. CRC Press Inc, USA, pp 1952–1966

Gerlach JC, Encke J, Hole O, Muller C, Courtney JM, Neuhaus P (1994) Hepatocyte culture between three dimensionally arranged biomatrix-coated independent artificial capillary systems and sinusoidal endothelial cell co-culture compartments. Int J Artif Organs 17: 301–306.

Gibbon JH Jr (1954) Application of a mechanical heart and lung apparatus to cardiac surgery. Minnesota Med 37: 71.

Higgins GM, Anderson RM (1931) Experimental pathology of the liver. Arch Pathol 31: 186–202.

Humes HD, Fissell WH, Weitzel WF, Buffington DA, Westover AJ, MacKay SM, Gutierrez JM. (2002) Metabolic replacement of kidney function in uremic animals with a bioartificial kidney containing human cells. Am J Kidney Dis 39: 1078–1087.

Iwata H, Murakami Y, Ikada Y (1999) Control of complement activities for immunoisolation. Ann NY Acad Sci 875: 7–23.

Kolff WJ, Berk HTJ (1943) De kunstmatige nier: een dialysator met groot oppervlak. Ned Tijdschr Geneeskd 87: 1684.

Levenspiel O (1972) Chemical Reaction Engineering, 2nd Ed. Wiley, New York, USA.

Lindner MD, Emerich DF (1998) Therapeutic potential of a polymer-encapsulated L-DOPA and dopamine-producing cell line in rodent and primate models of Parkinson's disease. Cell Transpl 7(2): 165–174.

Maki T, Monaco AP, Mullon CJP, Solomon BA (1996) Early treatment of diabetes with porcine islets in a bioartificial pancreas. Tissue Eng 2: 299–306.

Maxwell JC (1873) A Treatise on Electricity and Magnetism, vol 1. Clarendon Press, Oxford UK.

Middleman S (1972) Cardiovascular Transport Phenomena. Wiley-Interscience, New York.

Montesano R, Soriano JV, Pepper MS, Orci L (1997) Induction of epithelial branching tubulogenesis in vitro. J Cell Physiol 173: 152–161.

Millis JM, Maguire PJ, Cronin HC, Johnson R, Conlin CA, Brotherton J, O'Laughlin R, Triglia D (1999) Continuous human liver support as a bridge to transplantation. Hepatology 30: 168A.

Morsiani E, Brogli M, Gavalotti D, et al (2001) Long-term expression of highly differentiated functions by isolated porcine hepatocytes perfused in a radial-flow bioreactor. Artif Organs 25: 329–337.

Naruse K, Sakai Y, Nagashima I, Jiang GX, Suzuki M, Muto T (1996) Development of a new bioartificial liver module filled with porcine hepatocytes immobilized on non-woven fabric. Int J Artif Organs 19(6): 347–352.

Nyberg SL, Shatford RA, Peswa MV, White JG, Cerra FB, Hu WS (1993) Evaluation of a hepatocyte-entrapment hollow fiber bioreactor: a potential bioartificial liver. Biotechnol Bioeng 41: 194–203.

Orellana SA, Neff CD, Sweeney WE, Avner ED. (1996) Novel Madin Darby canine kidney cell clones exhibit unique phenotypes in response to morphogens. In vitro Cell Dev Biol Anim 32(6): 329–339.

Scharp DW, Swanson CJ, Olack BJ, Latta PP, Hegre OD, Doherty EJ, Gentile FT, Flavin MF, Ansara MF, Lacy PE. (1994) Protection of encapsulated human islets implanted without immunosuppression in patients with type I or type II diabetes and in nondiabetic control subjects. Diabetes 43: 1167–1170.

Schiodt FV, Atillasoy E, Shakil AO, Schiff ER, Caldwell C, Kowdley KV, et al (1999) Etiology and outcome for 295 patients with acute liver failure in the United States. Liver Transpl Surg 5: 86–89.

Silva AI, Norton de Matos A, Brons IG, Mateus M (2006) An overview of the development of a bio-artificial pancreas as a treatment of insulin-dependent diabetes mellitus. Med Res Review 26(2): 181–222.

Stockmann HB, JN IJ (2002) Prospects for the temporary treatment of acute liver failure. Eur J Gastroenterol Hepatol 14: 195–203.

Storrs R, Dorian R, King SR, Lakey J, Rilo H (2001) Preclinical development of the islet sheet. Ann NY Acad Sci 944: 252–266.

Sussman NL, Chong MG, Koussayer T, He D, Shang TA, Whissenand HH, Kelly JH (1992) Reversal of fulminant hepatic failure using an extracorporeal liver assist device. Hepatology 16: 60–65.

Suzuki K, Bonner-Weir S, Hollister-Lock J, Colton CK, Weir GC. (1998) Number and volume of islets transplanted in immunobarrier devices. Cell Transplant 7: 47–52.

Talbot NC, Caperna TJ, Wells KD. (2002) The PICM-19 cell line as an in vitro model of liver bile ductules: effects of cAMP inducers, biopeptides and pH. Cell Tissues Organs 171(2–3): 99–116.

Whitmore RL (1968) Rheology of the Circulation. Pergamon Press, Oxford.

Wilke CR, Chang P (1955) Correlation of diffusion coefficients in dilute solutions. AIChE J 1: 264–270.

Chapter 8
Plant Cell-Based Bioprocessing

R. Eibl and D. Eibl

Abstract Plant cell-based bioprocessing is the use of plant cell and tissue cultures for the production of biologically active substances (low molecular secondary metabolites and recombinant proteins). The most significant advantage of plant cell culture over the traditionally grown *whole wild plant* or engineered *transgenic plant* is the sterile production of metabolites under defined controlled conditions independent of climatic changes and soil conditions, which means that variations in product yield and quality can be better avoided. Furthermore, regulatory requirements such as the cGMP standards, which have to be adhered to in the early stages of pharmaceutical production, are more easily met.

Moreover, plant cells are capable of performing complex posttranslational processing, which is a precondition for heterologous protein expression. When compared with mammalian cells, which currently dominate in the commercial protein manufacture, plant cell cultures as alternative expression systems guarantee safer processes because there is a lower risk of contamination by mammalian viruses, pathogens, and toxins. In addition to this considerable advantage, the process costs can also be substantially reduced. This is due to the fact that plant cell culture medium is very simple in composition and therefore relatively inexpensive.

This chapter provides an overview of culture types, techniques, and suitable bioreactors used to produce secondary metabolites and recombinant proteins in plant cells. We describe plant cell culture basics, discuss key topics relevant to plant cell bioreactor engineering with application examples, and give an overview of approaches to improving productivity of plant cell-based processes.

R. Eibl, D. Eibl
Institute of Biotechnology, Zurich University of Applied Sciences, Department of Life Sciences and Facility Management, Wädenswil, Switzerland
regine.eibl@zhaw.ch

R. Eibl et al., *Cell and Tissue Reaction Engineering: Principles and Practice,*
© Springer-Verlag Berlin Heidelberg 2009

8.1 Plant Cell Culture Basics

8.1.1 Characteristics of Plant Cells and Culture Conditions

Plant cells are higher eukaryotic systems with the ability to produce secondary metabolites and glycoproteins, which are more similar to their mammalian counterparts than those from bacteria, yeasts, and fungi. Comparable to the typical cell growth processes of microorganisms and animal cells, those of plant cell cultures are ideally represented by a sigmoid curve with lag, exponential, delay, stationary growth, and lethal phases (Endress 1994; Takebe et al. 1971; Zenk et al. 1975). However, plant cells show morphological and chemical totipotency (potential to form all cell types and/or to regenerate a plant). Their large size of maximum $100\,\mu m$ results in doubling times of several days and corresponding lower growth rates than microbial cells. The robustness of plant cells, which varies with culture species, culture type, and culture age, is moderate and can be attributed to the presence of a relatively inflexible cellulose wall. It has been described that plant cells can be cultivated under reasonable agitation and aeration conditions without significant responses to hydrodynamic stress, such as loss of viability, cell death, or cell lysis. Less dramatic shear-related effects observed include changes in cell morphology, and reductions in metabolite yield and biomass productivity (Kieran et al. 1997).

The plant cell vacuole, which increases in size over the growth phase, is the main site of product accumulation. While plant cell-based secondary metabolites are largely accumulated in the intracellular environment, recombinant proteins are occasionally secreted into the culture medium due to their composition and structure (Fischer et al. 2004; Misawa 1994). Schillberg and Twyman (2004) report that molecules of 20–30 kDa generally pass through plant cell walls and are thereby secreted into the culture medium. Product formation is growth-associated during active cell growth (Kreis et al. 2001; Misawa 1994) or nongrowth associated when cell growth has ceased (Guardiola et al. 1994; Hamill et al. 1986; Mantell and Smith 1983). Indeed, a large cell mass in the correct cell cycle stage has been found to be the most important precondition for a high product yield.

A temperature between 23 and 29 °C and a medium pH between 5.0 and 6.0 represent optimal parameters for plant cell growth (Endress 1994; Fowler 1988). Aeration is also an important factor for the regulation of cell growth and product accumulation and, as the case may be, secretion. Volumetric oxygen requirements of plant cells are drastically lower than those of microbial systems but comparable to those of animal cells. However, plant cell growth and metabolite production can be limited as a consequence of an inadequate oxygen supply (Zhong 2001). Kieran et al. (1997) and Taticek et al. (1994) describe average oxygen uptake rates (OUR) from 1 to 10 mmol $L^{-1}\,h^{-1}$. In small scale, the culture containers or flasks are plugged by the use of gas diffusion materials, ensuring oxygen penetration into the culture containers as a result of molecular diffusion. It is clear that this natural diffusion of oxygen is limited and, as culture volume and plant cell mass increase, plant cell growth is inhibited without continuous shaking or forced aeration (0.1–0.5 vvm).

Interestingly, besides the buffering capacity of carbon dioxide, its addition in concentrations of 0.1–5% has been shown to stimulate the cell growth and product formation of some plant cell lines (Bergmann 1967; DiIorio et al. 1992; Fischer et al. 1995; Fowler 1988; Valluri et al. 1991; Weathers and Zobel 1992; Weathers et al. 1997). According to Mirjalili and Linden (1995), ethylene (ppm range) is a further gaseous metabolite proven to be important for cell growth and/or synthesis of metabolites in individual plant cell cultures. The periodic (dark/light cycle of 8h:16h) or continuous introduction of light (0.6–10klux or 80.7–1,345 µmole m^{-2} s^{-1}) with different wavelengths and intensities promotes the culture growth and product synthesis of heterotrophic, photomixotrophic, and photoautotrophic plant cells, whereas animal cells have to be cultivated in the dark as a consequence of their light-sensitive media compounds. Finally, plant cell cultures can be successfully initiated with relatively high cell concentrations (10% of the culture volume) – unlike bacterial or animal cells – to eliminate long lag phases of 120h or more and to ensure that initial specific growth rates correlate with maximum growth rates (Endress 1994; Than et al. 2004).

8.1.2 Media

The nutrient supply provided by the culture medium is another critical element when culturing plant cells. The constituents of plant cell culture media summarized in Table 8.1 may be seperated into four groups: (1) doubled distilled and deionized water (95%), (2) the basal medium, which consists of the carbon source and organic as well as inorganic supplements, (3) the phytohormones or growth regulators, and (4) the support matrix such as agar, agarose, or gellan gums (e.g., Phytagel, Gelrite) if required, as in the case of solidification of the medium (Murashige 1973). About 80% (Evans et al. 1983) of all plant cell cultivations are realized in MS basal (MS standard) medium developed by Murashige and Skoog for tobacco tissue cultures (Murashige and Skoog 1962) and/or their modifications. It differs from other typical plant cell culture basal media, such as B5 medium established by Gamborg et al. (1968), in its very high concentration of nitrate, potassium, and ammonia. Plant cell growth and corresponding product formation are strongly influenced by modifications in carbon source, phosphate level, nitrate-to-ammonium ratio, carbon-to-nitrogen ratio, and phytohormones. It is therefore common to optimize the medium by modifying one or two kinds of the basal culture media listed by Endress (1994), or by varying phytohormone types and their concentrations (Kreis et al. 2001; Saito and Mizukami 2002).

Generally, sucrose (2–5%) is the most used carbon source for plant cells. Extracellular invertase hydrolyzes sucrose to monosaccharides such as glucose and fructose. During plant cell growth invertase is excreted into the medium, and the monosaccharide concentrations change according to cultivation time. In contrast to phosphate (phosphate anions or potassium dihydrogen phosphate) and nitrogen (nitrate anions, ammonium cations, amino acids, or protein hydrolysates), which support

Table 8.1 General composition of plant cell culture media

Water	Basal medium			Phytohormones (growth regulators)	Support matrix
	Carbon source	Organic supplements	Inorganic supplements		
	Sucrose	Amino acids: Ala, Arg, Asn, Cys, His, Ile, Leu, Met, etc.	Microelements in µM concentrations: Fe, Mn, Zn, Cu, Mo, I, B, Co	Auxins: IAA, 2,4-D, NAA, IBA, 2,4,5-T	Agar
	Glucose	Vitamins and cofactors: myo-Inositol, thiamine, pyridoxine, folic acid, ascorbic acid, tocopherol, yeast extracts[a]	Macroelements in mM concentrations: N, S, P, K, Mg, Ca	Cytokinins: kinetin, BAP, zeatin, purine, adenine	Agarose
	Fructose			Coconut milk or coconut water[a]	Gellan gums
	Sorbitol			Gibberellines[a]: GA_3, GA_4, GA_7 ABA[a]	

[a] Occasionally used

cell growth and lower product synthesis (Nettleship and Slaytor 1974; Do and Cormier 1991), an increase in the initial concentration of sucrose can lead to an improved product concentration, as described by Petersen et al. (1992). However, opposite results have been reported by Oksman-Caldentey et al. (1994) and Chattopadhyay et al. (2003), who showed that maximum product content was obtained with a lower medium sucrose concentration than that required for maximum plant cell growth. They speculated that high sugar concentrations may result in plasmolysis of the plant cells and lead to decreased biosynthesis of secondary metabolites. Furthermore, physiological studies indicate that ammonium is generally used before nitrate in plant cells (Hilton and Wilson 1995; Wilhelmson et al. 2006).

Growth regulators decisively affect the growth process through the ratio of auxins, such as indole-3-acetic acid (IAA) or its most common alternatives 2,4-dichlorophenoxyacetic acid (2,4-D), naphthaleneacetic acid (NAA), indole-3-butyric acid (IBA), and 2,4,5-trichlorophenoxyacetic acid (2,4,5-T), to cytokinins, such as kinetin, benzylaminopurine (BAP), zeatin, purine, and adenine. While low auxin and high cytokinin concentrations typically stimulate cell growth, high auxin and low cytokinin concentrations promote cell division. In fact, it is important to select the most appropriate phytohormones (as they differ in effect) and to determine their optimal concentration by considering the desired tissue or cell culture type. For example, 2,4-D used as a dedifferentiating hormone in higher concentrations supports rapid callus induction, but its application may cause severe growth abnormalities or even result in culture mutations during long-term cultivations. It also often inhibits the regeneration process (Hess 1992). Recombinant protein production and secretion is influenced by supplements that stabilize the proteins (Sect. 8.4), such as polyvinylpyrrolidone, gelatin, bovine serum albumin, and sodium chloride added to the frequently used standard MS- or B5- medium (Shadwick and Doran 2004). Some attention should also be paid to medium sterility (which is achieved by classical standard- or overkill-sterilization treatment in autoclaves (121 °C, 15–20 min) or sterile filtration with 0.1 or 0.2-μm filters) as heat-sensitive constituents are contained in the Newtonian fluid-like culture medium.

8.1.3 Plant Cell Culture Types and Their In Vitro Initiation

8.1.3.1 Callus Cultures

In the in vitro culture of plant cells, the induction of callus is a fundamental step. The term callus describes an amorphous mass of unorganized parenchyma cells formed by the proliferation of plant cells as a protective response at the surface of explant tissue cut with a sterile scalpel. In this process, sterile organs or pieces of tissue such as seeds, embryos, stem sections, leaf- or root- pieces are referred to as explants. Achieving sterility of the plant material requires surface sterilization with chemical agents to kill bacteria and fungi living on the plants while ensuring minimum damage to the treated tissue. Solutions of calcium hypochlorite and sodium hypochlorite in concentrations of 0.5 up to 10% have usually proved to be the most suitable during exposure times of between 5 and 30 min (Bhojwani and Razdan 1983; Endress 1994). A further cheap, ready-made sterilant is a 5–7% solution of the toilet disinfectant "Domestos." The three-stage sterilization process depicted in Table 8.2 has become the accepted rule. For lignified plant material, such as stem tissues or leaves with a waxy surface, an additional ultrasonic treatment is required (Bentebibel 2003).

The formation of the callus is usually achieved by placing the explant on an appropriate solid growth medium with phytohormones in a closed Petri dish incubated at 25 °C in darkness or low light. The running process of cell dedifferentiation is characterized by cell division as well as by elongation and changes in metabolic activity.

After a period of 3–6 weeks, the primary callus formed is normally transferred to a fresh medium after sterile elimination of necrotic parts, which are brown in color (attributed to oxidation of phenolic ingredients). To ensure callus growth, the inoculum should be of uniform size and sufficiently large. As Endress (1994) outlined, a diameter of 5–10 mm for 20–100 mg fresh weight should be provided. Rapid growth with doubling times of between 7 and 10 days and a white, light yellow, green, or red color generally indicate a healthy callus culture. Besides color, two different types of callus can be distinguished: the friable callus (crumbled or fragmented) and the lignified callus (nonfriable callus). Figure 8.1 shows cells of friable callus of *Vitis vinifera* vs. Uva Italia induced from berry skin. These cells were grown on standard MS medium with additives (Calderón et al. 1994) and maintained at 25 °C in the dark.

The successful establishment of a callus culture and its appearance depend on the donor tissue, the surface sterilization method, the culture conditions, the age of the callus (aging of callus can be characterized by a growing number of lignified cells), and the medium. Sometimes it is advisable to optimize the callus medium by varying auxin and cytokinin concentrations and to work with a callus induction and callus growth medium separately (Endress 1994; Evans et al. 2003; Hess 1992). Above all, it is well known that callus cultures are predisposed to genetic instability

Table 8.2 General sterilization procedure for plant fragments

Stage of procedure	Description
Presterilization (1)	Scrub carefully, clean or rinse under running tap water
	Rinse or submerge briefly in absolute ethanol
Sterilization (2)	Submerge for 5–30 min in 0.5–10% sodium or calcium hypochlorite
Poststerilization (3)	Wash three times with sterile water
	Dry with sterile tissue paper

Fig. 8.1 Cells of friable callus of *Vitis vinifera* vs. Uva Italia (established by S. Cuperus, Zurich University of Applied Sciences, Switzerland)

and loss of their morphogenetic characteristics during long passage (subculture) periods. Therefore, it is recommended that callus cultures are passaged every 3–4 weeks depending on the species and the growth rate. The important steps in callus induction and propagation are schematically illustrated in Fig. 8.2.

8.1.3.2 Plant Cell Suspension Cultures

To initiate plant cell suspension cultures, the callus produced is dispersed by inoculating the fragments into a liquid culture medium (Fig. 8.2). The callus friability and its inoculation rate (50–100 g L^{-1} fresh weight) are key issues in this process. To obtain friable callus, Evans et al. (2003) describe a number of procedures including the application of a special callus medium with rising auxin concentration, transfer in liquid medium, and repeated passages in shake flasks (25 °C, 100 rpm) until the callus reaches friability, as well as the addition of pectinase to nonfriable callus growing in liquid medium. After the initial passage, which includes cells breaking off from the friable callus and a suspension beginning to form, the culture is usually filtered with a sieve (300–500 μm) to remove larger aggregates and is finally subcultured to fresh medium in a ratio of 1:1.

This whole procedure (Fig. 8.2) is called homogenization. It aims to produce a homogeneous suspension and is repeated over 2–4 passages. Established plant cell suspension cultures are generally maintained in shake flasks (100 mL flasks with 20 mL medium) at 25 °C and between 100 and 120 rpm, and serially subcultured in the late exponential phase. This is performed every 7–10 days for fast-growing cells that reach high cell masses, and every 14–21 days for slower growing cells characterized by moderate or low cell masses. It is obvious that plant cell suspension cultures grow faster than their callus cultures. Doubling times of between 0.6 and 5 days along with growth rates of 0.24–1.1 d^{-1} are reported in the literature (Hess 1992). Maximum biomass concentrations typically range between 10 and 18 g dw L^{-1} or 200 and 350 g fw L^{-1} (James and Lee 2001).

Plant cell suspension cultures display a high degree of culture heterogeneity as well as variability in terms of cell morphology (including cell size, cell shape, and cell aggregation), rheological characteristics, growth, and metabolic pattern (Hall and Yeoman 1987; Hess 1992; Yanpaisan et al. 1998, 1999). The changes are mainly associated with the chemical and hydrodynamic environment of the cells (Kieran et al. 1997, 2000). The size of single plant cells is typically in the range of 10–100 μm. The most common shapes of plant cells in suspension are the spherical morphology (e.g., the majority of carrot suspension cells, taxus suspension cells, and grape suspension cells) and the rod morphology (e.g., most tobacco suspension cells). Lengthy subculture intervals may slow down cell division, activate cell elongation (which occurs after cell division), and result in changed cell morphology. In this case, a change from an original spherical shape to an elongated shape may occur (Curtis and Emery 1993).

Generally speaking, plant suspensions cells very rarely grow as single cells. They form a few large (or even huge) aggregates, which reach many millimeters

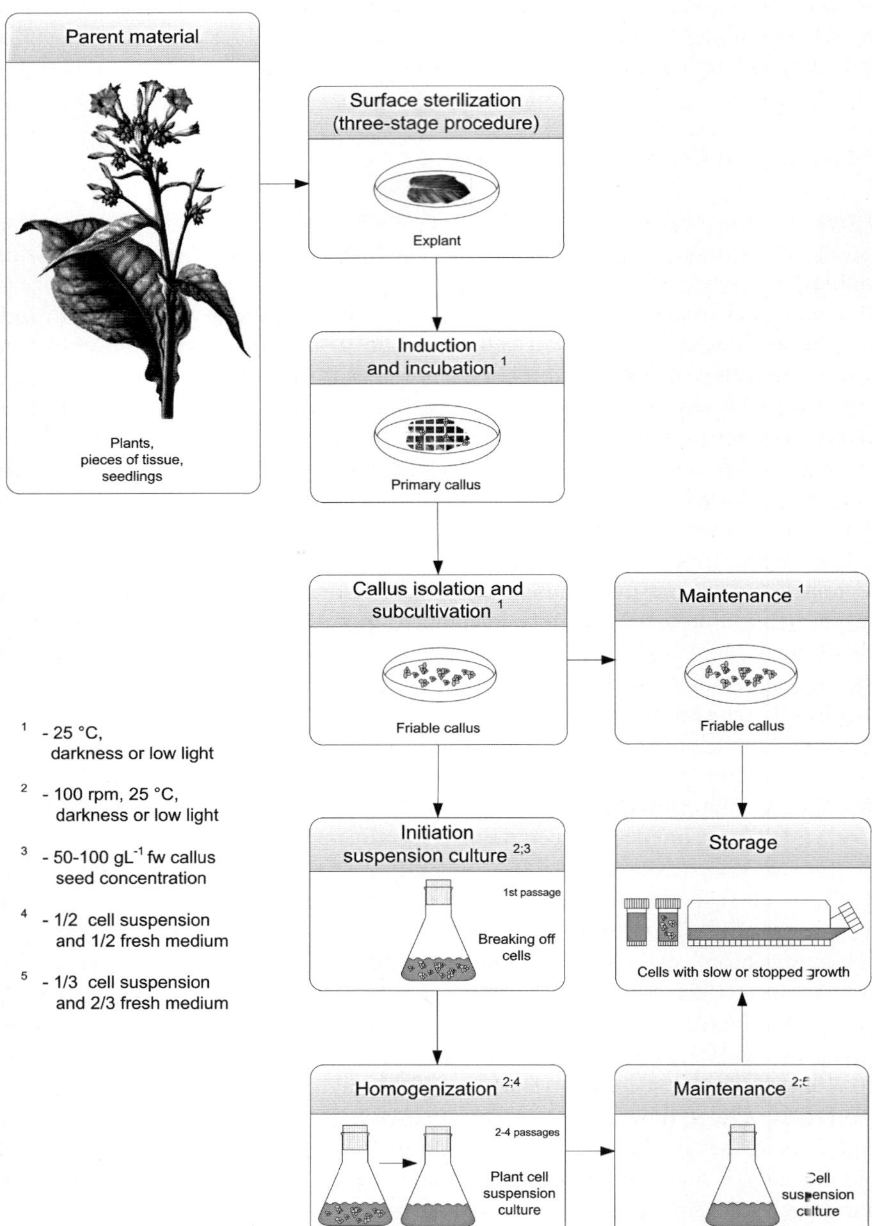

Fig. 8.2 Schematic representation of the procedure for plant suspension culture establishment and maintenance

in diameter and consist of hundreds of highly mitotic and less mitotic cells. The aggregation (clumping) is based on cell adhesion and results from the secretion of cell wall extracellular polysaccharides, which prevent cell separation after cell division (Glicklis et al. 1998). Chattopadhyay et al. (2002) reported an enhancement of cell clumping especially in the later growth stages. While cell aggregation is viewed as desirable for secondary metabolite production due to its promoting of cellular organization and differentiation, this feature may involve oxygen or nutrient gradients, which complicate the cultivation procedure. For plant cell suspension cultures, the critical dissolved oxygen level, below which growth is seen as a result of oxygen limitation, has been commonly quoted as ~15–20% air saturation (Kieran et al. 1997; Yu et al. 1997).

In addition, it appears that cell elongation and aggregation can influence rheological properties of plant cell suspension cultures. Most plant cell lines growing to low or moderate cell concentrations behave like a Newtonian fluid. Non-Newtonian behavior (Bingham plastics characteristics, pseudoplastics characteristics, Casson fluids) associated with higher culture viscosity has been described for high plant cell concentrations (high biomass concentrations, high cell density cultures) and elongated, filamentous cells entangled in a network of cells (Curtis and Emery 1993; Su 2006). Finally, it should be mentioned that suspension cultures often tend to be unstable and unproductive over time (Deus-Neumann and Zenk 1984). Possible explanations for this phenomenon are genotypic and phenotypic variations (i.e., somaclonal variations: changes in chromosome numbers, chromosome structure, and DNA sequence) during cultivation time. Nevertheless, suspension cultures are the most used plant cell culture type in the research and production of secondary metabolites and r-proteins.

8.1.3.3 Hairy Root Cultures

Hairy roots (or transformed roots, as they are more accurately called) are generated by the transformation of plants or explants with agropine- and mannopine-type strains (A4, ATCC, 15834, TR7, TR101, etc.) of *Agrobacterium rhizogenes*, a Gram-negative soil bacterium. When the bacterium infects the plant or explant, the T-DNA, that is, transfer DNA originating from root-inducing plasmid (Ri plasmid), is transferred and integrated into the genome of the host plant. This transformation process produces two by-products: hairy roots and opines (transferred genes, which have been involved in hairy root formation, including genes leading to opine production) (Chilton et al. 1982). In most cases, wounding a sterile leaf (midrib and major veins) or stem tissue is carried out with the sterile tip of a needle attached to a syringe before infection takes place. Two main infection methods are reported in the literature. In method 1, which is most suitable for leaf explants, one to two drops of fresh (two days old) undiluted bacterial suspension is transferred to each wound site (Hamill and Lidgett 1997). In contrast, method 2 works with liquid diluted infection medium in which the explants (leaves, callus, and seedlings) are submerged for about 3 min (Hamill and Lidgett 1997; Yoshikawa 1997; Komari et al. 2004; Vervliet et al. 1975).

Cocultivation follows, in which plant cells divide and dedifferentiate, and bacteria divide and infect the wounded plant tissue. The resulting transformed cells acquire the capacity to develop into a root system at the infection site. This system branches out to a greater extent than the ordinary roots of plants, and is covered with tiny multiple root hairs. Generally, cocultivation in Petri dishes or other suitable cultivation containers is accomplished by the incubation of the infected explants for 2–5 days at 25 °C in low light or darkness on solid MS- or B5- medium. After cocultivation, the infected explants are cleared of excessively growing *Agrobacteria* by transferring them to a culture medium, which also contains antibiotics such as carbenicillin, cefotaxime, or ampicillin. Because of the appearance of bacterial infections, repetition of the transfer to fresh medium with antibiotics at intervals of 2–4 days is stringently necessary. The neoplastic hairy roots may be maintained indefinitely in solid or liquid culture by subculturing the root tips as illustrated in Fig. 8.3. During the process of adapting hairy roots growing on solid culture medium to liquid medium, a change in root thickness (thinning) involving more rapid growth is not unusual (Carvalho et al. 1997).

Indeed, it is not surprising that root growth is not homogeneous because the rapidly dividing meristematic cells are restricted to root tips. After the cells stop dividing, they grow firstly by elongating and subsequently by branching, which results in lateral root production. Consequently, Hjortso (1997) distinguished between two types of cells in hairy roots: (1) dividing tip cells in the apical meristem and (2) nondividing cells elongating, differentiating, and representing the bulk of hairy roots. Even in shake flasks, a densely packed root mass with heterogeneous structure and the afore mentioned root hairs play a detrimental role for mass transfer of fluid and oxygen because the root hairs induce fluid flow stagnation and high levels of liquid entrainment as well as oxygen limitation (Bordonaro and Curtis 2000; Shiao and Doran 2000; Yu et al. 1997). In general, hairy roots respond to changes in their cultivation environment. For example, the number of root hairs, root length, and root tip viability can be reduced by drought or shear stress.

A clear indication of excessive shear stress is callus formation at root tips (Yu et al. 1997). On the one hand improved oxygen transfer efficiency increases the formation of root hairs and their length (Hofer 1996; Carvalho et al. 1997). On the other hand, Bates and Lynch (1996) describe an increase in root hair formation under nutrient-limiting conditions. Of course, the morphology of hairy roots is affected by many further factors including the plant species and the *Agrobacterium rhizogenes* strain used in hairy root induction.

Although the morphological character of hairy roots complicates their optimized in vitro cultivation, other properties make hairy roots very attractive for the commercial production of valuable metabolites. A significant advantage of hairy roots is that they are generally easy to isolate and grow in a defined medium. Because auxin metabolism is altered in plant cells after transformation with *Agrobacterium rhizogenes*, the addition of exogenous growth regulators or phytohormones becomes unnecessary. Most importantly, fully differentiated hairy roots have the strong tendency to be genetically and biochemically stable. They show a rapid

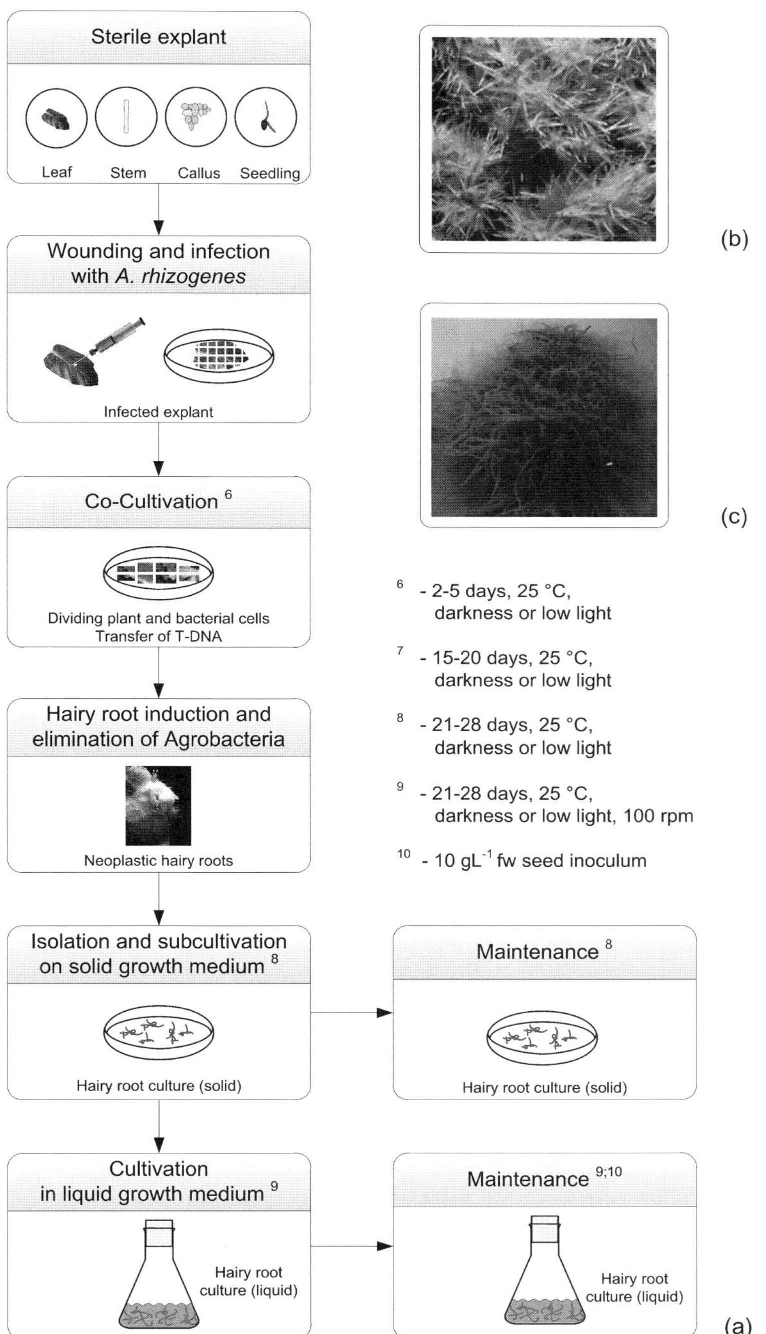

Fig. 8.3 Hairy root cultures: (**a**) Typical techniques to initiate hairy roots on solid and in liquid medium, (**b**) Hairy roots of *Hyoscymus muticus* growing on solid B5 medium

growth with average doubling times resembling those of disorganized cell suspensions. For example, Hess (1992) and Doran (2002) present doubling times of 1 day for hairy root cultures of *Atropa belladonna* and *Nicotiana tabacum*, and 8 days for hairy roots of *Arabidopsis thaliana* and *Solanum aviculare*.

Furthermore, hairy roots have been found to synthesize secondary metabolites at similar or higher levels to those found in whole plants (Flores 1987; Hu and Du 2006; Oksman-Caldentey and Hiltunen 1996; Rhodes et al. 1994; Sevón and Oksman-Caldentey 2002) and to have light-guiding properties (Towler et al. 2006). In addition to their recombinant protein expression ability, transgenic hairy root cultures exhibit a significantly greater long-time stability than transgenic suspension cultures (Sharp and Doran 2001).

8.1.3.4   Embryogenic and Shoot Cultures

Like hairy roots, embryogenic and shoot cultures belong to the group of differentiated organ cultures. But today, embryogenic and shoot cultures are mainly used for micropropagation and plant breeding. For both the in vitro production of secondary metabolites and recombinant proteins, these culture types are of minor importance in contrast to hairy root and plant cell suspension cultures. Therefore, the characteristics and the initiation procedure of embryogenic and shoot cultures is not included here. More information on this topic and on the process of developing embryos from somatic cells and tissue (somatic embryogenesis) is provided by Hess (1992), Endress (1994), Evans et al. (2003), Ducos et al. (2007), Gupta and Ibaraki (2006), and Hvoslef-Eide and Preil (2005).

8.1.4   Routine Working Methods in Plant Cell Cultivation

8.1.4.1   Determination of Plant Cell Growth

To characterize and design bioprocesses based on plant cells other than secondary metabolite production or protein accumulation/secretion and nutrient utilization (e.g., sucrose, glucose), the cell growth must be determined. Six direct and indirect methods, which provide information on plant cell growth, have been reported. These methods involve measurement of biomass accumulation (fresh weight, dry weight), cell mass, and cell number (cell concentration) as well as viability, conductivity, osmolarity, and pH (Endress 1994; Naill and Roberts 2005; Tanaka et al. 1993; Widholm 1972).

Plant cell growth is normally measured either by fresh weight or by dry weight. Its determination relies on population averages and does not take plant cell culture heterogeneity into account. Fresh weight values (expressed as g) are obtained by weighing freshly harvested cells. The increase in fresh weight that occurs is due to both cell growth and expansion (indicated by cell size). The dry weight (expressed as g) excludes errors caused by endogenous water content and is a more useful tool for biomass quantification of plant cells than fresh weight. For dry weight measurement, a known weight of fresh plant cells is dried in an oven at a temperature of between 50 and 60 °C to the point of constant weight (ca. 24–48 h). Alternatively, fresh cells are lyophilized.

For fine plant cell suspensions, the cell mass can be recorded as packed cell volume (PVC) immediately after gentle centrifugation at low speed. In fact, PCV (expressed as %) is defined as the ratio of the volume occupied by solid matter in the form of plant cells or cell aggregates to the volume of the whole sample (aliquots of about 15 mL). By comparing the packed cell volume to the biomass accumulation, a definite correlation can be determined. A further, less common method of measuring the biomass growth of suspension cultures is manual cell counting using an improved Neubauer-type hemocytometer with a depth of 0.1 mm to determine the number of plant cells per unit volume (expressed as cells mL^{-1}). The cell growth in aggregates and the large size of the plant cells presuppose cell dispersion with chemical agents such as hydrochloric acid, chromic acid (Reinert and Yeoman 1982), EDTA, or pectinase before counting. Nevertheless, if fluorescein diacetate, trypan blue, or Evan's blue is used as a dye for staining suspended cell counts, plant cell viability can be roughly determined (Steward et al. 1999; Widholm 1972). Viability is defined as the ratio of viable cells to total cells, and has values between 0 and 1, that is, between 0 and 100%.

Measurements with a conductivity meter in liquid culture medium readily allow the indirect monitoring of biomass growth. There is an inverse relationship between electrical conductivity (expressed as $mS\ cm^{-1}$) and cell weight. In the case of liquid plant cell cultures, a number of authors have shown that an increase in cell dry weight can be correlated linearly to a decrease in medium conductivity (Bais et al. 2002; Ryu and Lee 1990; Taya et al. 1989a; Than et al. 2004). This is probably caused by an uptake of medium salts in the form of ions by the cultured cells and resulting biomass growth during cultivation. But medium conductivity depends mainly on electrolyte (mineral) concentration and not on the sugars that are major components of the culture medium. Whereas small changes in biomass level are more difficult to estimate by conductivity measurement, relatively wide osmolarity changes correlate well with small changes in biomass of several glucose-consuming plant cell suspensions (Tanaka et al. 1993). Thus, Madhusudhan et al. (1995) recommend osmolarity ($mOsmol\ kg^{-1}$ units) as a sensitive measure of the growth of plant cell cultures with rapid sucrose-hydrolyzing capability.

The pH measurement is routinely made during the cultivation of liquid cultures. A gradual drop in pH to a value around 4 reflects the initial ammonium uptake and acidification caused by cell lysis within 20–48 h, whereas the pH returns to a stable value of about 5 related to the uptake of nitrates after a few days of cultivation (Ziv 2000). If the growth at the single-cell level should be quantified, flow cytometry may be utilized besides traditional measures of cell growth such as biomass accumulation for suspended plant cell populations (Evans et al. 2003; Naill and Roberts 2005; Yanpaisan et al. 1998, 1999). Parameters frequently calculated (specific growth rate and doubling time excepted) to characterize plant cell growth on a biomass basis are summarized in Table 8.3.

Table 8.3 Parameters characterizing plant cell growth on a biomass basis

Parameter	Description and formula	Used for
Fresh biomass concentration (FCM_{tx})	Ratio of cell fresh weight at time tx to initial volume of medium: $FCM_{tx} = fw_{tx}/V_i$ [g fw L^{-1}]	All types
Dry biomass concentration (DCM_{tx})	Ratio of cell dry weight at time tx to initial volume of medium: $DCM_{tx} = dw_{tx}/V_i$ [g dw L^{-1}]	
Dry substance (DS_{tx})	Ratio of cell dry weight at time tx to cell fresh weight at time tx: $DS_{tx} = dw_{tx}/fw_{tx}$ [none]	
Biomass productivity (BMP)	Ratio of cell weight increase to maximum cell weight and the cultivation time during which the maximum cell weight was obtained (FCM_f and DCM_f = maximum values): $BMP_{dw} = (DCM_f - DCM_i)/t$ [g dw L^{-1} d^{-1}]; $BMP_{fw} = (FCM_f - FCM_i)/t$ [g fw L^{-1} d^{-1}]	
Growth index or growth ratio (GI)	Ratio of final (maximum) cell dry weight to initial cell dry weight: $GI = dw_f/dw_i$ [none]	Organ cultures (e.g. hairy roots)

Mathematical models developed for callus and suspension cultures describe cell growth by modified Monod kinetics and reactions of the first order (Guardiola et al. 1994; Val 1993; Xing et al. 2001). To predict increase in biomass of hairy root cultures, branching models (Hjortso 1997; Kim et al. 1995; Taya et al. 1989c) incorporating the important features of hairy root growth (Sect. 8.1.3.3) and a population model (Han et al. 2004) have been proposed. The population balance model shows biomass growth resulting from root elongation caused by both cell division and the formation of new lateral branches. It also shows that the formation of a new branch depends on the age of the parent branch.

8.1.4.2 Genetic Transformation

Nowadays two indirect *Agrobacterium*-mediated transformation methods based on *Agrobacterium rhizogenes* or *Agrobacterium tumefaciens* are routinely used to transfer DNA into plant cells of both monocotyledonous and dicotyledonous species. While *A. rhizogenes* causes hairy root disease (Sect. 8.1.3.3), *A. tumefaciens* induces a tumor formation process, commonly known as crown gall disease, via an identical process. Genes involved in this tumor formation accumulate on the tumor-inducing plasmid, Ti plasmid, which encodes the virulence genes *vir* and transfer DNA into the plant genome. The most critical factors in efficient *Agrobacterium*-mediated transformation procedures are the vectors and strains, the type and quality of the starting material, the concentration of the *Agrobacterium* inoculum, the composition of the cocultivation medium, the temperature and pH range of cocultivation, and finally the antibiotics added to the plant cells to remove the *Agrobacterium* cells (Komari et al. 2004).

A more modern transformation method, developed in the 1980s, is particle bombardment (also called biolistic method, gene gun, or particle gun method), which has no intrinsic limitations with respect to target tissue, species, or genotype as described for *Agrobacterium*-mediated transformation. It remains a direct transformation method in which high-velocity microprojectiles from gold or tungsten breach the cell wall and membrane, and are thus used to introduce engineered DNA into plant cells and tissue of any type (leaves, shoot tips, embryogenic cultures, callus cultures, cell suspensions, whole plants). In this way, the DNA diffuses from the microprojectiles and is then transiently or permanently incorporated into the plant genome (Twyman and Christou 2004). Since the 1990s, particle bombardment has also been used successfully in combination with other transformation techniques (Bidney et al. 1992; Hansen and Chilton 1996; Rasmussen et al. 1994), for example, *Agrobacterium*-mediated transformation. As demonstrated, e.g., for mosses, the longstanding polyethyleneglycol (PEG)-mediated DNA transfer is the method of choice if Agrobacterium-mediated transformation or biolistic transformation is not possible (Gorr and Wagner 2005).

Studies have revealed that transgenic plant cell suspensions can be developed from transgenic callus cultures produced with a particle gun or explants from transformed plants, as well as from *Agrobacterium tumefaciens*-mediated transformation of wild-type suspensions, and the biolistic delivery of plasmid DNA into plant suspension cells. Currently, transgenic hairy roots can be initiated by infecting transgene-containing plants or explants with *A. rhizogenes*, performing root initiation and transformation using engineered *A. rhizogenes* strains containing modified T-DNA, and direct *A. tumefaciens*-mediated transformation of established hairy root cultures from wild-type (Shadwick and Doran 2004). As described for mammalian cells, doubling time and growth rate of plant cells can also be increased by genetic engineering (Su 2006). For more detailed information on the genetic transformation of plant cells, the reader is referred to Christou and Klee (2004) and Fischer and Schillberg (2004).

8.1.4.3 Storage

Plant cell cultures are maintained either in slow-growth storage or, for long-term conservation, in the form of cryopreservation. While preserving all the characteristics of the cells, both methods reduce the frequency of subcultivation and thereby reduce contamination risk, labor, as well as media costs. Slow growth can be achieved by various simple methods including a temperature shift to a range between 2 and 10 °C, a decrease in light (dim light at about 50 lux) or oxygen, the application of growth medium with stabilizers promoting cell survival, or combinations of these methods. According to Schumacher et al. (1994), slow-growth storage for more than 4 months in static tissue culture flasks with growth medium plus 0.01% charcoal and 1.0% gelatin at a reduced temperature of 10 °C in the dark may be preferable for suspension cultures of *Agrostis tenuis, Nicotiana tabacum, Nicotiana chinensis, Oryza sativa,* and *Solanum marginatum.*

In contrast to slow-growth storage, it is known from animal cell storage that cryopreservation, which finally occurs in liquid nitrogen and at $-196\,°C$, effectively stops all cellular processes. It is achieved by use of cryoprotectant solutions containing DMSO, glycerol, or sucrose in standard culture medium with controlled cooling or dehydration followed by rapid freezing (Withers 1991; Benson et al. 1998). A cooling rate in the range of $-0.25\,°C\ min^{-1}$ to $-2.0\,°C\ min^{-1}$ down to not less than $-40\,°C$ maintained for 30–60 min and followed by rapid cooling to liquid nitrogen temperature has proven to be suitable for many plant species and culture types (Moldenhauer 2003). Successful cryopreservation of cell suspension cultures by slow prefreezing is described by Seitz (1987), and Menges and Murray (2004). Vitrification using highly concentrated plant vitrification solution 2 (PVS2) represents another cryopreservation procedure with extended applicability (Kobayashi et al. 2006). In spite of the long list of cryopreserved plant cells, which includes callus cultures, suspension cultures, isolated protoplasts, embryogenic cultures, hairy roots, and shoot tips (Endress 1994; Kartha 1987; Moldenhauer 2003; Yoshimatsu et al. 1996), cryopreservation is still not widely used in processes that use plant cells as production organisms.

8.2 Bioreactors for Plant Cell Cultures

8.2.1 General Considerations

For plant cell suspensions and hairy roots, suitable bioreactors can be roughly divided into three main types according to their continuous phase (Kim et al. 2002a): liquid-phase bioreactors, gas-phase bioreactors, and hybrid bioreactors (Table 8.4). In liquid-phase bioreactors, the plant cells are immersed continuously and oxygen is usually supplied by bubbling air through the culture medium. The term "submerged bioreactor" is also used for liquid-phase reactors. Mechanically driven bioreactors, pneumatically driven bioreactors, and hydraulically driven bioreactors belong to this category (Sects. 8.2.2 and 8.2.3). Because of the low solubility of gases in liquid-phase systems, gas-exchange limitations and insufficient nutrient transfer (as a consequence of gentle mixing to lower hydrodynamic stress) frequently result in growth inhibition (Singh and Curtis 1994).

By using gas-phase reactors such as the spray reactor or the mist reactor, which represent typical emerged bioreactors (Sect. 8.2.3), oxygen transfer limitation can be reduced or even eliminated. Whereas plant cells, especially organ cultures, are exposed to humidified air or other gas mixtures, the medium containing the nutrients is delivered to cells as droplets produced by spray nozzles or ultrasonic transducers. As readily identifiable from the bioreactor name, the droplet size of mist bioreactors is smaller (usually $0.01–10\,\mu m$) than the potentially much greater droplet size of

Table 8.4 Plant cell bioreactor categorization

Plant cell bioreactors				
Liquid-phase bioreactor/ submerged bioreactor			Gas-phase bioreactor/ emerged bioreactor	Hybrid bioreactor
Mechanically driven bioreactor	Pneumatically driven bioreactor	Hydraulically driven bioreactor	Trickle bed reactor (droplet reactor or spray reactor)	Combinations of bubble column and gas-phase bioreactor, for example: bubble column-trickle bed
Stirred reactor	Airlift reactor	Radial flow bioreactor	Mist reactor	
Rotating drum reactor	Bubble column Plastic-lined reactor			Bubble column-mist reactor
BioWave	Slug Bubble bioreactor			Wilson Reactor (bubble column-spray reactor)
Wave & Undertow bioreactor				

spray bioreactors ($10–10^3$ μm). Finally, the hydrodynamic stress is low in gas-phase bioreactors (Towler et al. 2006; Weathers et al. 1999; Whitney 1992; Wilson 1997).

Hybrid bioreactors represent combinations of submerged and emerged bioreactors, such as the famous 500 L Wilson Reactor for hairy roots (Sect. 8.2). They have been developed to circumvent manual loading of the bioreactor growth chamber in order to uniformly distribute the production organisms, for example, roots. The hybrid bioreactor switches from liquid-phase to gas-phase operation after the inoculation, distribution, attachment to immobilization points, and short growth phase of the cells.

Small biomass concentrations of plant cell cultures can be grown in virtually any bioreactor configuration. Excellent plant cell growth with biomass productivity > 1 g dry weight L^{-1} day^{-1} requires an optimized and well-characterized bioreactor configuration. For plant cells tending to high cell density growth and/or aggregate formation, their trouble-free inoculation, transfer, and harvest presuppose specially designed bioreactor elements. For example, transfer pipes, inoculation pipes, harvest pipes, and ports all have to be sufficiently dimensioned. Moreover, inoculation should be performed by application of gravity, pressure, or vacuum, and high shear stress from pumping should be avoided.

Process monitoring and control are generally facilitated by standard bioreactor instrumentation, which uses pressure-, temperature-, pH-, and air sensors as well as gas flow rate-, pO_2- (dissolved oxygen), pCO_2- (dissolved carbon dioxide), and conductivity sensors. The integration of a gas analyzer allows the oxygen uptake rate (OUR) and carbon dioxide evolution rate (CER) to be determined.

As previously mentioned, the yield of bioactive substances is determined both by biomass production and by the level of bioactive substances produced per unit biomass (Sect. 8.1.1). Potentially, the importance of the bioreactor type used lies not in maximum biomass production, but rather in stimulating product formation by guaranteeing an optimal environment (Flores and Curtis 1992; Kim et al. 2002b). Therefore, the bioreactor should be selected and optimized with the culti-vation goal in mind (biomass or bioactive substance) taking into account the cell line's specific morphology, rheology, shear tolerance, growth behavior, and metabolism.

8.2.2   Suitable Bioreactors for Plant Cell Suspension Cultures

Cell suspension cultures in general are regarded as the most suitable plant cell cul-ture type for large-scale biotechnological applications and are cultured in mechanically or pneumatically driven aerated submerged bioreactors. Stirred reactors, bubble column reactors, and airlift reactors directly derived from microbial fermenters were initially used with only minor modifications to grow plant suspension cells (Eibl and Eibl 2002). In aerated/submerged plant cell suspension bioreactors, engineering analysis demonstrates that hydrodynamic-related stress adversely affecting the cultivation course is generally a result of the aeration and mixing sys-tem, aeration rate, and/or impeller tip speed used (Kieran et al 2000; Su 2006; Zhong 2001).

In most plant cell cultivations, the air is directly introduced via a sparger (ring, pipe, plate, frit) positioned in the lower part of the bioreactor (Eibl and Eibl 2002). According to Präve et al. (1994), such direct aeration guarantees the highest possible aeration efficiency, this being evaluated by volumetric oxygen transfer coefficients, $k_L a$ values, which are above $20\,h^{-1}$ for cell cultures. When pressure, temperature, and medium are fixed, $k_L a$ values of bioreactors are affected by the aeration rate, gas holdup, and size of the bubbles produced by the sparger. The higher the aeration rate, the larger the bubble size, speed between bubbles and cells, collision risk, and damaging ability become. If sensitive cells come into contact with air bubbles, cells cultivated in bubble-aerated bioreactors may be damaged as a consequence of cells and bubbles rising to the surface and subsequent bubble bursting (Storhas 1994). As mentioned in Sect. 8.1.1, typical oxygen demands and the resulting aeration rates for high plant cell growth do not lead to permanent cell damage for most plant cell lines. In general, $k_L a$ values of about $10\,h^{-1}$ are necessary for high plant cell culture growth (Takayama and Akita 2006) in large-scale culture systems. However, Pan et al. (2000) reported a growth stimulating effect of $k_L a$ values above $15\,h^{-1}$ in stirred *Taxus chinensis* suspension cell cultivations.

Another point to consider is that the presence of extracellular polysaccharides, fatty acids, and high sugar concentrations in the plant cell culture medium combined with bubble aeration promotes foaming at the culture broth surface, which is a serious problem particularly for the pneumatically driven plant cell bioreactors such as airlift and bubble column reactors. The result of extensive foaming is a crust of foam-entrapped cells adhering to the inside of the bioreactor's vessel (called wall growth phenomenon) as well as clogging of the air exhaust filter, which constitutes a contamination risk. Ceramic or sintering steel porous spargers (Kim et al. 1991; Zhong et al. 1993) generating fine bubbles with higher $k_L a$ values, bubble-free aeration via tubes of silicone (Abdullah et al. 2000; Ziv 2000), external aeration (Carvalho and Curtis 1998; Kino-Oka et al. 1999), and indirect aeration (spinfilter, eccentric motion stirrer, cell-lift impeller) (Eibl et al. 1999; Su 1995; Valluri et al. 1991) can be used to overcome the disadvantages of sparger rings to plant cells at high bubble aeration rates. However, with the exception of external aeration and ceramic or sintering steel porous spargers, sparger systems have a limited scale-up potential. It is evident that, as a general rule, direct aeration is more effective and advantageous than indirect medium aeration conducted in the outer loop. Paek et al. (2005) successfully reduced foaming and cell growth on the vessel wall in bubble columns by designing balloon-type bubble bioreactors. Similar effects, including minimized wall growth, can be achieved by oxygen enrichment, which causes low gas flow rates as found in the Plastic-lined bioreactor (Fig. 8.4g) by Curtis for *Hyoscyamus muticus* suspension cells. This low cost bioreactor type has a cultivation bag of plastic film, which is inserted in a bubble column vessel. Its 150 L system (100 L culture volume) produced biomass with over 15 g dw L^{-1} during the 33-day culture period (Curtis 2004).

In investigations carried out in our lab we observed that, in the course of different plant suspension cell cultivations producing stable foam layers, flotation effects increased, which led to rising cell entrapment in the foam layer and nutrient limitations for these cells during long-term cultivations. A reduction of foaming can be accomplished by the addition of antifoam agents such as silicone-based agents, polypropylene glycol antifoam agents, and/or mechanical foam disruption. It should be noted that in plant suspension cell cultivations an overdose of such antifoam agents may reduce oxygen transfer (Kawase and Moo-Young 1990), and that mechanical foam disruption is theoretically possible: unfortunately, applications with hydrocyclones or mechanical turbines do not exist. At high cell mass or biomass concentrations >30 g dw L^{-1}, a further disadvantage of using bubble columns and airlift bioreactors to cultivate plant cell suspension cultures is insufficient mixing. This results in poor oxygen transfer (air bubbles cannot be dispersed satisfactorily $\rightarrow k_L a$ decreases) and heterogeneous biomass distribution. Therefore, these bioreactor types are not recommended for culturing plant cells at high cell densities, the conventional stirred bioreactor modifications predominating in the literature being more suitable (Doran 1993; Tanaka 2000).

For stirred plant cell bioreactors, the aeration system type and arrangement should be taken into account when choosing the impeller system and its pumping mode. To guarantee mass and temperature homogeneity as well as optimal gas dispersion without causing excessive foaming and shear damage to the plant cell suspension, the sparger of the stirred bioreactor (which usually operates below the flooding point) is located in the flow direction of the impeller. For instance, surface aeration requires the use of an impeller type with downward pumping mode as this allows better oxygen transfer for the increased culture broth viscosities that result from high cell densities. Long-term experience made with mammalian cells has suggested the advantageous combination of upward pumping axial-flow and radial-flow impellers in stirred bioreactors equipped with frits for aeration. We assume that this will be used in plant suspension cell stirred bioreactors in the near future. Until now, impellers considered suitable have included large slow-moving axial flow impellers with low tip speeds of up to a maximum of $2.5\,m\ s^{-1}$, such as marine impellers or special pitched blade impellers, and impellers positioned near the vessel wall, for example, spiral stirrer, helical-ribbon impellers, and anchor impellers. Several descriptions of these impeller types for plant suspension cells can be found in the literature (Eibl and Eibl 2002; Joliceur et al. 1992; Kieran et al. 1997).

In addition, alternative impeller systems have been developed and installed in conventional stirred vessels. Prominent examples are the cell-lift impeller and the centrifugal-pump impeller, which function both as a fluid pump and aerator (Kim et al. 1991; Roberts and Shuler 1997; Wang and Zhong 1996a, 1996b; Zhong et al. 1999). For example, a maximum dry cell mass concentration of $26\,g$ dw L^{-1}, biomass productivity of $0.9\,g$ dw $L^{-1}\ d^{-1}$, and a ginsenoside productivity of $29\,mg\ L^{-1}\ d^{-1}$ were achieved in cultivations of *Panax notoginseng* cells in a centrifugal impeller bioreactor (Zhong 2001). Radial flow impellers, such as the Rushton impeller, achieve high-energy inputs and are therefore suitable for limited applications with plant cell suspensions. In certain cases, the design has been improved, as with concave blades (Eibl and Eibl 2002; Pan et al. 2000; Schiermeyer et al. 2004).

There have been many studies on the development and use of other mechanically driven bioreactors, characterized by their superior performance in terms of suspension homogeneity and low-shear environment. Tanaka et al. (1983), Shibasaki et al. (1992), and Takahashis and Fujita (1991) successfully used a rotating drum bioreactor for the cultivation of suspension cultures of *Catharanthus roseus, Nicotiana tabacum*, and *Lithospermum erythrorhizon*. In comparison to the other bioreactor types described earlier, the rotating drum bioreactor has a significantly larger surface area in relation to volume. Furthermore, in this bioreactor, mass transfer is achieved with comparatively less specific power input and therefore low hydrodynamic shear.

Eibl and Eibl (2006, 2007), Cuperus et al. (2006), and Krüger (2005) have pointed out the excellent biomass growth of tobacco, grape, and apple suspension cells (maximum biomass productivity in the range of $20\text{--}40\,g$ fw $L^{-1}\ d^{-1}$ and doubling times between 2 and 4 days) in the BioWave reactor (Fig. 8.4d) operating with 1 and $10\,L$ culture volume. Furthermore, reduced shearing (indicated by higher viabilities and no significant change in cell morphology) and reduced foaming were found in the disposable BioWave when compared with cultivations in $2\,L$ stirred bioreactors performed in our lab. These observations can be explained by the hydrodynamic characteristics of the cultivation environment and oxygen transport efficiency ensured by the BioWave. The energy input is caused by rocking the platform, which induces a wave in the culture bag. As a consequence of this wave induced motion, oxygenation and mixing are realized. The surface of the medium is continuously renewed and bubble-free surface aeration takes place. Provided batch mode and cultivation parameters in the BioWave and the applied stirred bioreactor were comparable, we found k_La values to be more than twice as high in the BioWave. Assuming a constant rocking angle and culture volume, as the rocking rate is increased, BioWave's power input rises until it reaches a stationary value, which may be followed by a slight decrease resulting from the occurring phase shift of the wave. For a 2 L culture bag containing 1 L culture volume, the phase shift took place at a rocking rate greater than 20 rpm (Chap. 5). The latest disposable bioreactor developments, including the Wave & Undertow bioreactor (WU bioreactor) and the Slug Bubble bioreactor (SB bioreactor), were successfully applied to grow tobacco and soya cells expressing isoflavones and monoclonal antibodies (anti-rabies) up to 100 L culture volume (Girard et al. 2006; Terrier et al. 2006). The WU bioreactor is based on the WIM principle like BioWave, while the SB bioreactor represents a bubble column that allows an easy increase of size by using multiple units.

If illumination is required, particular bioreactors made of glass vessels or plastic bags can be used in illuminated rooms or with external illumination lamps (i.e., fluorescent lamps) installed around them. There are only a few reports on and patents (Takayama and Akita 2006) for internal illumination systems made from tubes (Dubuis 1994) and glass fibers, for example, in respect of the 15 L bioreactor with eccentric motion stirrer and integrated illumination cage (Eibl et al. 1999; Eibl and Eibl 2002). As these systems are costly and fail to introduce light efficiently, they have not been commercialized.

Finally, while the majority of transgenic suspension cells that have been used to express recombinant proteins are ideally cultivated in stirred laboratory bioreactors (Hellwig et al. 2004), photoautotrophic moss suspension cells from *Physcomitrella patens*, which secreted 30 mg L^{-1} d^{-1} human vascular endothelial growth factor (VEGF), are cultivated in a tubular bioreactor. This photobioreactor type is well known from microalga cultivation where light is one of the main factors influencing biomass growth and also scale-up (Decker and Reski 2004; Gorr and Wagner 2005).

8.2.3 Suitable Bioreactors for Hairy Roots

From Sect. 8.1.3.3, it will have become clear to the reader that it is more difficult to cultivate hairy root cultures than plant suspension cells in bioreactors. Without design modifications supporting root inoculation, growth, sampling, and harvest, the stirred bioreactor is generally not suitable for hairy roots despite its dominance in biotechnology. Exceptions reported in the literature are the cultivation of *Catharanthus trichophyllus* L. hairy roots (Davioud et al. 1989), hairy root cultures of *Catharanthus roseus* (Nuutila et al. 1994), strawberry (Nuutila et al. 1997), and beetroot (Georgiev et al. 2006). In general, the impeller can damage the roots even at low specific power inputs. To rule out this possibility, a cage or mesh of stainless

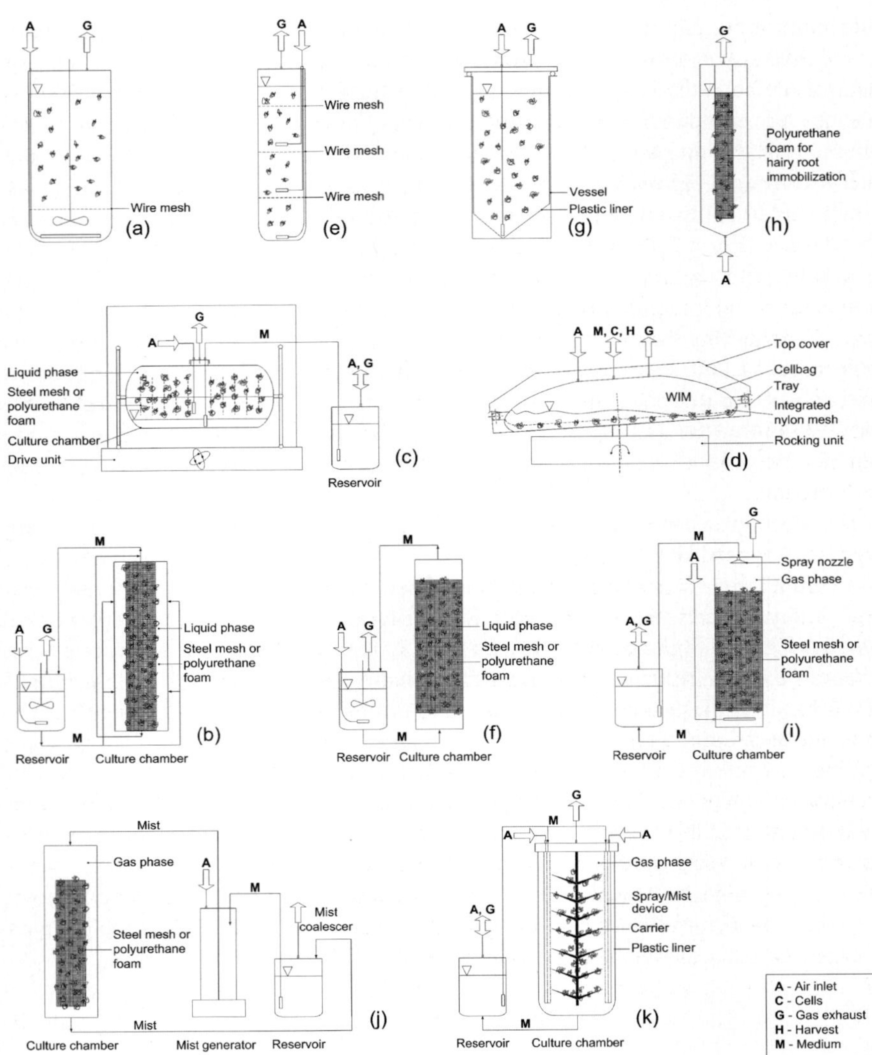

Fig. 8.4 Bioreactors for hairy root cultivation: (**a**) modified stirred bioreactor, (**b**) radial flow bioreactor, (**c**) Inversina, (**d**) BioWave, (**e**) segmented bubble column, (**f**) convective flow bioreactor, (**g**) Plastic-lined bioreactor, (**h**) airlift bioreactor, (**i**) trickle bed or spray bioreactor, (**j**) mist bioreactor, (**k**) LCMB Bioreactor

steel isolating the roots from the impeller was fitted inside the culture vessel of stirred bioreactors. The cage or mesh acts as a matrix and supports the self-immobilization of the roots. This can improve the growth of hairy roots in stirred bioreactors and further bioreactor types as shown by Hilton and Rhodes (1990), Lee et al. (1999), Shadwick and Doran (2004), and also by studies with different hairy root lines in our lab. In gas-phase bioreactors, there is an imperative need to immobilize the hairy roots since the continuous phase is gas. For these reasons, independent of the bioreactor type, immobilization of hairy roots by horizontal or vertical meshes as well as by cages or polyurethane foam meets the current standards (Wilson 1997). Reported productivities obtained from bioreactor cultivations of hairy roots are summarized in Table 8.5 and their schematics are depicted in Fig. 8.4. In addition to the previously mentioned modified stirred bioreactors, hairy roots were also cultivated in rotating and shaking representatives of mechanically driven bioreactors, different types of bubble column bioreactors as well as airlift bioreactors, the hydraulically driven radial flow bioreactor, and gas-phase bioreactors. Considering the bioreactor cultivation of hairy roots from an engineering standpoint, the most significant problem is the nonuniform distribution of biomass and, in particular, of high local root densities in the form of clumps. The latter is closely connected with mass transfer limitations and, in terms of efficient mass transfer (i.e., fluid mixing and oxygen transfer), presents considerable difficulties in bioreactor scaling-up. For laboratory scale, it is recognized that at hairy root concentrations up to $10\,\mathrm{g}$ dw $\mathrm{L^{-1}}$ the cultivation can be realized in essentially any bioreactor type listed in Table 8.5 (Curtis 2000).

Scale-up limitations due to high fluid flow resistance in dense root biomass at concentrations between 10 and $40\,\mathrm{g}$ dw $\mathrm{L^{-1}}$ were detected by Carvalho and Curtis (1998) in liquid-phase bioreactors. Yu et al. (1997) explained the growth-limiting effect of oxygen deficiency for hairy roots due to mass transport limitations under submerged cultivation conditions. Here gas–liquid oxygen transfer is hindered by reduced turbulence as a result of the roots present and also by coalescence of gas bubbles in the root clumps causing gas flow channeling around them. As a result, localized complete depletion of oxygen can arise. It has been suggested that liquid–solid mass transfer rather than gas–liquid mass transfer is the dominant influence on the rate of oxygen delivery to hairy roots. To improve the supply of oxygen to the growing roots in the bioreactor, different pneumatically driven bioreactor types have been developed such as the segmented bubble column operating with three spargers (Fig. 8.4e) and the convective flow bioreactor (Fig. 8.4f). Although the use of these bioreactors leads to higher biomass productivity for the configurations tested, scale-up problems can be expected (Carvalho and Curtis 1998; Kwok and Doran 1995).

The highest biomass productivity ($>2\,\mathrm{g}$ dw $\mathrm{L^{-1}}$ $\mathrm{d^{-1}}$) in the cultivation examples summarized in Table 8.5 was achieved by the radial flow bioreactor (Fig. 8.4b) and the BioWave (Fig. 8.4d). The radial flow bioreactor consists of a medium vessel (in which the culture medium is oxygenated externally), a peristaltic pump (which recirculates the aerated medium between the medium vessel and culture chamber), and a cylindrical culture chamber with an integrated wire mesh for root immobilization. After 12 days of cultivation, $51\,\mathrm{g}$ dw $\mathrm{L^{-1}}$ of beetroot hairy roots was harvested (Kino-Oka et al. 1999). In our lab, the application of the BioWave operating with Cellbag 2L (total volume) at specific power input values between 30 and $50\,\mathrm{Wm^{-3}}$ yielded final dry biomass concentrations of $60\,\mathrm{g}$ dw $\mathrm{L^{-1}}$ $\mathrm{d^{-1}}$ for hyoscyamine ($312\,\mathrm{mg}$ $\mathrm{L^{-1}}$), which produces hairy roots of Egyptian henbane, and about $46.4\,\mathrm{g}$ dw $\mathrm{L^{-1}}$ $\mathrm{d^{-1}}$

Table 8.5 Hairy root growth and metabolite production in suitable bioreactors

Product	Plant species	Bioreactor type	DCM_{final} (g dw L^{-1})	BMP_{dw} (g dw L^{-1} d^{-1})	Maximum metabolite productivity (mg L^{-1})	References
Hyoscyamine	*Datura stramonium*	Stirred bioreactor, Fig. 8.4a	8.4	0.24[#]	66.5	Hilton and Rhodes (1990)
Atropine	*Atropa belladonna*	Stirred bioreactor, Fig. 8.4a	60.2	1.81	325.1	Lee et al. (1999)
Antibody, murine IgG_1	*Nicotiana tabacum*	Stirred bioreactor, Fig. 8.4a	n.m.	n.m.	1.9	Shadwick and Doran (2004)
Atropine	*Atropa belladonna*	Bubble column, Fig. 8.4e	9.9	0.23	14	Kwok and Doran (1995)
Artemisin	*Artemisia annua*	Bubble column (*), feeding	15.3	0.4	0.025	Kim et al. (2002b)
Biomass	*Hyoscyamus muticus*	Bubble column, 8.4f	24.7	0.79	n.m.	Carvalho and Curtis (1998)
	Armoracia rusticana	Airlift bioreactor, Fig. 8.4h	11	0.35	n.m.	Taya et al. (1989b)
	Beta vulgaris	Radial flow bioreactor, Fig. 8.4b	51	3.73	n.m.	Kino-Oka et al. (1999)
	Hyoscyamus muticus	Rotated drum bioreactor, type Inversina, Fig. 8.4c	46	0.48	n.m.	Eibl et al. (1996)
	Hyoscyamus muticus	Plastic-lined bioreactor, Fig. 8.4g	8.8	0.28 (#)	n.m.	Curtis (2004)
	Hyoscyamus muticus	Spray bioreactor, Fig. 8.4i	31	1.1	n.m.	Not published
	Datura stramonium	Bubble column-spray bioreactor*	8 (#)	0.2	n.m.	Wilson (1997)
Hyoscyamine	*Hyoscyamus muticus*	BioWave, Fig. 8.4d, feeding	60	2.1	312	Eibl and Eibl (2006)
Ginsenosides	*Panax ginseng*	BioWave, Fig. 8.4d, feeding/exchange	46.4	0.51	145.6	Eibl and Eibl (2006)
Ginsenosides	*Panax ginseng*	Spray bioreactor Fig. 8.4i, exchange	9.9	0.17	39.2	Palazón et al. (2003)
L-DOPA	*Stizolobium hassjoo*	Mist bioreactor, Fig. 8.4j, feeding	11.5	0.72	644 (#)	Huang et al. (2004)

Nonexisting DCM_{final}: assumed to be 10% of FCM_{final}
n.m. – not measured
[#]BMP was calculated from DCM_{final} as a consequence of unavailable inoculum
*Wire-mesh cage or steel mesh

for ginsenosides (145.6 mg L^{-1}), which contain hairy roots of ginseng (Eibl and Eibl 2006). In this context, it is important to note that cultivations were carried out in feeding mode (200 mL → 500 mL), guaranteeing ebb-and-flow conditions for the growing hairy roots by changing the position of the BioWave rocker unit at constant rocking rate (6 rpm) and constant rocking angle (6°). The mat-like root tissue produced was uniformly distributed in the culture bag, which was completely filled with roots at the end of the cultivation. We are therefore of the opinion that the resulting reduction in mass transfer limitations caused biomass productivity to be ~40% higher than that achieved in the BioWave operating with culture bag 20 L (total volume). The root biomass was highly localized at three points in the culture bag 20 L. As far as root growth is concerned, a key to minimizing mass transfer limitations seems to be uniform distribution of root tissue in the bioreactor. Similar to BioWave, the Inversina (Fig. 8.4c), a rotating drum bioreactor characterized by oloid movement and low hydrodynamic shear, was run with alternate cycles of liquid- and gas phase, that is to say, in ebb-and-flow mode. When additional hairy root immobilization was guaranteed by meshes positioned at three levels in the Inversina bioreactor vessel, a final *Hyoscyamus muticus* dry biomass concentration of 46 g dw L^{-1} was obtained (Eibl et al. 1996).

Given that there are no limitations on the availability of oxygen to roots in a spray bioreactor (Fig. 8.4i), as described by Weathers et al. (1999), and that uniform biomass distribution is guaranteed, the question is why biomass productivity in a spray bioreactor does not exceed that of the BioWave for henbane and ginseng hairy roots. A possible explanation is that the lower biomass growth of both hairy root clones is caused by a nonoptimal spray cycle, which results in inadequate nutrient availability to the roots. It is clear that the availability of culture medium to the roots in spray and also mist bioreactors depends on the amount of droplets deposited on the root surface. This is a function of the spray or mist cycle (continuous/periodic/intermittent) as well as mist rate and the efficiency with which the hairy roots capture the droplets. We could not corroborate the hypothesis of Kim et al. (2002b) and Flores and Curtis (1992) that the importance of spray and mist bioreactors lies more in stimulating specific secondary metabolite production than in biomass production. In addition, the BioWave was superior to the spray reactor in ginsenoside production (Eibl and Eibl 2006).

Henbane and ginseng hairy roots grown in spray bioreactors possessed longer root hairs and were more abundant than those cultivated in the BioWave and the Inversina. These observations are consistent with the findings of Hofer as well as those of Williams and Doran. Doran also describes a much thicker layer of mucilage making mass transfer more difficult for hairy roots grown in liquid phases than for roots grown in gaseous phases (Williams and Doran 1999; Hofer 1996). However, it is also important to note that root hairs presenting a further significant barrier for mass transfer may affect the drainage of medium in gas-phase bioreactors (Shiao and Doran 2000). Ramakrishnan and Curtis (1994) demonstrated that profuse hairy roots entrain considerably more liquid than roots with fewer hairs. Because the number of hairs enhances spray/mist capture but may hinder liquid drainage in spray bioreactors, the mist bioreactor is preferable for the growth of profuse hairy roots (Kim et al. 2000a).

Uniform loading of hairy root matrices in the cultivation chamber and, most notably, achieving maximum use of the chamber are general problems for hairy root culture bioreactors. Whereas small-scale laboratory bioreactors are inoculated as open systems in a laminar flow, larger bioreactors necessitate the use of specific loading procedures and/or pharmaceutical isolator techniques for safe inoculation. For example, a pharmaceutical isolator is an integral part of the 50 L LCMB Bioreactor, a spray bioreactor of plastic film (Fig. 8.4k), which was introduced by Wink et al. (2005) for the production of podophyllotoxin. To allow hairy root attachment to the immobilization matrix, ebb-and-flow mode operation was included. The 500 L Wilson Bioreactor, which was specially designed for the investigation of hairy root growth in large scale, is a hybrid bioreactor (Table 8.5). It was used to cultivate hairy roots of *Datura stramonium* under submerged conditions for 21 days and subsequently in spray mode for 40 days. Besides the root immobilization assembly of wire chains and bars, it is specially constructed to allow mechanical root inoculation after growth of the root inoculum in a seed vessel. A total biomass of 39.8 kg fresh weight corresponding to 79.6 g fw L^{-1} was harvested (Wilson 1997). Although achieved biomass productivity was low it was demonstrated that large-scale cultivation of hairy roots is realizable. In the same way as hairy root loading requires opening the bioreactor, so does the hairy root harvest after the process has finished. In the end, the hairy roots have to be detached from their immobilization matrices.

In summary, it can be stated that optimum bioreactor design for hairy root cultures and their products is a balance between, firstly, meeting the biological needs for efficient root tissue growth without inducing undesirable responses and/or supporting the secondary metabolite or recombinant protein expression in sufficient quantity and, secondly, quality.

8.3 Approaches to Improving Productivity in Plant Cell-Based Bioprocessing

Besides bioreactor type (Sect. 8.2), bioreactor operating mode is another appropriate strategy with which to pursue high biomass productivity and/or high bioactive compound (secondary metabolite, recombinant protein) content in plant cell-based production processes. As a matter of course, it is essential that a highly productive cell line is used, the inoculum density is sufficiently high, and that culture medium and cultivation parameters including pH, temperature, aeration rate, composition of the bioreactor's gas phase (O_2, CO_2, C_2H_4) and, if necessary, light duration and intensity are optimized (Sects. 8.1.1 and 8.1.2). To select the most suitable operating mode, the product formation pattern (growth association or nongrowth association and product secretion or intracellular accumulation) has to be taken into account. Furthermore, the strong dependence of this product formation pattern on the promoter used (constitutive or inducible) in recombinant protein production should be borne in mind (Shadwick and Doran 2004; Su 2006).

Productivity of growth-associated intra and extracellular products can be improved by increasing plant cell growth, prolonging the exponential growth phase, and increasing final biomass concentration in the bioreactor. Therefore, fed batch cultivations operating with intermittent feeding of sucrose or maltose are frequently used to enhance the productivity of biomass and intracellular products in plant cell cultivation processes (Choi et al. 2000; Huang et al. 2004; Wang et al. 2000). However, perfusion performed continuously at high perfusion rates (by cell retention devices such as spinfilters or by cell immobilization) yields considerably higher biomass concentrations and secondary metabolite or protein levels than batch or fed batch operations (Bentebibel et al. 2005). Indeed, this also depends on the efficient and easy removal of undesirable by-products and the resulting prolongation of active cell growth at high metabolic culture activity. However, it should be noted that, if product inhibition occurs, a product harvest unit (e.g., affinity column, adsorption or extraction column) may be coupled to the bioreactor to avoid cell growth rate reduction, cell decease, and product level lowering in growth-associated, extracellular products (Su 2006). For nongrowth associated, intracellular products, advantages can be found in two-stage batch cultivations in which growth phase and production phase have been decoupled by the application of a growth and production medium (Fujita et al. 1981; Raval et al. 2003; Zenk et al. 1977). In contrast, as a consequence of potential product inhibition, cultivations in perfusion mode should be chosen for nongrowth associated, extracellular products (Seki and Furusaki 1999).

It should be borne in mind that the most effective approach to significantly increasing the productivity of secondary metabolites is elicitation (Reuben and Croteau 2004). Elicitation (nontransgenic technique) has commercial potential and can result in metabolite secretion (Sevón 1997). It is defined as induction of gene expression associated with stimulation of secondary metabolite production by agents and means, the so-called elicitors. Elicitors are able to activate plant defense mechanisms, for example, synthesis of phytoalexins or active oxygen species such as H_2O_2 (known as the oxidative burst), in response to pathogenic attacks (Wu and Ge 2004). Biotic elicitors, which are microbe-derived molecules (polysaccharides, glycoproteins, bacterial and fungal cell walls, and low molecular weight compounds), and abiotic elicitors (ultraviolet radiation, ultrasonic pulsed electric fields, salts of heavy metals and various chemicals such as MeJA or JA) were successfully tested (Endress 1994; Lin et al. 2001; Lin and Wu 2002; Ye et al. 2004; Yeoman and Yeoman 1996). Results revealed that effective elicitation requires two-stage cultivation (because elicitors are often added to the culture medium after cell growth is complete), an optimum elicitor dosage and exposure time (Kim et al. 2002a; Rhijwani and Shanks 1998; Sevón 1997; Singh 1997). We would like to point out that in many cases the combination of enhancement techniques (multiple elicitors or combination of elicitation with immobilization, fed batch mode, perfusion mode, etc.) was advantageous not only for the quantity but also for the quality of the desired secondary metabolite (Bentebibel et al. 2005; Qian et al. 2005; Zhang et al. 2002).

Efforts have also been made to increase the accumulation and secretion of recombinant proteins in plant cell cultures by their retention (in situ adsorption) and especially their stabilization (with medium additives such as gelatin, bacitracin,

BSA, PVP, NaCl, DMSO, KNO_3, amino acids, or by osmolarity or pH shift) after protein has been produced (Shadwick and Doran 2004; Su 2006). However, further improvements in recombinant protein yields are necessary to increase the economic competitiveness of plant cell cultures when compared with their mammalian cell counterparts.

8.4 Application Examples and Potential Active Agent Candidates

Whereas maximum protein levels in plant cell cultures reach values of around 4% of total soluble protein (TSP) or 200 mg L^{-1} volumetric productivity (Hellwig et al. 2004), maximum volumetric productivity titers of recombinant proteins secreted by mammalian cells range from 500 to 5,000 mg L^{-1} (Wurm 2005). Nowadays, this means that between 2.5- and 25-fold higher protein levels can be guaranteed if mammalian cells, particularly CHO cells, are used as production organisms instead of plant cells. The first recombinant protein made by cultured tobacco suspension cells was human serum albumin (HSA) (Sijmons et al. 1990). To date, 20 different recombinant proteins have been successfully produced in proof-of-principle studies with plant cell cultures (Hellwig et al. 2004; Hülsing 2005; Kreis et al. 2001; Marshall 2006; Yazaki 2004; Vanisree et al. 2004; Voedisch et al. 2005). An updated summary of such PMPs (plant-made pharmaceuticals = protein-based medicines produced in transgenic plants and plant cells) is given in http://www.molecularfarming. com. In addition to cell lines derived from the tobacco cultivars Bright Yellow 2 (BY-2, doubling time of 13 h according to Nagata et al. 1992) and *Nicotiana tabacum* 1 (NT-1), well-characterized fast-growing transformed rice, alfalfa, tomato, soybean, and moss (*Physcomitrella*) cells are frequently applied to express important therapeutic proteins including antibodies, for example, AAT (for lung diseases), EPO (for anemia), GM-CSF (to stimulate bone marrow to produce white cells more quickly after chemotherapy), HBsAg (for hepatitis B prevention), and VEGF (for tumor diagnostics, and circulatory disorder) (Gorr and Wagner 2005; Ma et al. 2003; Schillberg et al. 2005). The first registration of such a PMP (vaccine against *Newcastle Disease Virus*) was announced from engineered plant cell cultures in February 2006 (Evans 2006). In order that recombinant proteins can also be manufactured by plant cell cultures in the future, cell banking of plant cells including their cryopreservation has to be established as both a master cell bank (MCB) and a working cell bank (WCB), which are prerequisites for GMP production.

While recombinant protein expression is still in its infancy, plant cell-based secondary metabolite production has a tradition of more than 45 years. Maximum secondary metabolite levels in plant cell cultures vary from 10 to 20% of the cell dry weight, which corresponds to productivities between 1 and 2 g L^{-1} d^{-1}. The plant cell-based product with the highest productivity (2 g L^{-1} d^{-1}) reported in the literature is rosmarinic acid (RA), a caffeic acid dimer described as antimicrobial, antiviral, and antioxidative (Parnham and Kesselring 1985). Since RA is most notably used as a food additive and not as a high purity pharmaceutical, it is still commercially produced in the original way (i.e., by extraction from collected plant material), which is comparatively cheaper than the two-stage process with *Coleus blumei* suspension cells in stirred bioreactors (Kreis et al. 2001). It is worthy of note that the first industrial secondary metabolite production using *Lithospermum erythrorhizon* suspension cells was realized by Mitsui Petrochemical Industries, now Mitsui Chemicals, and concerned a naphtoquinone derivate called shikonin, a rich reddish-purple pigment for lipsticks that is also antimicrobial in activity. Under optimized conditions (two-stage mode), shikonin productivity averages out at $0.15\,g\,L^{-1}\,d^{-1}$ (Kreis et al. 2001). The shikonin production process is similar to ginsenoside (triterpenoid saponin) production (Nitto Denko Corporation), which has been routinely carried out in Japan since the 1980s. In the commercial ginsenoside production process, activities of 0.5 up to $0.7\,g\,L^{-1}\,d^{-1}$ were achieved in $25\,m^3$ stirred bioreactors operating with suspended *Panax ginseng* cells (Vanisree et al. 2004). We would like to point out that biotechnologically produced ginsenosides have been used as food additives (i.e., nutritional but not medical use) and whitening substances (cosmetics) since 1988 (Hibino and Ushiyama 1999). A landmark success in secondary metabolite manufacture was the industrial production of paclitaxel. Taxol is the generic name for the active ingredient of paclitaxel, an anti-cancer (ovarian, breast, and small-cell lung cancer) drug stopping cell division while stabilizing microtubule complexes (Wink et al. 2005). Cultivation of yew suspension cells in $70\,m^3$ stirred bioreactors (Phyton Biotech, Germany) ensured paclitaxel amounts exceeded those in intact plants, which constitute 0.01% of dry weight (Yazaki 2004). Further potential anticancer candidates being studied intensively in respect of their suitability for plant cell-based bioprocessing are camptothecin, podophyllotoxin, vincristine, vinblastine, and colchinine (Wink et al. 2005).

8.5 Conclusions

Plant cell-based bioprocessing aims to produce substances with antitumor, antiviral, hypoglycemic, anti-inflammatory, antiparasite, antimicrobial, tranquilizing, and immunmodulating activities, which can be categorized either as secondary metabolites or as recombinant proteins. Although to date only a few secondary metabolites and no PMPs have been manufactured commercially using plant cell cultures, their production in plant cell suspension cultures and hairy roots represents an alternative approach that may be advantageous when defined validated production conditions in compliance with GMP are required. When basic techniques are used to establish highly productive plant cell lines and successful cultivation in small-scale as well as large-scale bioreactors (m^3 range), basic techniques have been found to achieve moderate product yields for most target substances. An improved understanding of the manifold interactions between cultivated cells, product formation and bioreactor design, its features (fluid flow, mixing time, distribution time, oxygen transfer efficiency) affecting mass transfer and therefore product yield, will help not only to enhance and sustain high productivity but also to reduce existing high process costs and exploit the potential of plant cell cultures up to large scale. By using scalable, disposable, and/or low cost bioreactors such as the BioWave, the WU bioreactor, or the SB bioreactor, a further reduction in production costs is possible. Close cooperation of molecular biologists, pharmacists, bioprocess development specialists as well as bioreactor and regulatory guidance specialists will contribute to innovative new solutions. Plant cell-based bioprocessing can therefore be expected to increase in importance in the future.

8.6   Questions and Problems

1. Discuss the main operational parameters for optimal plant cell growth on the basis of plant cell characteristics.
2. Specify the constituents of plant cell culture media and describe their functions in mass propagation, cell differentiation, and metabolite production.
3. Compare the establishment, maintenance, and in vitro growth as well as secondary metabolite and recombinant protein production of callus cultures, plant cell suspension cultures, and hairy root cultures.
4. Explicate and review common methods of cell growth determination in plant cell-based processes and compare them with those of animal cell-based production processes.
5. You have the task of determining growth kinetics, maximum BMP, and doubling time of grape suspension cells growing in 250-mL shake flasks (25 °C, 100 rpm, darkness, batch) over 21 days. Per shake flask 50-mL modified MS medium has to be inoculated with 2.5 g fw of exponentially growing grape cell biomass with viability over 90%. Plan the experiment and discuss the experimental set up and analysis methods.
6. Describe possible signs of shear damage in plant cells cultivated in bioreactors.
7. Give reasons for the predominance of stirred bioreactors for plant cell suspension cultivation at high cell densities.
8. A genetically engineered BY-2 tobacco cell line has been used to produce hemoglobin in BioWave 20 SPS (CultiBag 20 L). Growth in batch culture has been monitored and the results summarized in Table 8.6. Determine maximum BMP and culture doubling time. Assess the growth behavior and discuss relevant approaches to improving it.
9. Hairy roots of *Harpagophytum procumbens* producing harpagosides grow in Petri dishes with doubling times of 7 days. At 25 °C and darkness, the roots are characterized by long hairs and intense hairiness on modified solid B5 medium. Develop and elucidate scale-up procedures for the application of suitable laboratory bioreactors. Decide which bioreactor type is theoretically most suitable for harpagoside expression that has been detected to have occurred due to growth association. In addition, discuss where and under what conditions you would expect mass transfer limitations, which result in low growth and harpagoside production.

Table 8.6 Experimental data

Time (d)	Biomass dry weight (g dw L^{-1})
0	0.02
3	0.05
5	1.47
7	10.2
9	12.1
11	11.9
13	9.6
15	9.55
17	8.35
19	7.2

10. Explain the most important strategies used for the enhancement of secondary metabolite production in plant cell cultures for growth- and nongrowth-associated processes.
11. Give examples of secondary metabolites and recombinant proteins produced by plant cell cultures, and describe their characteristics. What conditions are required for a high productivity process based on plant cell cultures?

List of Abbreviations and Symbols

AAT	alpha-1-antitrypsin
ABA	abscisic acid
Ala	alanine
Arg	arginine
Asn	asparagine
B	boron
BAP	6-benzylaminopurine
BMP	biomass productivity
BSA	bovine serum albumin
BY-2	Bright Yellow 2
CER	carbon dioxide evolution rate
Co	cobalt
CO_2	carbon dioxide
Cu	copper
Cys	cysteine
C_2H_4	ethylene
DCM	dry cell mass (dry biomass concentration)
DMSO	dimethyl sulphoxide
DS	dry substance
d	day
dw	dry weight

                                                                R. Eibl, D. Eibl

EDTA	ethylenediamine tetraacetic acid
EPO	erythropoietin
FCM	fresh cell mass (fresh biomass concentration)
cGMP	current Good Manufacturing Practice
f	final
fw	fresh weight
g	gram
GA_3	gibberellin A_3
GA_4	gibberellin A_4
GA_7	gibberellin A_7
GI	growth index
GM-CSF	granulocyte macrophage colony stimulating factor
His	histidine
h	hour
HBsAg	Hepatitis B surface antigen
HSA	human serum albumin
H_2O_2	hydrogen peroxide
I	iodine
i	initial
IAA	indole-3-acetic acid
IBA	indole-3-butyric acid
Ile	isoleucine
IPC	In-Process control
JA	jasmonic acid
kDa	kilodalton
$k_L a$	oxygen transfer coefficient
klux	kilolux
KNO_3	potassium nitrate
L	Liter
L-DOPA	3,4-dihydroxyphenylalanine
Leu	leucine
MCB	master cell bank
MeJA	methyl jasmonate
Met	methionine
mg	milligram
min	minute
mm	millimeter
Mn	manganese
Mo	molybdenum
MS	medium Murashige and Skoog medium
NAA	naphthaleneacetic acid
NaCl	sodium chloride
NT-1	Nicotiana tabacum 1
OUR	oxygen uptake rate

O_2	oxygen
pCV	packed cell volume
pCO_2	part of dissolved carbon dioxide in the medium
pO_2	part of dissolved oxygen in the medium
PEG	polyethyleneglycol
PMP	plant-made pharmaceutical
PVP	polyvinylpyrrolodine
PVS2	plant vitrification solution 2
r	recombinant
RA	rosmarinic acid
Ri	root-inducing
rpm	revolution per minute
T-DNA	transfer DNA
t	cultivation time
TSP	total soluble protein
tx	point x in time
V	volume
VEGF	vascular endothelial growth factor
vir	virulence
vvm	volumes per volume per minute
W	watt
WCB	working cell bank
Zn	zinc
2,4-D	2,4-dichlorophenoxyacetic acid
2,4,5-T	2,4,5-trichlorophenoxyacetic acid
μm	micrometer
μmole	micromole
°C	degree Centrigrade

Acknowledgement We wish to thank Dr. Kirsi-Marja Oksman-Caldentey for critical reading of the manuscript and her helpful comments.

References

Abdullah M, Ariff A, Marziah M, Ali A, Lajis N (2000) Strategies to overcome foaming and wall growth during the cultivation of *Morinda elliptica* cell suspension culture in a stirred-tank bioreactor. Plant Cell Organ Cult 60:205–212

Bais HP, Suresh B, Raghavarao KSMS, Ravishankar GA (2002) Performance of hairy root cultures of *Cichorium intybus L.* in bioreactors of different configurations. In Vitro Plant 38:573–580

Bates TR, Lynch JP (1996) Stimulation of hairy root elongation in *Arabidopsis thaliana* by low phosphorous availability. Plant Cell Environ 19:529–538

Benson EE, Lynch PT, Stacey GN (1998) Advances in plant cryopreservation technology: Current applications in crop plant biotechnology. AgBiotech News Inf 5:133N–142N

Bentebibel S (2003) Estudio de la produccion de taxanos por cultivos de células en suspensio e immovilizadas de *Taxus baccata*. PhD Thesis, University of Barcelona

Bentebibel S, Moyano E, Palazón J, Cusidó RM, Bonfill M, Eibl R, Pinol MT (2005) Effects of immobilization by entrapment in alginate and scale-up on paclitaxel and baccatin III production in cell suspension cultures of *Taxus baccata*. Biotechnol Bioeng 89:647–655

Bergmann L (1967) Wachstum grüner Suspensionskulturen von *Nicotiana tabacum* var. "Samsun" mit CO_2 als Kohlenstoffquelle. Planta 74:243–249

Bhojwani SS, Razdan MK (1983) Developments in Group Science 5. Elsevier, Amsterdam

Bidney D, Scelonge C, Martich J, Burrus M, Sims L, Huffmann G (1992) Microprojectile bombardment of plant tissues increases transformation frequency by *Agrobacterium tumefaciens*. Plant Mol Biol 18:301–313

Bordonaro JL, Curtis WR (2000) Inhibitory role of root hairs on transport within root culture bioreactors. Biotechnol Bioeng 70:176–86

Calderón AA, Zapata JM, Ros Barceló A (1994) Differential expression of a cell wall-localized peroxidase isoenzyme capable of oxidizing 4-hydroxystilbenes during the cell culture of grapevine (*Vitis vinifera cv*. Airen and Monastrell). Plant Cell Tissue Organ Cult 37:121–127

Carvalho EB, Holihan S, Pearsall B, Curtis WR (1997) Effect of root morphology on reactor design and operation for production of chemicals. In: Doran PM (ed) Hairy Roots: Culture and Application. Harwood Academic, Amsterdam, pp 151–167

Carvalho EB, Curtis WR (1998) Characterization of fluid-flow resistance in root cultures with a convective flow tubular bioreactor. Biotechnol Bioeng 60:375–384

Chattopadhyay S, Mehra RS, Srivastava AK, Bhojwani SS, Bisaria VS (2003) Effect of major nutrients on podophyllotoxin production in *Podophyllum hexandrum* suspension cultures. Appl Microbiol Biotechnol 60:541–546

Chattopadhyay S, Farkya S, Srivastava AK, Bisaria VS (2002) Bioprocess considerations for production of secondary metabolites by plant cell suspension cultures. Biotechnol Bioproc Eng 7:138–149

Chilton MD, Tepfer D, Petit A, David C, Casse-Delbart F, Tempe J (1982) *Agrobacterium rhizogenes* inserts T-DNA into the genomes of host plant cells. Nature 295:432–434

Choi HK, Kim SI, Son JS, Hong SS, Lee HS, Chung IS, Lee HJ (2000) Intermittent maltose feeding enhances paclitaxel production in suspension culture of *Taxus chinensis* cells. Biotechnol Lett 22:1793–1796

Cuperus S, Rouvinez V, Eibl R, Kneubühl M, Müller J, Hühn T, Amadó R (2006) Potential of the BioWave reactor for plant cell-based bioprocesses: *V. vinifera* cell suspension culture and their metabolites as a case study. In: Sorvari S, Toldi O (eds) Proceedings for Second International Conference on Bioreactor Technology in Cell, Tissue Culture and Biomedical Applications, Saariselkä, Lapland, 27–31 March 2006

Curtis WR (2004) Growing cells in a reservoir formed of a flexible sterile plastic liner. United States Patent, 6,709,862 B2

Curtis WR, Emery A (1993) Plant cell suspension culture rheology. Biotechnol Bioeng 42:520–526

Curtis WR (2000) Hairy roots, bioreactor growth. In: Spier RE (ed). Encyclopedia of Cell Technology, vol 2. Wiley VCH, New York, pp 827–841

Davioud E, Kan C, Hamon J, Tempé J, Husson HP (1989) Production of indole alkaloids by in vitro root cultures from *Catharanthus trichophyllus*. Phytochem 28:2675–2680

Decker EL, Reski R (2004) The moss bioreactor. Curr Opin Plant Biol 7:166–170

Deus-Neumann B, Zenk HM (1984) Instability of indole alkaloid production in *Catharanthus roseus* cells in suspension cultures. Planta Med 50:427–431

DiIorio AA, Cheetham RD, Weathers PJ (1992) Carbon dioxide improves the growth of hairy roots cultured on solid medium and in nutrient mists. Appl Microbiol Biotechnol 37:463–467

Do CB, Cormier F (1991) Effects of low nitrate and high sugar concentrations on anthocyanin content and composition of grape (*Vitis vinifera*) cell suspensions. Plant Cell Rep 9:500–504

Doran PM (2002) Properties and applications of hairy root cultures. In: Oksman-Caldentey KM, Barz WH (eds) Plant Biotechnology and Transgenic Plants. Marcel Dekker, New York, pp 143–161

Doran PM (1993) Design of reactors for plant cells and organs. Adv Biochem Eng 48:115–168

Dubuis D (1994) Design and scale-up of bubble-free fluidised bed reactors for plant cell cultures. PhD Thesis, Swiss Federal Institute of Technology, Zurich

Ducos JP, Lambot C, Pètiard V (2007) Bioreactors for coffee mass propagation by somatic embryogenesis. Int J Dev Biol 1:1–12

Eibl R, Eibl D (2002) Bioreactors for plant cell and tissue cultures. In: Oksman-Caldentey KM, Barz WH (eds) Plant Biotechnology and Transgenic Plants. Marcel Dekker, New York, pp 163–199

Eibl R, Hans D, Warlies S, Lettenbauer C, Eibl D (1999) Einsatz eines Taumelreaktorsystems mit interner Beleuchtung. BioWorld 2:10–12

Eibl R, Eibl D (2006) Design and use of the Wave Bioreactor for plant cell culture. In: Dutta Gupta S, Ibaraki Y (eds) Plant Tissue Culture Engineering, Series: Focus on Biotechnology, vol 6. Springer, Dordrecht, pp 203–227

Eibl R, Eibl D (2008) Design of bioreactors suitable for plant cell and tissue cultures. Phytochem Rev, DOI 10.1007/s11101-007-9083-z

Eibl R, Hans D, Eibl D (1996) Vergleichende Untersuchungen zur Kultivierung pflanzlicher Zellen in Laborbioreaktoren. In: iba (ed) Proceedings zum 8. Heiligenstädter Kolloquium "Technische Systeme für Biotechnologie und Umwelt", Heiligenstadt, Germany, pp 100–105

Endress R (1994) Plant Cell Biotechnology. Springer, Berlin, Heidelberg New York

Evans DA, Sharp WR, Ammirato PV, Yamada Y (1983) Handbook of Plant Cell Culture, vol 1. Macmillan, New York

Evans D, Coleman J, Kearns A (2003) Plant Cell Culture. BIOS Scientific, Abingdon

Evans J (2006) Plant-Derived Drug. http://www.rsc.org/chemistryworld/News/2006/February/07020602.asp. Cited 10 Apr 2007

Fischer U, Santore UW, Hüsemann W, Barz W, Alfermann AW (1995) Semicontinuous cultivation of photoautotrophic cell suspension cultures in a 20 L airlift-reactor. Plant Cell Tissue Organ Cult 38:123–134

Fischer R, Stoger E, Schillberg S, Christou P, Twyman RM (2004) Plant-based production of biopharmaceuticals. Curr Opin Plant Biol 7:152–158

Flores HE (1987) Use of plant cells and organ culture in the production of biological chemicals. In: LeBaron HM, Mumma RO, Honeycutt RC, Duesing JH (eds) Biotechnology in Agricultural Chemistry. ACS Symp Ser 334, Washington, pp 66–86

Flores HE, Curtis WR (1992) Approaches to understanding and manipulating the biosynthetic potential of plant roots. Ann NY Acad Sci 665:188–209

Fowler MW (1988) Problems in commercial exploitation of plant cell cultures. In: Bock B, Marsh J (eds) Application of Plant Cell and Tissue Culture. Ciba Foundation Symposium 137. Wiley, Chichester, pp 239–253

Fujita H, Hara Y, Suga C, Morimoto T (1981) Production of shikonin derivates by cell suspension cultures of *Lithospermum erythrorhizon* II: A new medium for the production of shikonin derivates. Plant Cell Rep 1:61–63

Gamborg OL, Miller RA, Ojimak K (1968) Nutrient requirements of suspension cultures of soybean root cells. Exp Cell Res 50:151–158

Georgiev M, Pavlov A, Bley T (2006) Betalains by transformed *Beta vulgaris* roots in stirred tank bioreactor: batch and fed-batch process. In: FinMed (ed) Proceedings for Second International Conference on Bioreactor Technology in Cell, Tissue Culture and Biomedical Applications, Saariselkä, Lapland, p 35

Girard LS, Fabis MJ, Bastin M, Courtois D, Pétiard V, Koprowski H (2006) Expression of a human anti-rabies virus monoclonal antibody in tobacco cell culture. BBRC 345:602–607

Glicklis R, Mills D, Sitton D, Stortelder W, Merchuk JC (1998) Polysaccharide production by plant cells in suspension: Experiments and mathematical modelling. Biotechnol Bioeng 57:732–740

Gorr G, Wagner S (2005) Humanized glycosylation: Production of biopharmaceuticals in a moss bioreactor. In: Knäblein J (ed) Modern Biopharmaceuticals, vol 3. Wiley VCH, Weinheim, pp 919–929

Guardiola J, Iborra JL, Cánovas M (1994) A model that links growth and secondary metabolite production in plant cell suspension cultures. Biotechnol Bioeng 46:291–297

Hall R, Yeoman M (1987) Intercellular and intercultural heterogeneity in secondary metabolite accumulation in cultures of *Catharanthus roseus* following cell line selection. J Exp Bot 38:1391–1398

Hamill JD, Lidgett AJ (1997) Hairy root cultures: Opportunities and key protocols for studies in metabolic engineering. In: Doran PM (ed) Hairy Roots: Culture and Applications. Harwood, Amsterdam, pp 1–29

Hamill JD, Parr AJ, Robins RJ, Rhodes MJC (1986) Secondary product formation by cultures of *Beta vulgaris* and *Nicotiana rustica* transformed with *Agrobacterium rhizogenes*. Plant Cell Rep 5:111–114

Han B, Linden JC, Gujarathi NP, Wickramasinghe SR (2004) Population balance approach to modelling hairy root growth. Biotechnol Prog 20:872–879

Hansen G, Chilton MD (1996) "Agrolistic" transformation of plant cells: Integration of T-strands generated in planta. Proc Natl Acad SCI USA 94:7469–7474

Hellwig S, Drossard J, Twyman RM, Fischer R (2004) Plant cell cultures for the production of recombinant proteins. Nature Biotech 22:1415–1422

Hess D (1992) Biotechnologie der Pflanzen. Eugen Ulmer, Stuttgart

Hibino K, Ushiyama K (1999) Commercial production of ginseng by plant tissue culture technology. In: Fu TJ, Curtis WR (eds) Plant Cell and Tissue Culture for the Production of Food Ingredients. Kluwer, New York, pp 215–224

Hilton MG, Wilson PDG (1995) Growth and uptake of sucrose and mineral ions by transformed root cultures of *Datura stramonium, Datura candida x aurea, Datura wrightii, Hyoscyamus muticus* and *Atropa belladonna*. Planta Med 61:345–350

Hilton MG, Rhodes MJC (1990) Growth and hyoscyamine production of hairy root cultures of *Datura stramonium* in a modified stirred tank reactor. Appl Microbiol Biotechnol 33:132–138

Hjortso MA (1997) Mathematical modelling of hairy root growth. In: Doran PM (ed) Hairy Roots: Culture and Applications. Harwood, Amsterdam, pp 169–178

Hofer R (1996) Root hairs. In: Waisel Y, Eshel A, Kafkafi U (eds) Plant Roots: The Hidden Half. Marcel Dekker, New York, pp 111–126

Huang SY, Hung CH, Chou SN (2004) Innovative strategies for operation of mist trickling reactors for enhanced hairy root proliferation and secondary metabolite productivity. Enzyme Microb Technol 35:22–32

Hu ZB, Du M (2006) Hairy root and its application in plant genetic engineering. J Integr Plant Biol 48:121–127

Hülsing B (2005) Biopharmazeutika aus transgenen Pflanzen. BioWorld 3:10–12

James E, Lee JM (2001) The production of foreign proteins from genetically modified plant cells. In: Tscheper T (ed) Advances in biochemical engineering: Plant cells. vol 72. Springer Berlin Heidelberg, pp 127–156

Joliceur M, Chavarie C, Carreau PJ, Archambault J (1992) Development of a helical-ribbon impeller bioreactor for high-density plant cell suspension culture. Biotechnol Bioeng 39:511–521

Kartha KK (1987) Advances in the cryopreservation technology of plant cells and organs. In: Green CE, Somers DA, Hackett WP, Biesboer DD (eds) Plant Tissue and Cell Culture. Plant Biology 3. Liss, New York, pp 447–458

Kawase Y, Moo-Young M (1990) The effect of antifoam agents on mass transfer in bioreactors. BioProcess Eng 5:169–173

Kieran PM, MacLoughlin PF, Malone DM (1997) Plant cell suspension cultures: Some engineering considerations. J Biotechnol 59:39–52

Kieran PM, Malone DM, MacLoughlin PF (2000) Effects of hydrodynamic and interfacial forces on plant cell suspension systems. In: Tscheper T, Schügerl K, Kretzmer G (eds) Advances in Biochemical Engineering Biotechnology: Influence of Stress on Cell Growth and Product Formation, vol 67. Springer, Berlin Heidelberg New York, pp 139–177

Kim DI, Cho GH, Pedersen H, Chin CK (1991) A hybrid reactor for high density cultivation of plant cells. Appl Microbiol Biotechnol 34:726–729

Kim S, Hopper E, Hjortso MA (1995) Hairy root growth models: Effect of different branching patterns. Biotechnol Prog 11:178–186

Kim Y, Wyslouzil BE, Weathers PJ (2002a) Secondary metabolism of hairy root cultures in bioreactors. In Vitro Cell Dev Biol Plant 38:1–10

Kim Y, Wyslouzil BE, Weather PJ (2002b) Growth of *Artemisia annua* hairy roots in liquid- and gas-phase reactors. Biotechnol Bioeng 80:454–464

Kino-Oka M, Hitaka Y, Taya M, Tone S (1999) High-density culture of red beet hairy roots by considering medium flow conditions in a bioreactor. Chem Eng Sci 54:3179–3186

Kobayashi T, Niino T, Kobayashi M (2006) Cryopreservation of tobacco BY-2 suspension cell cultures by vitrification with encapsulation. Plant Biotech 23:333–337

Komari T, Ishida Y, Hiei Y (2004) Plant transformation technology: Agrobacterium-mediated transformation. In: Christou P, Klee H (eds) Handbook of Plant Biotechnology, vol 1. Wiley, Chichester, pp 233–261

Kreis W, Baron D, Stoll G (2001) Biotechnologie der Arzneistoffe. Deutscher Apotheker Verlag, Stuttgart

Krüger A (2006) Untersuchungen zur in vitro Kultivierung von Zell- und Gewebekulturen des *Malus domestica*. Diploma Thesis, University of Applied Sciences Köthen/Anhalt

Kwok KH, Doran PM (1995) Kinetic and stoichiometric analysis of hairy roots in a segmented bubble column reactor. Biotechnol Prog 11:429–435

Lee KT, Suzuki T, Yamakawa T, Kodama T, Igarashi Y, Shimomura K (1999) Production of tropane alkaloids by transformed root cultures of *Atropa belladonna* in stirred bioreactors with stainless steel net. Plant Cell Rep 18:567–571

Lin LD, Wu JY, Ho KP, Qi SY (2001) Ultrasonic-induced physiological effects and secondary metabolite (saponin) production in *Panax ginseng* cell cultures. Ultrasound Med Biol 27:1147–1152

Lin LD, Wu JY (2002) Enhancement of shikonin production in single- and two-phase suspension cultures of *Lithospermum erythrorhizon* cells using low-energy ultrasound. Biotechnol Bioeng 78:81–88

Ma JKC, Drake PMW, Christou P (2003) The production of recombinant pharmaceutical proteins in plants. Nat Rev Genet 4:794–805

Madhusudhan R Ramachandra Rao S, Ravishankar GA (1995) Osmolarity as a measure of growth of plant cells in suspension cultures. Enzyme Microb Technol 17:989–991

Mantell SH, Smith H (1983) Plant Biotechnology. Cambridge University Press, Cambridge

Marshall B (2006) MolecularFarming.com. http://www.molecularfarming.com/PMPs-and-PMIPs.html. Cited 10 Apr 2007

Menges M, Murray JAH (2004) Cryopreservation of transformed and wild-type *Arabidopsis* and tobacco cell suspension cultures. Plant J 37:635–644

Mirjalili N, Linden JC (1995) Gas phase composition effects on suspension cultures of *Taxus cuspidata*. Biotechnol Bioeng 48:123–132

Misawa M (1994) Plant tissue culture: An alternative for production of useful metabolites. FAO Agricultural Services Bulletin N 108. Food and Agriculture Organization of the United Nations, Rome, p 95

Moldenhauer JR (2003) Cell culture preservation and storage for industrial bioprocesses. In: Vinci VA, Parekh SR (eds) Handbook of Industrial Cell Culture: Mammalian, Microbial and Plant Cells. Humana Press, Totowa, New Jersey, pp 498–501

Murashige T, Skoog F (1962) A revised medium for rapid growth and bioassays with tobacco tissue cultures. Physiol Plant 15:473–479

Murashige T (1973) Nutrition of plant cells and organs in vitro. In Vitro 9:81–85

Nagata T, Nemoto Y, Hasezawa S (1992) Tobacco BY-2 cell line as the "HeLa" cell in the cell biology of higher plants. Int Rev Cytol 132:1–30

Naill MC and Roberts SC (2005) Cell cycle analysis of *Taxus* suspension cultures at the single cell level as an indicator of culture heterogeneity. Biotechnol Bioeng 90:491–500

Nettleship L, Slaytor M (1974) Adaption of *Peganum harmala* callus to alkaloid production. J Exp Bot 25:1114–1123

Nuutila AM, Toivonen L, Kauppinen V (1994) Bioreactor studies on hairy root cultures of *Catharanthus roseus*: Comparison of three bioreactor types. Biotechnol Techniques 8:61–66

Nuutila AM, Linquist AS, Kauppinen V (1997) Growth of hairy root cultures of strawberry (*Fragaria ¥ ananassa Duch.*) in three different types of bioreactors. Biotechnol Techniques 6:363–366

Oksman-Caldentey KM, Sevón KM, Vanhala L, Hiltunen R (1994) Effect of nitrogen and sucrose on the primary and secondary metabolism of transformed root cultures of *Hyoscyamus muticus*. Plant Cell Tissue Org Cult 38:263–272

Oksman-Caldentey KM, Hiltunen R (1996) Transgenic crops for improved pharmaceutical products. Field Crops Res 45:57–69

Paek KY, Chakrabarty D, Hahn EJ (2005) Application of bioreactor systems for large scale production of horticultural and medicinal plants. Plant Cell Tiss Org Cult 81:287–300

Palazón J, Mallol A, Eibl R, Lettenbauer C, Cusidó RM, Piñol MT (2003) Growth and ginsenoside production in hairy root cultures of *Panax ginseng* using a novel bioreactor. Planta Med 69:344–349

Pan ZW, Wang HQ, Zhong JJ (2000) Scale-up study on suspension cultures of *Taxus chinensis* cells for production of taxane diterpene. Enzyme Microb Tech 27:714–723

Parnham MJ, Kesselring K (1985) Rosmarinic acid. Drugs Future 10:756–757

Petersen M, Dombrowski K, Gertlowski C, Haeusler E, Kawatzki B, Meinhard J, Alfermann AW (1992) The use of plant cell cultures to study natural product biosynthesis. In: Oono K, Hirabayashi T, Kikuchi S, Handa H, Kajiwara K (eds) Plant Tissue Culture and Gene Manipulation for Breeding and Formation of Phytochemicals. NAIR, Tsukuba, pp 293–296

Präve P, Faust U, Sittig W, Sukatsch DA (1994) Handbuch der Biotechnologie. Oldenbourg Verlag, München, p 189

Qian ZG, Zhao ZJ, Xu Y, Qian X, Zhong JJ (2005) Highly efficient strategy for enhancing taxoid production by repeated elicitation with a newly synthesized jasmonate in fed-batch cultivation of *Taxus chinensis* cells. Biotechnol Bioeng 90:516–521

Ramakrishnan D, Curtis WR (1994) Fluid dynamic studies on plant root cultures for application to bioreactor design. In: Ryu DD, Furusaki S (eds) Studies in Plant Science, 4. Advances in Plant Biotechnology: Production of Secondary Metabolites. Elsevier, Amsterdam, pp 281–305

Rasmussen JL, Kikkert JR, Roy MK, Sanford JC (1994) Biolistic transformation of tobacco and maize suspension cells using bacterial cells as microprojectiles. Plant Cell Rep 13:212–217

Raval KN, Hellwig S, Prakash G, Ramos-Plascencia A, Srivastava A, Büchs J (2003) Necessity of two-stage process for the production of Azadirachtin-related limonoids in suspension cultures of *Azadirachata indica*. J Biosc Bioeng 96:16–22

Reinert J, Yeoman MM (1982) Plant Cell and Tissue Culture: A Laboratory Manual. Springer, Berlin Heidelberg New York

Reuben JP, Croteau RB (2004) Metabolic Engineering of plant secondary metabolism. In: Christou P, Klee H (eds) Handbook of Plant Biotechnology, vol 1. Wiley, New York, pp 609–627

Rhijwani SK, Shanks JV (1998) Effect of elicitor dosage and exposure time on biosynthesis of indole alkaloids by *Catharanthus roseus* hairy root cultures. Biotechnol Prog 14:442–449

Rhodes MJC, Parr AJ, Giulietti A, Aird ELH (1994) Influence of exogenous hormones on the growth and secondary metabolite formation in transformed root cultures. Plant Cell Tissue Organ Cult 38:143–151

Roberts SG, Shuler ML (1997) Large-scale plant cell culture. Curr Opin Biotech 8:154–159

Ryu DDY, Lee SO (1990) Determination of growth rate for plant cell cultures: Comparative studies. Biotechnol Bioeng 35:305–311

Saito K, Mizukami H (2002) Plant cell cultures as producers of secondary metabolites. In: Oksman-Caldentey KM, Barz WH (eds) Plant Biotechnology and Transgenic Plants. Marcel Dekker, New York, pp 77–109

Schiermeyer A, Dorfmüller S, Schinkel H (2004) Production of pharmaceutical proteins in plants and plant cell suspension cultures. In: Fischer R, Schillberg S (eds) Molecular Farming. Wiley-VCH, Weinheim, pp 91–112

Schillberg S, Twyman RM (2004) Emerging production systems for antibodies in plants. In: Christou P, Klee H (eds) Handbook of Plant Biotechnology, vol 2. Wiley, Chichester, pp 801–810

Schillberg S, Twyman RM, Fischer R (2005) Biopharmaceutical production in cultured plant cells. In: Knäblein J (ed) Modern biopharmaceuticals, vol 3. Wiley VCH, Weinheim, pp 949–965

Schumacher HM, Malik KA, van Iren F (1994) Simple storage of plant cell cultures in liquid media. Publication No. 13, UNESCO/WFCC Technical Information Sheet:1–4

Seki M, Furusaki S (1999) Medium recycling as an operational strategy to increase plant secondary metabolite formation. In: Fu TJ, Curtis WR (eds) Plant cell and tissue culture for the production of food ingredients. Kluwer, New York, pp 157–163

Seitz U (1987) Cryopreservation of plant cell cultures. Planta Med 53:311–314

Sevón N (1997) Tropan alkaloids in hairy roots and regenerated plants of *Hyoscyamus muticus*. PhD Thesis, University of Helsinki

Sevón N, Oksman-Caldentey KM (2002) *Agrobacterium rhizogenes*-mediated transformation: Root cultures as a source of alkaloids. Planta Med 68:859–868

Shadwick FS, Doran PM (2004) Foreign protein expression using plant cell suspension and hairy root cultures. In: Fischer R, Schillberg S (eds) Molecular Farming. Wiley-VCH, Weinheim, pp 13–36

Sharp JM, Doran PM (2001) Strategies for enhancing monoclonal antibody accumulation in plant cell and organ cultures. Biotechnol Prog 17:979–992

Shiao TL, Doran PM (2000) Root hairiness: Effect on fluid flow and oxygen transfer in hairy root cultures. J Biotechnol 83:199–210

Shibasaki N, Hirose K, Yonemoto T, Takadi T (1992) Suspension culture of *Nicotiana tabacum* cells in a rotary-drum bioreactor. J Chem Technol Biotechnol 53:359–363

Sijmons PC, Dekker BMM, Schrammeijer B, Verwoerd TC, van den Elzen PJM, Hoekema A (1990) Production of correctly processed human serum albumin in transgenic plants. Bio/Technology 8:217–221.

Singh G, Curtis WR (1994) Reactor design for plant root culture. In: Shargool PD, Ngo TT (eds) Biotechnological Applications Plant Cultures. CRC Press. Boca Raton, FL, pp 185–206

Singh G (1997) Elicitation-manipulating and enhancing secondary metabolite production. In: Fu TJ, Curtis WR (eds) Plant cell and tissue culture for the production of food ingredients. Kluwer, New York, pp 101–127

Steward N, Martin R, Engasser JM, Goergen JL (1999) Determination of growth and lysis kinetics in plant cell suspension cultures from the measurement of esterase release. Biotechnol Bioeng 66:114–121

Storhas W (1994) Bioreaktoren und periphere Einrichtungen. Vieweg, Braunschweig/Wiesbaden

Su WW (2006) Bioreactor engineering for recombinant protein production using plant cell suspension culture. In: Dutta Gupta S, Ibaraki Y (eds) Plant Tissue Culture Engineering, Series: Focus on Biotechnology, vol 6. Springer, Dordrecht, pp 135–159

Su WW (1995) In situ filtration of *Anchusa officinalis* culture in a cell-retention stirred tank bioreactor. Biotechnol Techniques 9:259–264

Takahashis S, Fujita Y (1991) Production of shikonin. In: Komamine A, Misawa M, DiCosmo F (eds) Plant Cell Culture in Japan. CMC, Tokyo, pp 72–78

Takayama S, Akita M (2006) Bioengineering aspects of bioreactor application in plant propagation. In: Dutta Gupta S, Ibaraki Y (eds) Plant Tissue Culture Engineering, Series: Focus on Biotechnology, vol 6. Springer, Dordrecht, pp 83–100

Takebe J, Labib G, Melchers G (1971) Regeneration of whole plants from isolated mesophyll protoplasts of tobacco. Naturwiss 5:318–320

Tanaka H, Uemura M, Kaneko Y, Aoyagi H (1993) Estimation of cell biomass in plant cell suspensions by the osmotic pressure measurement of culture broth. J Ferment Bioeng 76:501–504

Tanaka H (2000) Technological problems in cultivation of plant cells at high density. Biotechnol Bioeng 67:1203–1217

Tanaka H, Nishijima F, Suwa M, Iwamoto T (1983) Rotating drum fermentor for plant cell suspension cultures. Biotechnol Bioeng 25:2359–2370

Taticek RA, Lee CWT, Shuler ML (1994) Large-scale insect and plant cell culture. Curr Opin Biotechnol 5:165–174

Taya M, Hegglin M, Prenosil JE, Bourne JR (1989a) On-line monitoring of cell growth in plant tissue cultures by conductometry. Enzyme Microb Technol 11:170–176

Taya M, Yoyoma A, Kondo O, Kobayashi T, Matsui C (1989b) Growth characteristics of plant hairy roots and their cultures in bioreactors. J Chem Eng Jpn 22:84–89

Taya M, Kino-Oka M, Tone S (1989c) A kinetic model of branching growth of plant hairy root. J Chem Eng Jpn 22:698–670

Terrier B, Courtois D, Hénault N, Cuvier A, Bastin M, Aknin A, Dubreuil J, Pétiard V (2006) Two new disposable bioreactors for plant cell culture: The wave & undertow bioreactor and the slug bubble bioreactor. Biotechnol Bioeng, Early View: 1–25

Than NT, Murthy HN, Yu KW, Jeong CS, Hahn EJ, Paek KY (2004) Effect of inoculum size on biomass accumulation and ginsenoside production by large-scale cell suspension cultures of *Panax ginseng*. J Plant Biotechnol 6:265–268

Towler MJ, Kim Y, Wyslouzil BE, Correll MJ, Weathers PJ (2006) Design, development and applications of mist bioreactors for micropropagation and hairy root culture. In: Dutta Gupta S, Ibaraki Y (eds) Plant Tissue Culture Engineering, Series: Focus on Biotechnology, vol 6. Springer, Dordrecht, pp 119–134

Twyman RM, Christou P (2004) Plant transformation technology: Particle bombardment. In: Christou P, Klee H (eds) Handbook of Plant Biotechnology, vol 1. Wiley, Chichester, pp 263–289

Val J (1993) Modelling the physiology of plant cells in suspension culture. PhD Thesis, Leiden University

Valluri JV, Treat WJ, Soltes E (1991) Bioreactor culture of heterotrophic sandalwood (*Santalum album* L.) cell suspensions utilizing a cell-lift impeller. Plant Cell Rep 10:366–370

Vanisree M, Lee CY, Lo SF, Nalawade SM, Lin CY, Tsay HS (2004) Studies on the production of some important secondary metabolites from medicinal plants by plant tissue cultures. Bot Bull Acad Sin 45:1–22

Vervliet G, Holsters M, Teuchy H, van Montagu M, Schell J (1975) Characterization of different plaqueforming and defective temperature phages in Agrobacterium strains. J Gen Virol 26:33–48

Voedisch B, Menzel C, Jordan E, El-Ghezal A, Schirrmann T, Hust M, Jostock T (2005) Heterologe Expression von rekombinanten Proteinpharmazeutika. Laborwelt 6:26–31

Wang HQ, Yu JT, Zhong JJ (2000) Significant improvement of taxane production in suspension cultures of *Taxus chinensis* by sucrose feeding strategy. Process Biochem 35:479–483

Wang SJ, Zhong JJ (1996a) A novel centrifugal impeller bioreactor I. Fluid circulation, mixing, and liquid velocity profiles. Biotechnol Bioeng 51:511–519

Wang SJ, Zhong JJ (1996b) A novel centrifugal impeller bioreactor II. Oxygen transfer and power consumption. Biotechnol Bioeng 51:511–519

Weathers PJ, Zobel RD (1992) Aeroponics for the culture of organisms, tissues and cells. Biotechnol Adv 10:93–115

Weathers PJ, Wyslouzil BE, Whipple M (1997) Laboratory-scale studies of nutrient mist reactors for culturing hairy roots. In: Doran PM (ed) Hairy Roots: Culture and Applications. Harwood Academic, Amsterdam, pp 191–199

Weathers PJ, Wyslouzil BE, Wobbe KK, Kim YJ, Yigit E (1999) The biological response of hair roots to O_2 levels in bioreactors. In Vitro Cell Dev Biol Plant 35:286–289

Whitney P (1992) Novel bioreactors for the growth of roots transformed by *Agrobacterium rhizogenes*. Enzyme Microb Technol 14:13–17

Widholm J (1972) The use of fluorescein diacetate and phenosafranine for determining viability of cultured plant cells. Stain Technol 47:189–194

Wilhelmson A, Häkkinen ST, Kallio PT, Oksman-Caldentey KM, Nuutila AM (2006) Heterologous expression of *Vitreoscilla* haemoglobin (VHb) and cultivation conditions affect the alkaloid profile of *Hyoscyamus muticus* hairy roots. Biotech Prog 22:350–358

Williams GRC, Doran PM (1999) Investigations of liquid–solid hydrodynamic boundary layers and oxygen requirements of hairy root cultures. Biotechnol Bioeng 64:729–740

Wilson PDG (1997) The pilot-scale cultivation of transformed roots. In: Doran PM (ed) Hairy Roots: Culture and Applications. Harwood Academic, Amsterdam, pp 179–190

Wink M, Alfermann AW, Franke R, Wetterauer B, Distl M, Windhoevel J, Krohn O, Fuss E, Garden H, Mohagheghzadeh A, Wildi E, Ripplinger P (2005) Sustainable production of phytochemicals by plant in vitro cultures: anticancer agents. Plant Gene Res 3:90–100

Withers LA (1991) Maintenance of plant tissue cultures. In: Kirsop BE, Doyle A (eds) Maintenance of Microorganisms, 2nd ed. Academic Press, London, pp 243–267

Wu J, Ge X (2004) Oxidative burst, jasmonic acid biosynthesis, and taxol production induced by low-energy ultrasound in *Taxus chinensis* cell suspension cultures. Biotechnol Bioeng 85:714–721

Wurm F (2005) Manufacture of recombinant biopharmaceutical proteins by cultivated mammalian cells in bioreactors. In: Knäblein J (ed) Modern Biopharmaceuticals, vol 3. Wiley VCH, Weinheim, pp 723–759

Xing HH, Ono A, Miyanaga K, Tanji Y, Unno H (2001) A kinetic model for growth of callus derived from *Eucommia ulmoides* aiming at mass production of a factor enhancing collagen synthesis of animal cells. Math Comput Simulat 56:463–474

Yanpaisan W, King N, Doran P (1998) Analysis of cell cycle activity and population dynamics in heterogeneous plant cell suspensions using flow cytometry. Biotechnol Bioeng 58:515–528.

Yanpaisan W, King N, Doran P (1999) Flow cytometry of plant cells with applications in large-scale bioprocessing. Biotechnol Adv 17:3–27

Yazaki K (2004) Natural products and metabolites. In: Christou P, Klee H (eds) Handbook of Plant Biotechnology, vol 2. Wiley, Chichester, pp 811–857

Ye H, Huang LL, Chen SD, Zhong JJ (2004) Pulsed electric field stimulates plant secondary metabolism in suspension cultures of *Taxus chinensis*. Biotechnol Bioeng 88:788–795

Yeoman MM, Yeoman CL (1996) Manipulating secondary metabolism in cultured plant cells. New Phytol 134:553–569

Yoshikawa T (1997) Production of ginsenosides in ginseng hairy root cultures. In: Doran PM (ed) Hairy Roots: Culture and Applications. Harwood Academic, Amsterdam, pp 73–79

Yoshimatsu K, Yamaguchi H, Shimomura K (1996) Traits of *Panax ginseng* hairy roots after cold storage and cryopreservation. Plant Cell Rep 15:555–560

Yu S, Mahagamasekera MGP, Willimas GRC, Kanokwaree K, Doran PM (1997) Oxygen effects in hairy root culture. In: Doran PM (ed) Hairy Roots: Culture and Applications. Harwood Academic, Amsterdam, pp 139–150

Zenk MH, El-Shagi H, Schulte U (1975) Anthraquinone production by cell suspension cultures of *Morinda citrifolia*. Planta Med Suppl 79–101

Zenk MH, El-Shagi H, Arenes H, Stockgit J, Weiler EW, Dues D (1977) Formation of indole alkaloids serpentine and ajmalicine in cell suspension cultures of *Catharanthus roseus*. In: Barz W, Reinhard E, Zenk MH (eds) Plant Tissue Culture and its Biological Application. Springer, Berlin Heidelberg New York, pp 27–44

Zhang W, Curtin C, Kikuchi M, Franco C (2002) Integration of jasmonic acid and light irradiation for enhancement of anthocyanin biosynthesis in *Vitis vinifera* suspension cultures. Plant Sci 162:459–468

Zhong JJ (2001) Biochemical engineering of the production of plant-specific secondary metabolites. In: Tscheper T (ed) Advances in Biochemical Engineering: Plant Cells, vol 72. Springer, Berlin Heidelberg New York, pp 1–26

Zhong JJ, Yoshida M, Fujiyama K, Seki T, Yoshida T (1993) Enhancement of anthocyanin production by *Perilla frutescens* cells in a stirred bioreactor with internal light irradiation. J Ferment Bioeng 75:299–303

Zhong JJ, Chen F, Hu WW (1999) High density cultivation of *Panax notoginseng* cells in stirred bioreactors for the production of ginseng biomass and ginseng saponin. Process Biochem 35:491–496

Ziv M (2000) Bioreactor technology for plant micropropagation. In: Janick J (ed) Horticultural Reviews, vol 24. Wiley, New York, pp 1–30

Complementary Reading

Christou P, Klee H (2004) Handbook of Plant Biotechnology, vol 1–2. Wiley, Chichester
Endress R (1994) Plant Cell Biotechnology. Springer, Berlin Heidelberg New York
Evans D, Coleman J, Kearns A (2003) Plant Cell Culture. BIOS Scientific, Abingdon
Doran PM (1997) Hairy Roots: Culture and Applications. Harwood, Amsterdam
Dutta Gupta S, Ibaraki Y (2006) Plant Tissue Culture Engineering, Series: Focus on Biotechnology, vol 6. Springer, Dordrecht
Fischer R, Schillberg S (2004) Molecular Farming: Plant-made Pharmaceuticals and Technical Proteins. Wiley, Weinheim
Hess D (1992) Biotechnologie der Pflanzen. Eugen Ulmer, Stuttgart
Hvoslef-Eide AK, Preil W (2005) Liquid Culture Systems for In Vitro Plant Propagation. Springer, New York
IAEA (2004) Low cost options for tissue culture technology in developing countries. Proceedings of a Technical Meeting Organized by the Joint FAO/IAEA Division of Nuclear Techniques in Food and Agriculture, Vienna, 26–30 August 2002
Oksman-Caldentey KM, Barz WH (2002) Plant Biotechnology and Transgenic Plants. Marcel Dekker, New York

Index

Printing: Krips bv, Meppel, The Netherlands
Binding: Stürtz, Würzburg, Germany